VoIP Performance Management and Optimization

Adeel Ahmed, CCIE No. 4574

Habib Madani

Talal Siddiqui, CCIE No. 4280

Cisco Press

800 East 96th Street

Indianapolis, IN 46240

VoIP Performance Management and Optimization

Adeel Ahmed, Habib Madani, Talal Siddiqui

Copyright© 2011 Cisco Systems, Inc.

Published by:
Cisco Press
800 East 96th Street
Indianapolis, IN 46240 USA

ISBN-13: 978-1-58705-528-7

ISBN-10: 1-58705-528-7

Printed in the United States of America

First Printing July 2010

Library of Congress Cataloging-in-Publication Data:

Ahmed, Adeel.
    VoIP performance management and optimization / Adeel Ahmed, Habib Madani, Talal Siddiqui.
       p. cm.
     ISBN 978-1-58705-528-7 (hardcover)
    1. Internet telephony—Management. 2. Computer network protocols. I. Madani, Habib, 1969- II. Siddiqui, Talal, 1973- III. Title.

    TK5105.8865.A38 2010
    621.385—dc22   2010023573

Warning and Disclaimer

This book is designed to provide information about managing and optimizing VoIP networks using a metrics-based approach that relies on collecting, analyzing, and correlating VoIP performance data from various network elements. Every effort has been made to make this book as complete and as accurate as possible, but no warranty or fitness is implied.

The information is provided on an "as is" basis. The authors, Cisco Press, and Cisco Systems, Inc., shall have neither liability nor responsibility to any person or entity with respect to any loss or damages arising from the information contained in this book or from the use of the discs or programs that may accompany it.

The opinions expressed in this book belong to the author and are not necessarily those of Cisco Systems, Inc.

Trademark Acknowledgments

All terms mentioned in this book that are known to be trademarks or service marks have been appropriately capitalized. Cisco Press or Cisco Systems, Inc., cannot attest to the accuracy of this information. Use of a term in this book should not be regarded as affecting the validity of any trademark or service mark.

Corporate and Government Sales

The publisher offers excellent discounts on this book when ordered in quantity for bulk purchases or special sales, which may include electronic versions and/or custom covers and content particular to your business, training goals, marketing focus, and branding interests. For more information, please contact: **U.S. Corporate and Government Sales 1-800-382-3419** corpsales@pearsontechgroup.com

For sales outside the United States please contact: **International Sales** international@pearsoned.com

Feedback Information

At Cisco Press, our goal is to create in-depth technical books of the highest quality and value. Each book is crafted with care and precision, undergoing rigorous development that involves the unique expertise of members from the professional technical community.

Readers' feedback is a natural continuation of this process. If you have any comments regarding how we could improve the quality of this book, or otherwise alter it to better suit your needs, you can contact us through email at feedback@ciscopress.com. Please make sure to include the book title and ISBN in your message.

We greatly appreciate your assistance.

Publisher: Paul Boger

Associate Publisher: Dave Dusthimer

Manager, Global Certification: Erik Ullanderson

Business Operation Manager, Cisco Press: Anand Sundaram

Executive Editor: Brett Bartow

Managing Editor: Sandra Schroeder

Development Editor: Ginny Bess Munroe

Senior Project Editor: Tonya Simpson

Copy Editor: John Edwards

Technical Editors: Brad Stewart, David L. Blair

Editorial Assistant: Vanessa Evans

Book Designer: Louisa Adair

Cover Designer: Sandra Schroeder

Composition: Mark Shirar

Indexer: Tim Wright

Proofreader: Apostrophe Editing Services

Americas Headquarters
Cisco Systems, Inc.
San Jose, CA

Asia Pacific Headquarters
Cisco Systems (USA) Pte. Ltd.
Singapore

Europe Headquarters
Cisco Systems International BV
Amsterdam, The Netherlands

Cisco has more than 200 offices worldwide. Addresses, phone numbers, and fax numbers are listed on the Cisco Website at **www.cisco.com/go/offices.**

CCDE, CCENT, Cisco Eos, Cisco HealthPresence, the Cisco logo, Cisco Lumin, Cisco Nexus, Cisco StadiumVision, Cisco TelePresence, Cisco WebEx, DCE, and Welcome to the Human Network are trademarks; Changing the Way We Work, Live, Play, and Learn and Cisco Store are service marks; and Access Registrar, Aironet, AsyncOS, Bringing the Meeting To You, Catalyst, CCDA, CCDP, CCIE, CCIP, CCNA, CCNP, CCSP, CCVP, Cisco, the Cisco Certified Internetwork Expert logo, Cisco IOS, Cisco Press, Cisco Systems, Cisco Systems Capital, the Cisco Systems logo, Cisco Unity, Collaboration Without Limitation, EtherFast, EtherSwitch, Event Center, Fast Step, Follow Me Browsing, FormShare, GigaDrive, HomeLink, Internet Quotient, IOS, iPhone, iQuick Study, IronPort, the IronPort logo, LightStream, Linksys, MediaTone, MeetingPlace, MeetingPlace Chime Sound, MGX, Networkers, Networking Academy, Network Registrar, PCNow, PIX, PowerPanels, ProConnect, ScriptShare, SenderBase, SMARTnet, Spectrum Expert, StackWise, The Fastest Way to Increase Your Internet Quotient, TransPath, WebEx, and the WebEx logo are registered trademarks of Cisco Systems, Inc. and/or its affiliates in the United States and certain other countries.

All other trademarks mentioned in this document or website are the property of their respective owners. The use of the word partner does not imply a partnership relationship between Cisco and any other company. (0812R)

About the Authors

Adeel Ahmed, CCIE No. 4574, is a senior manager in the Cisco Advanced Services group. He has been with Cisco for more than 11 years. His areas of expertise include Access/Dial, Broadband Cable, SP Voice, and IPv6. He has worked with major cable MSOs in North America, EMEA, and ASIAPAC in designing and troubleshooting cable networks. He has written several white papers and design guides used by customers, sales teams, and Cisco engineers in deploying multiservices over cable networks.

He has represented Cisco at industry technical forums such as IETF, APRICOT, SCTE, CableLabs, NCTA, NAv6TF, Networkers, and Global IPv6 Summit. He is the coauthor of *Deploying IPv6 in Broadband Access Networks*. Adeel holds a bachelor's and a master's degree in electrical engineering.

Habib Madani works as a network consulting engineer in the Cisco Advanced Services group. He has been with Cisco for the past five years; before that, he worked for more than ten years as a lead engineer at Yahoo!, CenterPoint Broadband Technology, and Motorola. His background is SP VoIP, Data Center, Network Management, and OSS solutions. He has worked with major enterprise customers and MSOs in the United States and abroad across these segments. He has represented Cisco at industry technical forums such APRICOT and Networkers. Habib holds a bachelor's degree in electrical engineering from the University of Arizona and a master's in electrical engineering and computer science from the University of Illinois.

Talal Siddiqui, CCIE No. 4280, is a senior manager in the Cisco Advanced Services group. He has been with Cisco for 13 years. He has been involved with beachhead customers who were deploying these solutions for the past 12 years. As a consulting engineer, he has worked with MSOs, CLECs, and ILECs who have deployed Cisco SS7 Interconnect Solutions through planning, designing, deploying, network expansion, and operational troubleshooting phases. As a technical leader in Unified Communications Practice, he worked with Cisco partners in planning, designing, and deploying IP telephony solutions for large enterprise customers. Currently he is working on developing service offerings for Communication/Collaboration Enabled Business Process (CEBP) solutions and service delivery tools. Talal holds BSEE and MBA degrees and is a CCIE.

About the Technical Reviewers

Brad Stewart is a manager of the Consulting group of the U.S. Service Provider (SP) Operations and Network Management (ONM) Team at Cisco based in Englewood, Colorado. He earned his Bachelor of Science, Master of Science, and Doctor of Philosophy degrees from the Systems Engineering Department at the University of Virginia. Brad has more than 25 years of experience in telecommunications, high technology, defense, and management consulting. He has expertise in systems engineering, network management processes and systems, and managing software and systems development, with particular strengths in operations optimization, program and project management, and strategic business and technical consulting. He has effectively led large project teams in the analysis, design, implementation, customization, and deployment of enterprise and service provider Business Support Systems (BSS), Operations Support Systems (OSS), and Network Management Systems (NMS). Brad has taught several courses in various aspects of systems engineering in both corporate and academic settings.

David L. Blair, CCVP, CIPTSS, CIPTOS, MCSE, Security+, is a customer support engineer with the Cisco Remote Operations Services (ROS), where he works as a voice engineer in the Go to Market (G2M) team. David has 21 years of information technology industry experience, including work with Cisco AAVVID solutions, WAN management and design, Cisco routers and switches, and the administration and design of WANs and LANs.

Dedications

I would like to dedicate this book to my parents for everything they have done for me. They helped me become a better person and always provided love, guidance, and support whenever I needed it. To my wife Seema for her unconditional love, support, and patience throughout the writing of this book. To my lovely children: Asad, Aashir, and Zeerak; you are the light of my eyes and my inspiration. To all the people suffering in this world and to those people who work endlessly to help others and make this world a better place. —Adeel Ahmed

I would like to dedicate this book to my ammi and abu for everything they have done for me. Their support has helped me become a better person, and they always provided love and guidance. To my wife Qamar: for her unconditional love, support, and patience. To my beloved children: A'na, Alia, Aisha, Aadil, and Yusuf; you inspire me and make me strong. To my brother Atiq: You have always been there for support and guidance. —Habib Madani

I would like to dedicate this book to my parents for their unconditional love and to the teachers who gave me discipline, curiosity, and motivation to undertake projects like this. I owe an enormous gratitude to my family: my wife Aisha for bearing with me when I had to seclude myself during nights and weekends, and to my children: Raiyyan, Nabeeha, Hadi, and Humza, who sacrificed time with their father in order for me to complete this manuscript. —Talal Siddiqui

Acknowledgments

The authors would like to thank the technical reviewers, Brad Stewart and David Blair, for their valuable feedback in improving the quality of this book. Without their critical analysis and detailed feedback, we would not have been able to add the technical depth and clarity in the text to better explain and illustrate our ideas.

A big thank-you goes out to the production team for this book. Brett Bartow, Andrew Cupp, Christopher Cleveland, and Ginny Munroe have been very professional and helpful throughout the writing of this book. They have been very patient, especially when we missed our deadlines. They are a pleasure to work with. We couldn't have asked for a better team.

Last but not least, we would like to thank our management for their support, understanding, and encouragement toward this "side project."

Special Acknowledgments from Adeel Ahmed:

I would like to acknowledge a few individuals who have influenced my life and personality, and have provided guidance and inspiration throughout my life. These individuals include Muhammad ibn Abdullah and his companions, Sanjeeda Ahmed, Aziz Ahmed, Khairunisa Begum, Aquila Khanum, Shaikh Suhaib Webb, Shaikh Saad Hassanin, Natasha Ahmed, Najia Ahmed, Farzana Khan, Ismail Saadiq, Nasir Ali, Akram Hussain, Imran Chaudhary, and Salman Asadullah. I would also like to thank my colleagues Tom Chang, Umair Siddiqui, and Jeff Riddel.

Special Acknowledgments from Habib Madani:

I would like to acknowledge my coauthors, Zeeshan Naseh and Adnan Ashraf, as they have influenced me both technically and personally, and my friend Akram Hussain, who has always been there for any kind of support. I also want to acknowledge my dear friend Man'Altaher, who has deeply influenced my life. Finally, I would also like to acknowledge my colleagues at Cisco and partners Naeem Akhter and Justin Thompson.

Special Acknowledgments from Talal Siddiqui:

I would like to acknowledge my esteemed colleagues at Cisco and our partners who provided invaluable technical expertise and professional guidance. These individuals include Luc Bouchard, Ramesh Kaza, Faisal Chaudhry, Niels Brunsgaard, Tito Dayans, Akram Hussain, Abderrahmane Mounir, Himanshu Desai, Rajesh Ramarao, Imran Chaudhary, Brian Goodwin, Robert Donze, and Michael Crane. Their knowledge, mentorship, and creative thoughts were a tremendous help for me.

Contents at a Glance

Contents

Icons Used in This Book

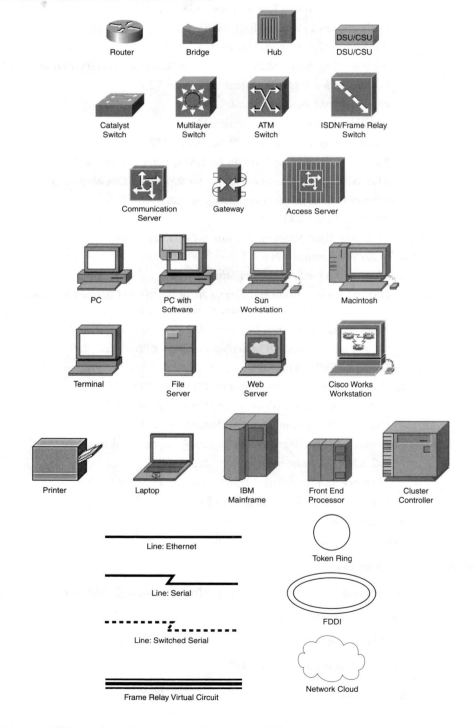

Command Syntax Conventions

The conventions used to present command syntax in this book are the same conventions used in the IOS Command Reference. The Command Reference describes these conventions as follows:

- **Boldface** indicates commands and keywords that are entered literally as shown. In actual configuration examples and output (not general command syntax), boldface indicates commands that are manually input by the user (such as a **show** command).

- *Italic* indicates arguments for which you supply actual values.

- Vertical bars (|) separate alternative, mutually exclusive elements.

- Square brackets ([ ]) indicate an optional element.

- Braces ({ }) indicate a required choice.

- Braces within brackets () indicate a required choice within an optional element.

Foreword

According to a Gartner market share analysis done for Enterprise Unified Communications on June 23, 2009, the total vendor revenue from the entire enterprise unified communications segment in 2008 was $5.1 billion. FierceVoIP quoted ISP-Planet study in its online newsletter on July 28, 2008, saying that the total subscribers in just the United States for the top 10 VoIP service providers are around 18 million and growing in double digits. Comcast came in on top at 5.2 subscribers followed by Time Warner (3.2 million) and Vonage (2.6 million), based on their first-quarter reporting in 2008. This puts great emphasis on managing VoIP for both enterprises and service providers.

Management of a VoIP network is a cyclic process that starts even before VoIP is deployed. The first stage is planning, which includes forming a team, defining the scope of deployment, requirements validation, and assessment of the IP network to determine whether the infrastructure is adequate to support media traffic. The design phase includes comprehensive design-based traffic engineering and validated requirements. It not only covers call-processing servers, remote gateways, and features implementation but also covers changes to the IP network in the form of quality of service design and provisions for network resiliency. It is followed by the implementation phase, which is governed by project management principles and ensures that best practices for deployment are followed for on-time completion.

Implementation also includes a test plan execution and transfer of information to network operations prior to commissioning. The correct execution of these phases ensures minimum problems and decreases the total cost of deployment. Implementation is followed by the operations phase, with continuous optimization to close the loop. This book briefly mentions planning, design, and implementation stages and emphasizes the operations and optimization phases.

First, the hand-off to operations needs to be complete, including remediation of any issues discovered when the postdeployment test plan was executed. All the deployed devices must be discovered by the network management systems. But most important, VoIP can no longer be managed in a silo that is separate from the data network management subteam. This book emphasizes correlating network problems with VoIP-related key performance indicators for faster problem resolution by isolating it and fixing the root cause.

Operational data provides critical feedback for continuous optimization of the network, including its performance and capacity. Optimization is not limited to fine-tuning the traffic engineering process for future growth but also for extending VoIP for the next evolution to collaboration-enabled business transformation.

What is presented in this book is the authors' collective experience and knowledge, working with several other colleagues from Advanced Services, Cisco Remote Operations Service, the product development teams, and most important, Cisco customers, whose feedback was critical in developing best practices for VoIP management and optimization.

Regards,

Talal Siddiqui, Senior Manager, Unified Communication/Collaboration Practice Cisco Advanced Services

Introduction

With the exponential growth of the Internet and an increasing number of VoIP deployments, customers are looking for new ways to manage and scale their networks to meet the growing needs of end users. Customers not only need to fix problems in a timely manner with minimal downtime, but they also need to proactively monitor their networks to fix potential problems before they become service and revenue impacting.

The complexity of an IP network increases with the addition of new services, and as these networks start to scale, managing them becomes a challenge. Customers are looking for new ways to manage their networks and effectively scale these services.

Customers are looking for new techniques and efficient ways to monitor multivendor products in the network and use tools/applications that can scale with the growth of their networks. We got feedback from our customers and VoIP SPs through forums such as SANOG, NANOG, APRICOT, and Cisco Live (formerly known as Networkers) about what they would like to see in a VoIP management book. This feedback can be boiled down to "We want a practical guide with specific details and examples that we can use right away...something that is a desk reference for NOC (Network Operations Center) staff and the network architects."

This book addresses some of the challenges associated with deploying and managing VoIP networks and also provides guidelines on how to optimize these networks.

Goals and Methods

The most important goal of this book is to help define a methodology and framework of collecting, analyzing, and correlating VoIP performance data from various network elements. When correlated in a meaningful way, this data can help network operators identify problematic trends in their VoIP networks, and isolate and fix problems before they become service impacting.

One key methodology in this book is to use a layered approach when troubleshooting VoIP network problems. This helps narrow the scope of the problem in an efficient manner and also helps find the root cause. By quickly identifying the root cause of the problem, the network operator can resolve issues in a timely manner and minimize customer impact.

This book also provides guidelines for optimizing VoIP networks by defining the following:

- What VoIP performance data should be collected from various network elements?

- How to collect VOIP performance data?

- How to use dashboards to analyze and correlate VoIP metrics?

- How to use the VoIP dashboard for trend analysis and capacity planning?

Who Should Read This Book

This book is meant to be used as a guide by network engineers, architects, and operations personnel in managing and optimizing their VoIP networks.

This book also helps network operators troubleshoot VoIP-related issues efficiently and identify root causes to fix problems in a timely manner. However, it does not focus on traces, logs, and debug messages but rather on analyzing trends and correlating network issues to address core issues. This book compliments other Cisco Press publications:

- Kaza, Ramesh and Asadullah, Salman. *Cisco IP Telephony: Planning, Design, Implementation, Operation, and Optimization.* Indianapolis, IN: Cisco Press, February 23, 2005.

- Halmmark, Addis, Giralt, Paul and Smith, Anne. *Troubleshooting Cisco IP Telephony.* Indianapolis, IN: Cisco Press, December 11, 2002.

- Clemm, Alexander. *Network Management Fundamentals.* Indianapolis, IN: Cisco Press, Nov 21, 2006.

How This Book Is Organized

This book discusses some of the challenges faced by service providers and enterprise customers in deploying, managing, and optimizing VoIP in their networks. It provides guidance on how to address voice quality issues and proactively monitor key performance indicators (KPI) to help gauge the health of the VoIP network.

The first part of the book provides an overview of VoIP and key network management concepts. It also discusses a metrics-based approach of managing and optimizing VoIP networks.

The second part of the book concentrates on different VoIP deployment models in SP and enterprise networks, and reviews the common VoIP-related problems in each deployment approach.

Note   The first and second parts of the book set the stage for how VoIP is deployed in enterprise and SP networks and discusses the challenges associated with such deployments. You might feel that both these parts of the book are brief and high-level; they do not cover in-depth technology and protocol details. For example, what is DOCSIS and how does it work? How does the Session Initiation Protocol (SIP) work, and what are the various SIP messages? This is by design; it is assumed that you already understand these basics as this information has already been covered in various other texts. The main focus of this book is on managing and optimizing VoIP networks; these concepts are covered in detail in the third part of the book. That is why chapters in the third part of the book are longer and more detailed than the chapters in the first and second parts of the book.

The third part of the book focuses on a proactive approach to diagnosing problems in VoIP networks and fixing these problems before they become service impacting. This part of the book also talks about what tools can be used by customers in gauging the health of their VoIP network and improve network performance. Using performance counters, Call Detail Records (CDR), and Call Agent trace logs, customers can utilize real-time data to gauge the health of their voice network and make capacity-planning decisions before network resources get congested.

Chapters 1 through 8 cover the following topics:

- Chapter 1, "Voice over IP (VoIP) and Network Management": This chapter talks about VoIP media transport and signaling protocols, some common voice quality issues, and their causes. The second half of the chapter discusses network management methodologies such as Telecommunications Management Network (TMN); Fault, Configuration, Accounting, Performance, and Security (FCAPS); and Information Technology Infrastructure Library (ITIL). It also talks about the strategic importance of managing VoIP networks.

- Chapter 2, "A Metrics-Based Approach for Managing the VoIP Network": This chapter highlights the key performance indicators that can be utilized to effectively manage a VoIP network. It encourages the use of a layered approach for isolating localized and systemic issues. It explains how performance data from various network segments and service flows can be used to manage SLAs in a VoIP network.

- Chapter 3, "VoIP Deployment Models in Service Provider Networks": This chapter discusses various VoIP solutions in an SP environment. The deployment models cover scenarios in which broadband SPs provide VoIP service to residential and business customers. These providers own the last-mile connection to end users; they use their infrastructure to not only provide Internet connectivity but also to offer VoIP services using the same infrastructure. Because they own the last-mile connection and the VoIP infrastructure, they can provide better QoS to VoIP traffic and offer high-quality VoIP services.

- Chapter 4, "Internet Telephony": This chapter describes how VoIP is deployed over a publicly shared infrastructure such as the Internet. In such deployment models, the company providing VoIP services might not own the entire network infrastructure, such as the last-mile connection to the end users, which is used for deploying this service. They might use infrastructure, owned by other entities, to provide VoIP as an overlay service by deploying some of their own network components that are required for offering the VoIP service. This deployment model is different from the models discussed in Chapter 3. The VoIP SP is faced with several challenges with providing QoS to VoIP traffic; these issues are also discussed in this chapter.

- Chapter 5, "VoIP Deployment Models in Enterprise Networks": This chapter explains various deployment models that are commonly used in typical enterprise networks, including the fundamental models: central call processing and distributed call processing. It also discusses large-campus deployment schemes.

This chapter discusses the differences in hosted and managed services around Unified Communications solutions. It also presents a brief overview of IP Contact Centers, which are essentially an extended functionality of a Unified Communications solution.

- Chapter 6, "Managing VoIP Networks": This chapter discusses the best practices for planning media deployment over IP networks starting from how to assess the readiness of the network, traffic engineering, high availability, and managing the IP network and its integrated components that process voice and other media transmissions. This chapter also covers the monitoring mechanism available to network administrators and their scope and effectiveness in managing VoIP networks.

- Chapter 7, "Performance Analysis and Fault Isolation": This chapter discusses an approach for proactive monitoring of the VoIP network for performance analysis and fault isolation of problems caused by anomalies in the network. It starts with explaining the VoIP network monitoring aspects including collection, categorization, and correlation of performance counters for both enterprise and service provider networks. It also discusses different ways of gauging the performance of a large-scale VoIP network by looking at various key performance indicators (KPIs).

- Chapter 8, "Trend Analysis and Optimization": This chapter explains the use of VoIP dashboards to monitor and trend performance data from different components in the VoIP network. This trend analysis can help network operators not only establish a baseline but also help with resource optimization and capacity planning by looking at problematic trends in the network, such as resource overutilization and changes in traffic patterns.

## Chapter 1

Voice over IP (VoIP) and Network Management

The Internet Protocol (IP) network was primarily designed for carrying best-effort (no assurance for delivery without any impairment) data traffic. Traditionally, voice communication was primarily carried over the public switched telephone network (PSTN), which is a global network. Because of the success of IP in becoming a worldwide standard and the ubiquity of IP networks, interconnectivity of IP networks provides global connectivity to its users. The global nature of IP networks provides alternative solutions to carrying media, including voice, as long as it can be carried as a payload in an IP packet.

The change from circuit-switched technology to packet-switched technology leverages some common technologies but also requires some unique methods for transporting voice packets. In the PSTN, the analog voice signal (human speech is analog voice) needs to be digitized using pulse code modulation (PCM), and the signaling happens transparently to the end user to provide dial tone, call routing, and other call features. The IP networks digitize voice signals at the telephone set by employing digital signal processors (DSP), and they put it in an IP packet. The IP packet containing the digitized voice traverses the IP network, leveraging its routing protocols using various available paths to reach the end destination. During the course of its journey, the voice packet shares links/paths with other packets carrying other data traffic, whereas in the PSTN, the circuit or path is established during the initial call setup and is dedicated for that voice call for its entire duration.

The PSTN has a dedicated call-signaling plane called Signaling System 7 (SS7). SS7 behaves like an IP routing and call-signaling protocol; however, in IP networks, call signaling shares common network resources with the data traffic.

The packet-switching mechanism in IP networks not only includes packetization of analog voice with the help of coders and decoders called CoDec but also encapsulation of this digitized voice data and voice signaling in packets as it is guided through the IP network based on routing protocols on the routers. This transport method is referred as

Voice over Internet Protocol (VoIP). VoIP introduces several new issues on the packet-switched network that a circuit-switched network does not have:

- Serialization delay

- Propagation delay with variance

- Packet loss during transmission

- Service degradation as the voice packets share a link on the LAN and Internet with data traffic

This chapter discusses some of these issues briefly.

VoIP Technology

VoIP technology was initially targeted for achieving toll savings called toll bypass because it uses the available bandwidth at the client's local-area network (LAN), wide-area network (WAN), and the Internet. VoIP technology turns analog voice into digital data packets that can be stored, searched, manipulated, copied, combined with other data, and distributed to virtually any device that connects to the IP network.

VoIP can interact seamlessly with other IP-based data systems because of the common standards defined by the International Telecommunications Union (ITU-T), the Internet Engineering Task Force (IETF), and the Institute of Electrical and Electronics Engineers (IEEE) that cover the VoIP call-signaling protocols, encryption, compression, encoding, and decoding techniques. This interconnectivity not only has made VoIP almost as ubiquitous as the time-division multiplexing (TDM)–based voice technology, but it now also facilitates enhanced collaboration experience by integrating with business process applications in manufacturing, health care, finance, government, education, and many other industry verticals. VoIP has provided opportunities to the service providers to offer new communication, enabling services to consumers and enterprises that contribute significantly to their revenue. The growing strategic importance of VoIP makes the management of the network extremely critical.

Network management's purpose is to maintain the quality and optimum delivery of services provided by the network. This is critical for service providers whose business is network-centric and enterprise or commercial segments because of business dependency on information technology. Network management has evolved to keep up with the technology advancements and convergence such as VoIP. This evolution is not limited to the technology involved in network management, including collection methods, configuration management, SLA management, trouble ticketing software, and so on. The advancements in network management include process improvement and best practices formulation into standards such as Telecommunications Management Network (TMN), Fault Configuration Accounting Performance Security (FCAPS), enhanced Telecom Operations Map (eTOM) framework, and Information Technology Infrastructure Library (ITIL). At a detailed level, we have seen continuous improvements and advancements in collection methods using technologies such as Simple Network Management Protocol versions

(SNMPv1, SNMPv2, SNMPv3), Remote network MONitoring (RMON), and eXtensible Markup Language (XML) that provide a central view of overall performance of the network by integrating the various network management tools and correlating data to provide intelligent and actionable reports.

This chapter lays the foundation for this book by explaining the fundamental workflow and process management concepts in network management and by providing an overview of TMN, FCAPS, and ITIL. We cover how these concepts apply to converged networks and then highlight the specific aspects that are fundamental to the topic of this book, that is, VoIP performance management and optimization based on analysis of key performance indicators or metrics. In the next section, we look at VoIP technology basics, its underlying protocols, common network problems in VoIP networks, and some voice quality–related problems in IP networks. We then cover the VoIP basics and network management concepts and discuss how they apply specifically to VoIP networks. We also discuss the strategic value of VoIP to service providers and enterprises and the importance of managing VoIP networks.

VoIP Overview

The PSTN is considered a *connection-oriented network*, in which a path from the source to the destination is established using TDM trunks between the intermediate central offices before the audio path is cut through, so end users can start their conversations over their phones. TDM is a *multiplexing scheme* in which two or more bit streams of digitized voice signal are transferred apparently and simultaneously as subchannels by taking turns on the common physical channel. The time domain is divided into several recurrent time slots of fixed length, one for each subchannel. This subchannel is referred to as Digital Signal 0 (DS0). DS0 has a bandwidth of 64,000 bits per second (bps), which is determined by the Nyquist theorem, which states that the minimum sampling rate of twice the frequency of the signal to be sampled will result in an accurate representation of the original signal. Because the human voice is limited to 4000 hertz, a sampling rate of 8000 samples per second, or every 125 microseconds, is used. The conversion process begins by analyzing each voice sample and converting it into an 8-bit word, also called an *octet*. If there are 8 bits per sample and 8000 samples per second, the product is 64,000 bits per second. Hence, the bandwidth of DS0 is 64 kbps.

The communication starts with the end user taking the telephone off the hook, which notifies the network that the service is requested. The network then returns a dial tone, and the end user dials the destination number. The call routing information is relayed by an overlay signaling network known as Signaling System 7 (SS7), which is a global network. All the central offices, including international gateways, connect to it. When the destination party answers, the end-to-end connection is confirmed through the various central offices along the path. When the conversation is complete, the two parties hang up, and their network resources can be reallocated for someone else's conversation. Also note that the voice signal can also come from data devices such as modems rather than just a phone.

Because of the connection-oriented nature of the PSTN, which holds the call path or "circuit" as constant for the duration of the call, the characteristics of that path, including serialization delay at different connection points, propagation delay, and information sequencing, remain constant for the duration of the call. Because these constants add to the reliability of the system, the term *reliable network* is often used to describe a connection-oriented environment. However, this reliability comes at a cost of a dedicated network meant for only one purpose, that is, to switch calls.

In contrast, traditional data networks, including intranets and the Internet, are considered *connectionless networks*, in which the full source and destination address are attached to a packet of information, and then that packet is passed through the network for delivery to the ultimate destination. As this packet traverses through the network routers and switches, they route or switch the packet based on the header information in the packet and the routing configuration and routing protocols on the Layer 3 devices in the IP network. The dynamic nature of IP routing might cause these packets for the same call to take different routes during the entire duration of the call. This increases the potential for packets arriving out of sequence and with varying delay with increased probability of drop. The network devices might have different processing power, capacity, and connections of varying speed and bandwidth, which make the serialization and propagation delay variable and potentially large.

There are some inherent problems with digitization of analog voice signals, such as noise and echo, which are introduced by impedance mismatch in the hybrid (2-wire to 4-wire conversion) coder/decoder (codec) circuits. Because of the additional delay in IP networks, these problems get exacerbated and are more noticeable. For these reasons, the terms *best effort* and *unreliable* are often used to describe a connectionless environment. These problems can be mitigated through proper planning and employing special techniques to guarantee quality of service (QoS) for media traffic and proactive network monitoring to continue to optimize the network characteristics for better end-user experience.

Note   VoIP is also referred to as IP telephony, or IPT. Both terms are used to refer to sending voice packets across an IP network. The distinction is based on the endpoints used in the communication. In a VoIP network, the TDM-based PSTN network interconnects with the IP network, usually through a voice gateway–enabling communication between an IP endpoint and another traditional endpoint in the PSTN. In an IP telephony environment, the communication typically takes place between two IP endpoints.

Unified Communications (UC) is a framework of hardware and software products that facilitate multiple communication means such as voice video, presence, instant messaging, and other collaboration technologies over an IP network that might integrate with other external networks. UC also encompasses management software for monitoring and configuring the devices in the network. The term *UC* is more commonly used in the enterprise context.

This book uses VoIP and IP telephony interchangeably. UC will be used in an enterprise context, where services other than just voice, such as voicemail and presence, are involved.

Before we elaborate on some of the common problems (and their root causes) encountered by service providers and network architects when deploying voice over an IP infrastructure, we briefly cover some protocol basics.

Media Transport Protocol for VoIP—RTP

Using Transmission Control Protocol (TCP) for transport on top of IP provides enhanced reliability (albeit with additional protocol overhead) as compared with User Datagram Protocol (UDP), although it is still not a true equivalent of a connection-oriented service. But a connection-oriented transport infrastructure such as PSTN is not absolutely necessary to support interactive communication. This is why Real Time Protocol (RTP) was introduced to carry real-time, interactive media traffic with greater efficiency and reliability on a fundamentally unreliable protocol (IP).

RTP runs on top of UDP to avoid the overhead associated with the otherwise more reliable TCP. RTP is currently the cornerstone for carrying real-time traffic across IP networks. To date, all VoIP signaling protocols utilize RTP/UDP/IP as their transport mechanism for voice traffic. Often, RTP packet flows are known as *RTP streams* or *media streams*. Therefore, you can use IP in conjunction with UDP and RTP to replace a traditional voice circuit.

Figure 1-1 illustrates the different fields within the RTP header.

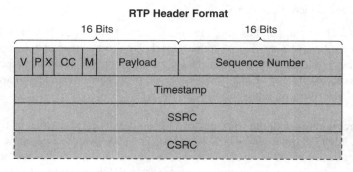

Figure 1-1 *Real-time Transport Protocol Header Fields*

The RTP header fields are described in the following list:

- **Version (V):** This field is 2 bits long and identifies the version of RTP. Most of the current applications use version 2. RFC 3550 covering RTP also defines version 2.

- **Padding (P):** This field is 1 bit long. If the padding bit is set, the packet contains one or more additional padding octets at the end that are not part of the payload.

- **Extension (X):** This field is 1 bit long. If the extension bit is set, the fixed header *must* be followed by exactly one header extension.

- **CSRC Count (CC):** This field is 4 bits long. The Contributing SouRCe (CSRC) count contains the number of CSRC identifiers that follow the fixed header.

- **Marker (M):** This field is 1 bit long. The interpretation of the marker is defined by a profile.

- **Payload Type (PT):** This field is 7 bits long and identifies the format of the RTP payload and determines its interpretation by the application.

- **Sequence Number:** This field is 16 bits long. The sequence number increments by one for each RTP data packet sent and can be used by the receiver to detect packet loss and to restore packet sequence. The initial value of the sequence number should be random (unpredictable).

- **Timestamp:** This field is 32 bits long. The timestamp reflects the sampling instant of the first octet in the RTP data packet. The initial value of the timestamp should be random. Several consecutive RTP packets have equal timestamps if they are (logically) generated at once, for example, if they belong to the same video frame. Consecutive RTP packets can contain timestamps that are not monotonic if the data is not transmitted in the order it was sampled, as in the case of MPEG-interpolated video frames.

- **SSRC:** This field is 32 bits long. The Synchronization SouRCe (SSRC) indentifier field identifies the synchronization source. This identifier is chosen randomly, with the intent that no two synchronization sources within the same RTP session have the same SSRC identifier.

- **CSRC:** This field is 32 bits long and contains 0 to 15 items, 32 bits each. The CSRC list identifies the contributing sources for the payload contained in this packet. The number of identifiers is given by the CC field. If there are more than 15 contributing sources, only 15 can be identified. CSRC identifiers are inserted by mixers using the SSRC identifiers of contributing sources.

The IP packet containing the voice payload has an IP packet header that is 40 bytes (IP = 20 bytes, UDP = 8 bytes, and RTP = 12 bytes). The IP packet containing the voice payload varies in size based on the codec type, its bit rate, and the codec packetization period, which is also known as the *codec sample interval*. For example, a G.711 codec with a bit rate of 64 kbps and a sample interval of 20 milliseconds (ms) would yield a voice payload of 160 bytes. This is calculated as follows:

Voice payload size = Codec bit rate * Codec sample interval
$$= \text{(64 kbps for G.711)} * \text{(20 ms)}$$
$$= \text{(8000 bytes per second)} * \text{(.02 seconds)}$$
$$= 160 \text{ bytes}$$

The total length of the voice packet is 200 bytes (40 bytes IP header + 160 bytes voice payload), and this does not include the Layer 2 header overhead. Figure 1-2 illustrates the RTP encapsulation format.

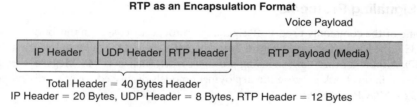

Figure 1-2 *Real-time Transport Protocol Encapsulation Format*

Similarly, for a G.729 codec with a bit rate of 8 kbps and a sample interval of 20 ms, the voice payload size comes out to 20 bytes. The total length of the voice packet in this case comes out to 60 bytes (40 bytes header + 20 bytes voice payload). Because voice packets are relatively small, a loss of one to two packets does not severely impact the quality of the voice conversation, and in most cases, the user might not experience a noticeable difference in the quality of the voice call.

In the case of packet loss, the receiving station waits for a period of time (per its jitter buffer) and then runs a *concealment strategy*. This concealment strategy replays the last packet received, so the listener does not hear gaps of silence. Because the lost speech is only typically 10 or 20 ms, the listener most likely does not hear the difference. You can accomplish this concealment strategy only if one packet is lost. If multiple consecutive packets are lost, the concealment strategy is run only once until another packet is received. In IP networks, it is common and normal for packet loss to occur. In fact, TCP/IP was built to utilize packet loss as a means of controlling the flow of packets. In TCP/IP, if a packet is lost, it is retransmitted. In most real-time applications, retransmission of a packet is worse than not receiving a packet because of the time-sensitive nature of the information. That is why real-time applications do not use TCP and instead use UDP.

The ITU-T recommends a one-way delay of no more than 150 ms. In a Cisco VoIP network, the unidirectional delay might be 120 ms (currently, 65 to 85 ms of that 120-ms delay is derived from two Cisco VoIP gateways when using G.729). If the receiving station must request that a packet be retransmitted, the delay is too large and large gaps and breaks in the conversation occur.

RTP has a timestamp field that records the exact time the packet was sent (in relation to the entire RTP stream). This information is used by the device terminating/receiving the audio flow. The receiving device uses the RTP timestamps to determine when a packet was expected, whether the packet was in order, and whether it was received when expected. All this information helps the receiving station determine how to tune its own settings to mask any potential issues such as delay, jitter, and packet loss, which are the result of the inherent nature of an IP network. These network-related problems are discussed in more detail later in this chapter.

VoIP Signaling Protocols

Some of the commonly used VoIP signaling protocols discussed in this book include H.323, the Media Gateway Control Protocol (MGCP), and the Session Initiation Protocol (SIP). Although other VoIP signaling protocols exist, such as H.248/Megaco, they are not covered in this book because the deployment models discussed are primarily based on H.323, MGCP, and SIP:

- H.323 is an ITU-T specification for transmitting audio, video, and data across an IP network, including the Internet. When compliant with H.323, vendors' products and applications can communicate and interoperate with each other. The H.323 standard addresses call signaling and control, multimedia transport and control, and bandwidth control for point-to-point and multipoint conferences. The H series of recommendations also specifies H.225 for connection establishment and termination between endpoints, H.245 for multimedia communications, H.320 for Integrated Services Digital Network (ISDN), and H.324 for plain old telephone service (POTS) as transport mechanisms. H.323-based deployment models are covered in detail in Chapter 5, "VoIP Deployment Models in Enterprise Networks."

- Media Gateway Control Protocol (MGCP) is defined in RFC 3435. MGCP is a protocol used by media gateway controllers (MGC), also known as call agents, to control media gateways (MG). MGCP is based on a master/slave relationship in which MGC is the master that issues commands to the MG (slave). The MG acknowledges the command, executes it, and notifies the MGC of the outcome (successful or not). In this architecture, the MG handles the media functions, such as conversion of TDM/analog signals into Real-time Transport Protocol (RTP)/Real-time Transport Control Protocol (RTCP) streams. MGC handles the call-signaling functions. MGCP-based call control is also referred to as a centralized switching deployment model. This is discussed in more detail in Chapter 3, "VoIP Deployment Models in Service Provider Networks."

- Session Initiation Protocol (SIP) is defined in RFC 3261. SIP is a signaling protocol that controls the initiation, modification, and termination of interactive multimedia sessions. The multimedia sessions can be as diverse as audio or video calls among two or more parties, chat sessions, or game sessions. SIP extensions have also been defined for instant messaging, presence, and event notifications. SIP is a text-based protocol that is similar to HTTP, Simple Mail Transfer Protocol (SMTP), and SDP.

SIP is a peer-to-peer protocol, which means that network capabilities such as call routing and session management functions are distributed across all the nodes (including endpoints and network servers) within the SIP network. This is in contrast to the traditional telephony model, where the phones or end-user devices are completely dependent on centralized switches in the network for call session establishment and services. SIP-based deployment models are covered in detail in Chapter 4, "Internet Telephony."

Common Network Problems in VoIP Networks

Deploying a VoIP infrastructure introduces a new set of challenges that do not exist in circuit-switched networks like the PSTN. Some of the common network problems encountered by providers deploying VoIP infrastructure include the following:

- Delay/latency
- Jitter
- Packet loss
- Voice Activity Detection (VAD)
- Other issues

These issues can affect the quality of voice service and result in a poor user experience when voice packets are transported over an IP infrastructure. Figure 1-3 categorizes the type of voice quality issues experienced by the users and shows their associated root causes.

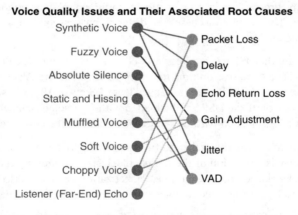

Figure 1-3   *Voice Quality Issues and Their Associated Root Causes*

Before we continue, it is important to understand the meaning and correlation of the previously mentioned issues and their impact on voice quality.

Delay/Latency

VoIP delay or latency is characterized as the amount of time it takes for speech to exit the speaker's mouth and reach the listener's ear. Three types of delay are inherent in today's packet-based voice networks:

- Propagation delay is caused by the length a signal must travel through light in fiber or electrical impulse in copper-based networks.

- Processing delay defines many different causes of delay (actual packetization, compression, and packet switching) and is caused by devices that forward the frame through the network.

- Serialization delay, also called queuing delay, is the amount of time it takes to actually place the voice packet onto an interface.

Propagation Delay

Light travels through a vacuum at a speed of 186,000 miles per second, and electrons travel through copper or fiber at approximately 125,000 miles per second. A fiber network stretching halfway around the world (13,000 miles) induces a one-way delay of about 70 ms. Although this delay is almost imperceptible to the human ear, propagation delays in conjunction with processing delays can cause noticeable speech degradation.

Processing Delay

As mentioned previously, devices that forward the frame through the network cause a processing delay. Processing delays can impact traditional phone networks and packet-based networks. This section discusses the different processing delays and describes how they affect voice quality.

The time taken by the DSP to compress a block of PCM samples is called *compression* or *coder delay*. The compression time for a Conjugate Structure Algebraic Code Excited Linear Prediction (CS-ACELP) process ranges from 2.5 ms to 10 ms based on the loading of the DSP processor. If the DSP is fully loaded with four voice channels, the coder delay is 10 ms. If the DSP is loaded with only one voice channel, the coder delay is 2.5 ms.

Packetization delay is the time taken to fill a packet payload with encoded/compressed speech. This delay is a function of the sample block size required by the coder and the number of blocks placed in a single frame. Packetization delay can also be called *accumulation delay*, as the voice samples accumulate in a buffer before they are released.

In the Cisco IOS VoIP product, the digital signal processor (DSP) can generate a speech sample every 10 ms when using G.729. Two of these speech samples (both with 10-ms speech data which causes a delay) are then placed within one packet. The packet delay is, therefore, 20 ms. An initial look-ahead of 5 ms occurs when using G.729, which gives an initial delay of 25 ms for the first speech frame. The compression algorithm relies on known voice characteristics to correctly process sample block N. The algorithm must have some knowledge of what is in block N+1 to accurately reproduce sample block N. This look-ahead is also known as the *algorithmic delay*.

Vendors can decide how many speech samples they want to send in one packet. In the previous example, G.729 uses 10-ms speech samples; each increase in samples per frame raises the delay by 10 ms. In fact, Cisco IOS enables users to choose how many samples to put into each frame.

Cisco gave DSP much of the responsibility for framing and forming packets to keep router/gateway overhead low. The RTP header, for example, is placed on the frame in the DSP instead of giving the router that task. So the primary task of the router is to forward the voice packets as fast as possible to reduce the switching or forwarding delay.

Serialization/Queuing Delay

A packet-based network experiences delay for several reasons. Two of these are the time necessary to move the actual packet to the output queue and queuing delay.

When packets are held in a queue because of congestion on an outbound interface, the result is queuing delay. Queuing delay occurs when more packets are sent out than the interface can handle at a given interval.

As mentioned previously, the ITU-T G.114 recommendation specifies that for good voice quality, no more than 150 ms of one-way, end-to-end delay should occur. With today's Cisco VoIP implementation, two routers with minimal network delay (back to back) introduce or require only about 60 ms of end-to-end delay. This leaves up to 90 ms of network delay to move the IP packet from source to destination.

Jitter

Jitter is the variation of packet interarrival time. Jitter is one issue that exists only in packet-based networks. While in a packet voice environment, the sender is expected to reliably transmit voice packets at a regular interval (for example, send one frame every 20 ms). These voice packets can be delayed throughout the packet network and not arrive at that same regular interval at the receiving station (for example, they might not be received every 20 ms). The difference between when the packet is expected and when it is actually received is jitter.

To compensate for jitter and conceal interarrival packet delay variation, VoIP endpoints are equipped with jitter buffers. Voice packets in IP networks have highly variable packet interarrival intervals. Recommended practice is to count the number of packets that arrive late and create a ratio of these packets to the number of packets that are successfully processed. You can then use this ratio to adjust the jitter buffer to target a predetermined, allowable late-packet ratio. This adaptation of jitter buffer sizing is effective in compensating for delays.

Note that jitter and total delay are not the same thing, although having plenty of jitter in a packet network can increase the amount of total delay in the network. This is because the more jitter you have, the larger your jitter buffer needs to be to compensate for the unpredictable nature of the packet network. Because more packets are accumulated in the jitter (or dejitter) buffer before play-out, it contributes to the overall delay budget because of larger accumulation time.

Most DSPs do not have infinite jitter buffers to handle excessive network delays. When network delay is predictable and has fewer variations, it is better to just drop packets or have fixed-length buffers instead of creating unwanted delays in the jitter buffers. As long as the packet drop because of jitter buffer overflow is minimal, it might not significantly affect the voice quality compared to excessive delay. If your data network is engineered well and you take the proper precautions, jitter is usually not a major problem, and the jitter buffer does not significantly contribute to the total end-to-end delay. Although in situations where network delay is not constant, especially where network path traverses the Internet, adaptive or variable-length jitter buffers will be beneficial.

RTP timestamps are used within Cisco IOS Software to determine what level of jitter, if any, exists within the network.

The jitter buffer found within Cisco IOS Software is considered a dynamic queue. As voice frames arrive too quickly, the queue fills up. Similarly, when voice frames arrive too slowly, the queue empties out. This way the voice frame play-out rate is constant.

Although many vendors choose to use static jitter buffers, Cisco found that a well-engineered dynamic jitter buffer is the best mechanism to use for packet-based voice networks. Static jitter buffers force the jitter buffer to be either too large or too small, thereby causing the audio quality to suffer, because of either lost packets or excessive delay. The Cisco jitter buffer dynamically increases or decreases based upon the interarrival delay variation of the last few packets. More details on configuring jitter buffer can be found in *Troubleshooting Cisco IP Telephony*, by Paul Giralt, Addis Hallmark, and Anne Smith.

Packet Loss

Packet loss in data networks is both common and expected. Many data protocols use packet loss to detect the condition of the network and can reduce the number of packets they are sending. Data can tolerate a packet being retransmitted. But retransmitting a packet containing voice traffic (RTP) is not an option because the packets might arrive at the destination out of order and with excessive delay. Voice traffic can tolerate a small amount of packet loss as long as the gap in voice is not perceivable to human ear, although excessive packet drop will cause a noticeable gap in voice, making it sound choppy. Therefore, when putting voice on data networks, it is important to use a mechanism like quality of service (QoS) to make the voice traffic somewhat resistant to periodic packet loss by prioritizing it over data traffic, although QoS might not completely prevent packet loss.

Cisco has developed many QoS tools that enable administrators to classify and manage traffic through a data network. If a data network is well engineered, you can keep packet loss to a minimum.

The Cisco VoIP implementation enables the voice-configured router to respond to periodic packet loss. If a voice packet is not received when expected (the expected time is variable), it is assumed to be lost and the last packet received is replayed. If the packet lost is only 20 ms of speech, the average listener does not notice the difference in voice quality.

Voice Activity Detection (VAD)

In normal voice conversations, someone speaks and someone else listens. Today's toll networks contain a bidirectional, 64,000-bps (bits per second) channel, regardless of whether anyone is speaking. This means that in a normal conversation, at least 50 percent of the total bandwidth is wasted. The amount of wasted bandwidth can actually be much higher if you take a statistical sampling of the breaks and pauses in a person's normal speech patterns.

When using VoIP, you can utilize this "wasted" bandwidth for other purposes when VAD is enabled. VAD works by detecting the magnitude of speech in decibels (dB) and deciding when to cut off the voice from being transmitted.

Typically, when the VAD detects a drop-off of speech amplitude, it waits a fixed amount of time before it stops putting speech frames in packets. This fixed amount of time is known as *hangover* and is typically 200 ms.

With any technology, trade-offs are made. VAD experiences certain inherent problems in determining when speech ends and begins and in distinguishing speech from background noise. This means that if you are in a noisy room, VAD is unable to distinguish between speech and background noise. This is also known as the *signal-to-noise threshold*. In these scenarios, VAD disables itself at the beginning of the call.

Another inherent problem with VAD is detecting when speech begins. Typically, the beginning of a sentence is cut off or clipped. This phenomenon is known as *front-end speech clipping*. Usually, the person listening to the speech does not notice front-end speech clipping.

Other Issues

Besides the common problems mentioned previously, other issues can impact voice quality. These issues can include the following:

- **Physical layer impairments:** Noise, interference in the line, loose connectors, badly terminated punch-down block, and so on

- **Last-mile connection bandwidth:** Low-speed connections, oversubscription of circuits resulting in congestion, and so on

- **Network resource overutilization:** High CPU and memory utilization on network devices, oversubscription of IP links resulting in congestion, high number of input/output drops under interfaces, lack of QoS for voice, and so on

- **VoIP application issues:** Poor software implementation on PC-based soft clients, lack of QoS and prioritization of resources for voice, and so on

The impact of these issues on voice quality and their fixes are discussed in detail in Chapter 6, "Managing VoIP Networks," and Chapter 7, "Performance Analysis and Fault Isolation."

However, we now look at some of the common voice quality problems that occur as a result of these issues.

Common Voice Quality Problems in IP Networks

The network problems listed in the previous section can seriously impact the quality of voice in an IP network. Some of the common voice quality issues experienced by providers include the following:

- **Noise:** This is typically any noise on the line introduced by an analog source in addition to the voice signal. Noise will typically leave the conversation intelligible but still far from excellent. Static, hum, crosstalk, and intermittent popping tones are examples where the calling and called parties can understand each other, but with some effort. Some noises are so severe that the voice becomes unintelligible.

- **Voice distortion:** This is typically any problem that affects the voice (RTP/media stream) itself. This category is further divided as follows:

  - **Echoed voice:** Echo voice is where the voice signal is repeated on the line. It can be heard at either end of the call, in varying degrees and with many combinations of delay and loss within the echoed signal.

  - **Garbled voice:** A garbled voice signal is one where the actual character of the voice is altered to a significant degree and often has a fluctuating quality. On some occasions, the voice becomes unintelligible.

  - **Volume distortion:** Volume distortion problems are associated with incorrect volume levels, whether constant or in flux.

Table 1-1 summarizes the different types of voice quality issues, their impact, and root causes.

In a nutshell, the IP network should be architected and configured in a way to transmit voice traffic the fastest way possible as a steady, smooth stream. Delay should be kept under 150 ms in one direction, and average jitter should be below 30 ms. Delay sources should be reduced, and dejitter buffers should be used carefully because compensating for jitter itself can create additional delay. Any out-of-order packets should be dropped because voice does not tolerate delay associated with packet retransmitting.

Quality of service best practices, traffic engineering for capacity planning, and proper network readiness assessment that takes into consideration delay budget planning in the predeployment phases for VoIP address these requirements. Echo- and noise-related issues can be compensated for by proper network wide-gain adjustment and use of echo cancellers. However, this must be done proactively in the design or pilot stages because changing the gain levels at the network boundary, where the IP and TDM network interface, can potentially affect the install base that did not originally have issues. Network transmission loss plan (NTLP) is a key step in VoIP implementation project. The concepts of network readiness assessment, traffic engineering for voice, delay budget planning, and NTLP are discussed in detail in Chapter 6.

Table 1-1   *Voice Quality Issues and Their Causes*

Voice Quality Issue	Symptom	Root Cause
Noise		
Absolute silence	This type of silence between speech can be understood if you have ever had the experience of not knowing whether the other person is still there because there is no sound on the line.	A common cause for this problem is VAD without comfort noise. To experience this symptom, the background noise is usually loud enough for the silence insertion to be noticeable but soft enough so that VAD will be engaged.
Clicking	Clicking is an external sound similar to a knock that is usually inserted at intervals.	A common cause is clock slips or other digital errors on the line.
Crackling	Crackling is an irregular form of very light static, similar to the sound a fire makes.	A common cause is poor electrical connections, in particular poor cable connections. Other causes are electrical interference and a defective power supply on the phone.
Crosstalk	Crosstalk is a familiar concept where you can hear someone else's conversation on the line. Commonly the other parties cannot hear you. There are also forms of crosstalk where all parties can hear each other.	Wires in close proximity, where the signal of one is induced into the other, is a common cause of this problem.
Hissing	Hissing is more driven and constant than static. White noise is a term often associated with strong hissing. Pink noise is a less constant hissing noise, and brown noise is even less constant.	A common cause of hissing is VAD. When VAD kicks in, comfort noise packets are introduced into the audio stream. The hissing sound is caused by the introduction of comfort noise into the conversation.
Static	Severe static is an example of static that in addition to creating background noise, affects the dial and ring tones and the voice itself. Another name for this symptom might be scratchy or gravel voice.	A common cause is A-law/Mu-law codec mismatch. A-law is a codec companding scheme used outside of the United States, whereas Mu-law is a U.S.-specific codec companding scheme. This is typically involved in international calls originating or terminating in the United States.

continues

Table 1-1   *Voice Quality Issues and Their Causes*   *(continued)*

Voice Quality Issue	Symptom	Root Cause
Echoed Voice		
Listener echo	Listener and talker echo sound similar, although the signal strength of listener echo might be lower. The essential difference between them is who hears the echo and where it is produced. Listener echo is the component of the talker echo that leaks through the near-end hybrid and returns again to the listener, causing a delayed softer echo. The listener hears the talker twice.	Common causes are Insufficient loss of the echo signal. The reduction in the echo level produced by the tail circuit without the use of an echo canceler is referred to as Echo Return Loss (ERL). So if a speech signal enters the tail circuit from the network at a level of X dB, the echo coming back from the tail circuit into the terminal of the echo canceller is (X − ERL). Long echo tail. Echo cancellers in the gateway adjacent to the near-end hybrid circuit not activating.
Talker echo	Talker echo is the signal that leaks in the far-end hybrid and returns to the sender (talker). The talker hears an echo of his own voice.	Common causes are Insufficient loss of the echo signal. Echo cancellers in the gateway adjacent to the far-end hybrid not activating. Acoustic echo caused by the listener's phone.
Tunnel voice	Tunnel voice sounds similar to talking in a tunnel or on a poor-quality mobile phone car kit.	A common cause is tight echo with some loss. For example, 10-ms delay and 50 percent loss on the echo signal.
Garbled Voice		
Choppy voice	Choppy voice describes the sound when there are gaps in the voice. Syllables appear to be dropped or badly delayed in a start-and-stop fashion. Note: Other terms used to describe this sound are *clipped voice* and *broken voice*.	Common causes are consecutive packets being lost or excessively delayed such that DSP predictive insertion cannot be used and silence is inserted instead, for example, delay inserted into a call through contention caused by a large data packet.

Table 1-1 *Voice Quality Issues and Their Causes*

Voice Quality Issue	Symptom	Root Cause
Synthetic (robotic) voice	The term *synthetic* means that the sound of the voice is artificial and with a quiver. Predictive insertion causes this synthetic sound by replacing the sound lost when a packet is dropped with a best guess from a previous sample. Synthetic voice and choppy voice commonly occur together.	A common cause is single packet loss or delay beyond the bounds of the dejitter buffer playout period. DSP predictive insertion causes the synthetic quality of the voice, for example, when a call was provided insufficient bandwidth (such as a G711 codec across 64 kbps).
Underwater voice	Unintelligible underwater voice describes a distortion that makes it impossible to understand the voice. Descriptions of this sound include the sound of a cassette tape being fast forwarded, a gulping sound, and a wishy-washy sound.	A common cause of this problem is a G729 IETF and pre-IETF codec mismatch.
Volume Distortion		
Fuzzy voice	Fuzzy voice sounds similar to the radio being turned up too loud and the voice is shaky. This can only occur at certain signal levels within the sentence depending on the level of gain applied.	This is often caused by too much gain on the signal, possibly introduced at one of a number of points in the network. It can also apply to IP phones when used in noisy environments or when the volume is set to the high end.
Muffled voice	Muffled voice sounds similar to speaking with your hand over your mouth.	A common cause is an overdriven signal or some other cause that eliminates or reduces the signal level at frequencies inside the key range for voice (between 440 and 3500 Hz).
Soft voice	Soft voice is like a low voice that is hard to hear.	Soft voice is usually caused by too much attenuation on the signal, possibly introduced at one of a number of points in the network such as a voice gateway when trying to reduce echo.
Tinny voice	Tinny voice is similar to listening to an old-fashioned wireless broadcast.	A common cause is an overdriven signal or some other cause that eliminates or reduces the signal level at frequencies outside the key range for voice (less than 440 Hz and greater than 3500 Hz) but important to the richness of the voice.

Before we get into the discussion of network management concepts and how they apply specifically to VoIP, we discuss the strategic value of VoIP in providing business-critical services to customers and hence the importance of managing VoIP networks.

Strategic Importance of VoIP and Management

Businesses and consumers have been able to find applications for the communication network and turn them into critical services. This phenomenon has been in effect since the invention of the telegraph in the mid-1800s, when stock and commodity traders used it for obtaining stock prices, and it became the de facto stock ticker for Wall Street. Telephone systems have been around for over 100 years now. You might lose power but the phone still works. Users have gotten to the point where they expect the phone system to always be working. The public telephony system has supported life support services and is the backbone of how businesses conduct day-to-day activities. Therefore, any interruption in services provided by a communications network can have dire consequences.

VoIP technology is no different when it comes to the business criticality of the services it provides. Initially, cost cutting might be the deciding factor for service providers in making the initial investment by leveraging their data networks. The cost savings aspect is rooted in this fundamental technology difference: PSTN networks use the SS7 protocol that runs on a dedicated signaling network TDM to set up a call path or circuit. This circuit requires 64 kbps of network bandwidth for a single voice channel throughout the network. Packet telephony or VoIP uses the network bandwidth more efficiently by using statistical analysis to multiplex voice traffic, including both call signaling and the voice stream, alongside data traffic and sharing it across multiple logical connections. This reduces the overall bandwidth requirement, especially if it is combined with compression techniques that are possible only with packet switching. In addition to optimized bandwidth utilization, service providers can relay the toll savings to their consumers by making their services more competitive by leveraging the Internet instead of paying for interconnection charges.

Similarly, enterprises are also interested in leveraging their intranet, which spans all their corporate footprint, especially if it has excess capacity, to save on long-distance toll charges. As smartphones penetrate the cellular phone industry with technologies such as Edge, 3G, and PDSN to access the Internet, the consumers of these devices have began experimenting with VoIP by leveraging clients such as Skype regardless of bandwidth constraints and the best effort for packet delivery nature of Internet. This has made mobile service providers seriously consider offering VoIP as they make the transition to 4G networks with more efficient IP transport to prevent (or compensate for) revenue erosion because of the diminishing adoption of traditional voice over the public land and mobile network (PLMN).

VoIP technology in general and Unified Communications in particular allow businesses to push additional applications to their employees, partners, and customers in their ecosystem to increase productivity by coupling them with their business processes to provide a comprehensive business-centric collaboration platform. This includes using unified messaging that makes the delivery of voice and voicemail through any IP-capable medium, including email clients, web browsers, or smartphones. Other advanced features include coupling IP phones with web services, tracking users' present status and location, integrated information systems, the ability to initiate a call on demand from a website, mobility over IP network, Single Number Reach (SNR) for greater access, and location awareness.

A few of these examples include a collaboration-enabled radiology system in a hospital that allows a radiologist to contact the referring physician from his or her workstation used to read the radiology images without the need to look up the contact information using the preferred way of communications. The radiologist can share images and other diagnostic details related to a particular study. This saves time, which increases productivity and additionally provides compliance with health-care standards because the entire transaction can be coordinated and recorded over an IP infrastructure for auditing purposes. Similarly, a nurse in a hospital setup can scan a list of available doctors when confronted with an urgent medical need, and with the click of a mouse, speak instantly with the most appropriate specialist, wherever he or she is located. If a VoIP call control system is integrated with hospital databases and monitoring systems, it can provide a real-time view of the patient's history and vital statistics during the call, expedite test orders, and keep track of prescriptions by converting system messages into alerts and notifications and immediately delivering them to the medical staff. The VoIP system can also tie into external systems such as medical insurance providers to make the entire care provider system more efficient.

Some collaboration-enabled business processes are applicable in other market segments, such as finance, manufacturing, education, and government, when a VoIP system is integrated with other productivity applications and business systems process management systems, including SAP, Oracle, PeopleSoft, SalesForce.com, and other Business Process Management (BPM) platforms.

Because IP packets carrying voice or any other payload can be rerouted, copied, stored, originated from different IP-capable interfaces such as a telephone, instant messaging, email, and web presence, it is changing the way customer relationship management (CRM) systems, including contact centers, are modeled. VoIP provides agility when designing CRM systems by providing contact center agents with consolidated and up-to-date contextual customer records along with a customer's preferred medium of communication. Contact center software typically uses either time or skill-level selection. Time selection is based upon agent free time. Skill-level selection is what is mentioned here. In the past, time selection was mainly used. Skill selection requires more intelligent software. Also, the contact center agents can be located anywhere in the world, which allows an organization to have access to different pools of expertise anytime from anywhere. This flexibility is cheaper over IP networks and allows easier expansion as business needs grow over the longer period and can bring agents online on demand on short notice.

Because VoIP provides more than a mechanism for voice transport and allows organizations to develop their critical business processes, its strategic value is significant. We have to keep in mind that VoIP networks are now required to provide the same level of service as the PSTN, including emergency services, and comply with the same regulations that were once meant for the PSTN only. This makes the management of a VoIP network a business-critical function. The next section discusses various management methodologies that can be used to devise a comprehensive plan to manage the entire life cycle of a VoIP network.

Network Management Methodologies

Because of the critical nature of a communications network, as discussed earlier, the Consultative Committee for International Telegraph and Telephone (CCIT), dating back to 1865, has provided guidance and structure for network management. Its focus was primarily telecommunications network management for stability and facilitating interoperability between national networks to enable global communication. The International Telecommunications Union (ITU-T), which was created in March 1993 and replacing CCIT, developed one of the first modern network management methodologies, including the Telecommunications Management Network (TMN), which was followed by a more comprehensive methodology that defines key management areas covering Fault, Configuration, Accounting, Performance, and Security (FCAPS). The United Kingdom's Central Computer and Telecommunications Agency (CCTA) created the Information Technology Infrastructure Library (ITIL), focusing on the service delivery and support methodologies. The TeleManagement Forum (TMF) defines a framework called the enhanced Telecommunications Operations Map (eTOM) to help its members reduce the costs and manage risks associated with creating and delivering services profitably.

Telecommunications Management Network

The International Telecommunications Union (ITU-T) introduced recommendation M.3010 in May 1996. This delivered the concept of the Telecommunications Management Network (TMN). Recommendation M.3010 provided a framework for service providers to manage their service delivery network. This framework defined four management architectures at different levels of abstraction: functional, physical, informational, and logical layers. The TMN-prescribed framework also provided a common methodology and logic that was applicable to the management of private corporate-owned IT networks. Figure 1-4 includes the four layers of abstraction defined in M.3010.

The Business Management Layer (BML) and the Service Management Layer (SML) provide a relationship between the IT and the business of the corporation. The Network Management Layer (NML) deals with fault and performance data for the network. The Element Management Layer deals with configuration management, fault, and performance at the device level.

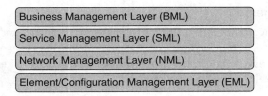

Figure 1-4 *TMN Logical Layers*

FCAPS Model

The ITU-T introduced another recommendation, M.34000, in April 1997, that further defined the general management functionality. Recommendation M.3400 breaks the management system into five functional key areas of FCAPS. The FCAPS model does not have a particular focus on the business-related role of a management system within the telecommunications network. However, this functional model does provide informational elements that can help in the business aspects of the telecommunications network, such as analyzing accounting and performance data to extract meaningful information to make business decisions about the network's capacity for new service offerings. This concept is discussed in detail in Chapter 8, "Trend Analysis and Optimization." The following sections describe the five functional areas in the FCAPS model.

Fault Management

Fault management is about recognizing the problem through continuously monitoring the entire network, correlating the fault data, and isolating the problem to the source. Fault management involves the entire life cycle, including fault detection, handling of alarms, fault isolation by the filtration and correlation process, fault correction for network recovery, tracking the incident by error logging, and managing the entire workflow through a trouble ticketing system.

Configuration Management

Configuration management deals with managing the configuration change control process, including commissioning and decommissioning of network devices, backing up and restoring the methodology for configurations, and overall workflow management for the administrators performing the configuration changes.

Accounting Management

Accounting management covers methods to track usage statistics and costs associated with time and services provided with devices and other network resources. Some aspects of accounting management overlap with fault management to provide comprehensive data to validate service-level agreement compliance.

Performance Management

This area covers the network management system's (NMS) capability to track system statistics to help identify network trends. In a sense, this provides feedback to the fault management layer to establish thresholds for determining the network and device-level faults proactively. It also provides data for capacity planning for network growth and new service offerings (by determining network readiness).

Security Management

Security management addresses access rights that include authentication and authorization, data privacy, and auditing security violations.

Figure 1-5 includes a list of the FCAPS functional elements.

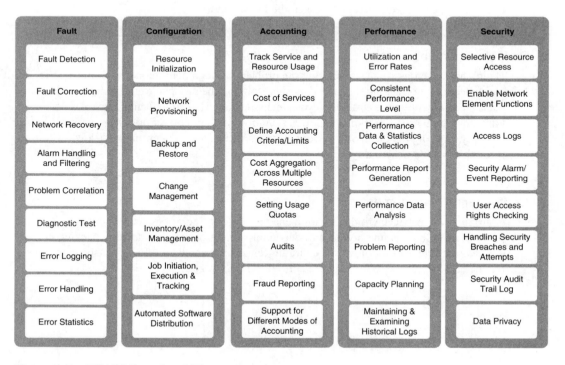

Fault	Configuration	Accounting	Performance	Security
Fault Detection	Resource Initialization	Track Service and Resource Usage	Utilization and Error Rates	Selective Resource Access
Fault Correction	Network Provisioning	Cost of Services	Consistent Performance Level	Enable Network Element Functions
Network Recovery	Backup and Restore	Define Accounting Criteria/Limits	Performance Data & Statistics Collection	Access Logs
Alarm Handling and Filtering	Change Management	Cost Aggregation Across Multiple Resources	Performance Report Generation	Security Alarm/ Event Reporting
Problem Correlation	Inventory/Asset Management	Setting Usage Quotas	Performance Data Analysis	User Access Rights Checking
Diagnostic Test	Job Initiation, Execution & Tracking	Audits	Problem Reporting	Handling Security Breaches and Attempts
Error Logging	Automated Software Distribution	Fraud Reporting	Capacity Planning	Security Audit Trail Log
Error Handling		Support for Different Modes of Accounting	Maintaining & Examining Historical Logs	Data Privacy
Error Statistics				

Figure 1-5 *FCAPS Functional Elements*

Information Technology Infrastructure Library (ITIL)

ITIL is a set of best practices for IT service management. It has become a worldwide de facto standard in service management since the late 1990s. ITIL provides network administrators and CIOs a customizable framework for best practice to achieve quality service and overcome difficulties associated with the growth of networks and the services it offers to its subscribers. Because of its customizable nature, ITIL standards are being adopted by organizations both big and small to improve their business processes related to IT.

These are the five fundamental processes defined by ITIL version 3 that cover the entire life cycle of an IT project, from starting its architectural planning, spanning through the design and implementation phases, and covering the operational phases to a continued loop with the services optimization.

Service Strategy

Service strategy provides guidance to all IT service providers about the following:

- What services should be offered
- Who the services should be offered to
- How the customer and stakeholders perceive and measure the value, and how this value is created
- How to evaluate and leverage partners for partial or complete sourcing of the services
- How visibility and control over value creation are achieved through financial management
- How the allocation of available resources is tuned to optimal effect across the portfolio of services
- How service performance is measured

The financial management aspect of service strategy covers the function and processes responsible for managing a provider's budgeting, accounting, and charging requirements. It provides IT with the quantification in financial terms about the value of network services and the infrastructure upon which they are delivered and the qualification of operational forecasting.

Service portfolio management is a continuous and proactive process that deals with defining services for the planning/concept phase, design, and transition pipeline, and maintaining them through these phases.

Demand management is targeted to understanding and influencing customer demand for services and the provision of capacity to meet these demands.

Service Design

Service design starts with a set of business requirements and ends with the development of a service solution designed to meet documented business requirements and outcomes for handover into service transition. It covers the following management aspects:

- Service Catalogue Management provides a single consistent source of information on all the agreed services and ensures that it is available to all the authorized users.

- The Service Level Management (SLM) process is intended to ensure that all operations services and their performances are measured in a consistent, professional manner throughout the IT organization, and that the services and the reports produced meet the needs of the business and customers. This includes service-level agreements (SLA), operational-level agreements (OLA), and the production of the Service Improvement Plan and the Service Quality Plan.

- Capacity Management includes business, service, and capacity management across the service life cycle. It is therefore important to consider capacity management at the onset of the design stage.

- Availability Management deals with availability-related issues pertaining to services, components, and resources to ensure that the availability targets in all areas are measured and achieved in a cost-effective manner.

- IT Service Continuity Management is targeted toward maintaining the appropriate ongoing recovery capability within IT services to match the agreed needs, requirements, and time scales of the business. It also ensures that all the activities are aligned with business continuity plans and business priorities.

- Information Security Management is part of the overall corporate governance framework. Its purpose is to ensure that complete information is available when required to the authorized personnel with focus on its authenticity and nonrepudiation.

- The Supplier Management process ensures that suppliers and the services they provide are managed to support IT service targets and business expectations while conforming to all the terms and conditions of their contracts and agreements.

Figure 1-6 illustrates the ITIL Services and Support delivery components. Service Design, Service Transition, and Service Operation are cyclic processes based on Service Strategy. Each of these phases provides feedback to the next step for continuous improvement, as shown by the curved arrows in the diagram.

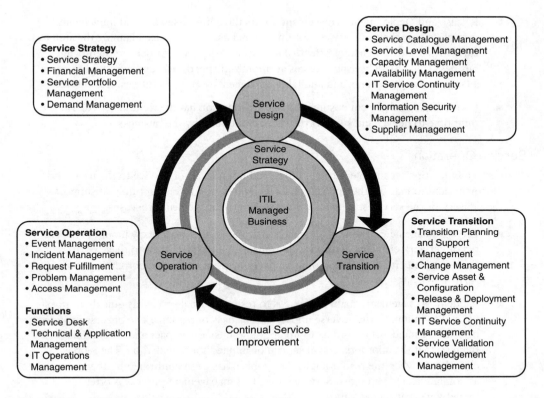

Figure 1-6   *ITIL Service and Support Delivery Components*

Service Transition

The main role of Service Transition is to deliver services that are required by the business into operational use. This is achieved through the following processes and activities:

■   Transition Planning and Support is geared toward planning and coordination of resources to ensure that the requirements of Service Strategy as encoded in Service Design are effectively realized in Service Operations. Transition Planning is done by identifying, managing, and controlling the risks of failure and disruption across transition activities.

■   Change Management ensures that changes are planned, prioritized, evaluated, tested, authorized, and implemented while being recorded and documented in a controlled manner.

■   Service Asset and Configuration Management lay a framework for providing accurate information and control across all assets and relationships that go up an organization's infrastructure. Its scope can also be extended to non-IT assets and to internal and external service providers where shared assets need to be controlled. Cloud computing, hosted services, and software as a service (SaaS) are some of the examples for external or hybrid resources.

- Release and Deployment management covers the entire assembly and implementation of new or changed services for operational use from release planning through to early life support. It ensures effective release, and deployment delivers significant business value by delivering changes at optimized speed, risk, and cost, and offering a consistent, appropriate, and auditable implementation of useful business services.

- Knowledge Management ensures that the right person has the right knowledge at the right time to deliver and support the services required by the business.

Service Operation

Service Operation's main role is to deliver agreed-upon levels of service to the users and customers and to manage the applications, technology, and infrastructure that support the delivery of the services. It includes the following management processes:

- Event Management generates, detects notifications, and proactively monitors the status of network elements. It also includes managing the responses to these notifications received either reactively or proactively by prioritizing and correlating them for further analysis.

- Incident Management's main purpose is to restore normal service as quickly as possible and to minimize the adverse impact on business operations. The process includes prioritizing incidents according to urgency and business impact and categorizing them so that the appropriate skill set can be applied for remediation. The loop is closed only when the resolution to the problem has been confirmed by the end user as communicated through a Service Desk. This management process is often aided by network management tools.

- The Request Fulfillment process enables users to request and receive standard services, to source and deliver these services, to provide information to users and customers about services and procedures for using them, and to assist with general information, complaints, and comments.

- The Problem Management process is geared toward preventing problems and resulting incidents from happening, to eliminate recurring incidents, and to minimize the impact of incidents that are inevitable. This is done by diagnosing causes of incidents, determining the resolution, and ensuring that the resolution is implemented and verified.

- The Access Management process manages the rights and permissions for users to be able to access a service or group of services. It also tracks access and role changes for auditing purposes. In that sense, access management ensures the confidentiality, availability, and integrity of data and intellectual property.

Service Operations relies on key functions, including Service Desk, which is a central point of contact for users of services; Technical and Application Management, including the people's technical expertise and management of the network infrastructure and the applications running on it; and IT Operations Management, which is staffed by operators carrying routine operational tasks.

Continual Service Improvement

Continual Service Improvement is considered a feedback loop that provides a way for an organization to identify and manage appropriate improvements by contrasting its current position and the value it is providing to the business with its long-term goals and objectives, and identifying any gaps that exist. This is a continuous process to address changes in business requirements, to refresh technology cycles, and to ensure that high quality is maintained. Continual Service is shown in Figure 1-6 by arrows representing the feedback between service phases transition.

Enhanced Telecom Operations Map (eTOM)

Enhanced Telecom Operations Map, or eTOM, defines a business process framework that is part of TeleManagement Forum's Next Generation Operations Systems and Software (NGOSS). It is the most commonly adopted framework among telecommunications service providers. The eTOM framework defines three major process areas that are fundamentally technology and industry segment agnostic. Each area has four levels of processes defined, with each level covering specific process details. These three process areas include the following:

■ **Strategy, Infrastructure and Product:** This area covers services offering a strategy and commit process and the entire infrastructure and product life cycle management. The four hierarchal levels in order include marketing and offer management, service development and management, resource development and management, and supply chain development and management.

■ **Operations:** This area addresses operations support and readiness and fulfillment, assurance, and billing (FAB). The corresponding four levels include customer relationship management, service management and operations, resource management, and partner relationship management.

■ **Enterprise Management:** This is an overlying process area that covers strategic planning, financial assets manangement, brand management and associated marketing activities, human resource management, reasearch and development, stakeholder and external relations management, disaster recovery, security and fraud management, and quality management.

Note   Because service providers offer products and services to enterprises relevant to VoIP and Unified Communications, including infrastructure such as Metro-Ethernet, MPLS, and other more comprehensive services such as UC as a Service (UCaaS), there was a growing need to map eTOM to ITIL. ITIL is widely adopted by enterprises. This mapping was targeted to better align SP's IT service management framework with its enterprise customers. In 2004, the TM forum formulated a team to focus on mapping eTOM to ITIL. The TM Forum's technical report is referred as "An Interpreter's Guide for eTOM and ITIL Practitioners." It is available at http://www.tmforum.org/community/groups/the_business_process_framework.

Comprehensive Network Management Methodology

The FCAPS model provides essential foresight and knowledge to optimize network performance through fault, performance, and configuration management by focusing on technology management aspects. It also addresses security, which is a major concern to network operators with respect to theft of service and data protection. ITIL provides the methodology and discipline to execute essentially all aspects of service management encompassing the network used for delivery of these services. The management framework and functions, as described earlier in the "Service Operation" section of the ITIL v3 foundation discussion in this chapter, overlap with specific steps laid out in the FCAPS framework. However, the ITIL methodology addresses certain tasks such as configuration under its Service Transition process definitions. TMN also possesses similarities with ITIL and FCAPS, where its definition of Business Management and Service Management layers match ITIL's Service Strategy and Service Design definitions. The Fault and Performance and Element/Configuration Management layer covers the operational aspect of the service delivery network in the same manner as the Service Transition and Service Operation definition in ITIL and all the aspects of FCAPS. However, ITIL covers the entire life cycle of the service delivery network starting from planning, designing, implementation/commissioning, operation, and optimization phases.

There has been an increase in services in the telecom world because of competition and technological maturity that has brought us services such as VoIP and Collaboration services, including web conferencing, video, presence, and single-number reach, to name a few. This situation becomes more complex with new service models such as cloud computing and hosted or remotely managed services. IT has to continuously evolve with the introduction of new service businesses by being more agile to support and grow them. At the same time, typically there is less focus on business-to-IT alignment. Sometimes there is a tendency to measure productivity, profit, and loss for individual departments in an organization. This is because of a lack of a service-centric approach, which often ignores the fact that these services are being delivered on the converged network. An absence of services-centric views increases the time to adapt to new business models, market needs and makes it difficult to efficiently manage them. This fragmented approach also affects budgeting decisions that are made in silos and might not optimally realize the potential of the converged network. The management processes defined under the ITIL v3 Foundation provide a comprehensive framework to bridge the gap between business processes and IT operations.

The ITIL methodology is flexible enough to build a corporate-specific framework to correlate discrete data collection points illustrated by FCAPS and present them in service-centric dashboards that are relevant to the core business of the corporation. For example, it can provide IT decision makers with a view of the collaboration application on the network, concentrating on their availability, adoption, and capacity to grow in a business context that shows the savings as a result of business transaction efficiencies achieved by using these tools and applications. Chapter 8 introduces the concept of dashboards that provide an overall network view by focusing on key performance indicators in the context of the services offered by the network.

Any comprehensive solution must begin at the planning and design stage. When a problem is identified after introducing a new technology by following ITIL management methodology and FCAPS-based metrics, it might not be easy to remediate it. A proactive planning and design methodology executed with foresight ensures that the network is fundamentally ready for the introduction of new technologies such as voice and video on converged networks. This entails performing network readiness assessment, traffic engineering (that might be covered through capacity management as prescribed in ITIL), network transmission loss plan establishment, a policy for quality of service classes, and corporate operational readiness assessment. We cover these proactive aspects related to overall network performance and specific to VoIP in Chapter 6.

Because of the comprehensive nature of ITIL, the customization of its various processes make the ITIL compliance a rather long project, especially for large service providers and corporations. Of course, the ITIL methodology adoption is easier for greenfield deployments, but the existing service delivery networks should employ the fundamental operations strategy as laid out in FCAPS.

This concept, showing a relationship among FCAPS, ITIL, TMN, Planning, and Design methodology, is shown in Figure 1-7.

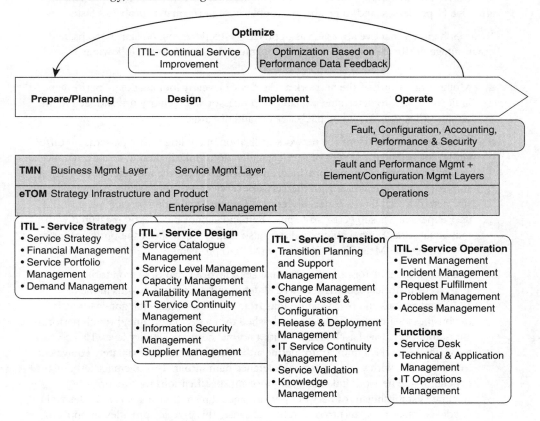

Figure 1-7 *Comprehensive Network Management Methodology*

Focusing on Performance Metrics

Since the advent of the service delivery networks, network operators have focused on managing availability. This includes proactive checks, including SNMP polling and reactive notifications through alarms and traps; for example, if the router is up, the link is down or congested, or the server needs to be restarted. However, these days, the service provider and enterprise networks have been able to achieve 99.9 percent uptime. The reasons include technology improvements, regulatory compliance with stricter penalties, and considerably heavier reliance on the networks where most of the enterprise applications are networks that motivate network managers to put more redundancy and other resiliency measures in the network.

A Yankee Group report titled "Performance is the New Mandate for Network Management" refers to a study on enterprise application management. This study found that enterprises report an average productivity drop of 14 percent when experiencing performance degradations in Oracle, SAP, PeopleSoft, Siebel, and custom .NET and J2EE applications. Ashton, Metzler & Associates conducted a survey of 176 IT professionals that indicated that over a quarter of network operations centers (NOC) do not meet their organizations' current needs. To fulfill current and emerging requirements, NOCs are being driven to do a better job of managing application performance, to implement more effective IT processes, and to be able to troubleshoot performance problems faster.

To adopt a more proactive approach as opposed to solely relying on alarms, traps, and other outage notifications, network operators must be able to do the following:

- Monitor the health of the network at the device level by looking at key performance indicators (KPI) to determine availability, processor and memory utilization, and interface utilization, including packet errors and discards.

- Maintain the visibility of the network utilization, including traffic patterns on different layers, segments, classes of service, and various links including delay, jitter and packet loss, application performance, and when users are the most active in relationship with their interaction with the network services and application. A comprehensive network visibility is especially critical for voice and video service because the user experience is subjective to large extent and cannot be directly measured with individual KPIs. Having contextual data also helps troubleshoot the problems faster and reduces the Mean Time To Resolve (MTTR) problems.

- Establish a baseline for normal performance. It involves real-time monitoring and analysis of historical records such as call detail and diagnostic records in the case of VoIP applications and performance logs from servers running vital applications for the business. After the baseline is established over a period of time through performance monitoring, the baseline represents a normal network activity level. This characterization helps the administrator define a threshold that is approximately equivalent to the baseline with some additional tolerance built into it. This tolerance level might depend on the network link and/or component specifications for safe operation. Accurate establishment of network performance thresholds can also aid in determining deviation as they start to occur. In that sense, this practice provides an opportunity to mitigate them before they cause service interruption. Network baseline also provides guidance in determining when and where there is inappropriate and wasteful use of network resources. Also, it helps network engineers understand where there is the potential for cost savings and efficiencies on the network. These concepts are discussed in more detail in Chapters 7 and 8.

To accomplish these three key tasks, the network operator needs a framework and a methodology to determine what is practical to monitor, where to monitor, how often to monitor (to avoid adverse effects on the monitored devices), and what information needs to be correlated to provide meaningful information. There is a famous network management idiom: "You cannot manage what you don't measure, and you cannot measure what you don't collect." Performance metrics are core to an FCAPS-based management scheme. The Fault and Performance Management layer in TMN, Service Operations and its functions in ITIL, and Accounting and Performance in FCAPS revolve around KPIs or key performance metrics (KPM). In that sense, performance metrics have a pivotal role in overall network management in the operational stage as well as in network optimization and growth planning.

The main focus of this book is to define and illustrate an approach of collecting, analyzing, and trending data using performance counters. This approach is based on defining and measuring KPIs at different layers (the physical layer, data link layer, IP layer, and application layer) from various devices in the VoIP network and using this data to gauge the health of the VoIP network using a consolidated dashboard view. Figure 1-8 shows elements of network management methodology that contribute to KPIs for performance measurement and proactive fault detention.

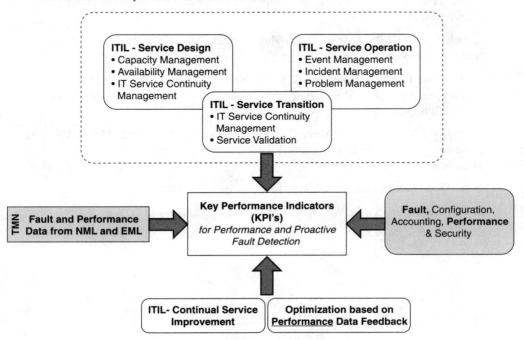

Figure 1-8 *Role of KPIs in Network Management Methodology*

Summary

Some of the voice quality problems such as noise, echoed voice, and volume distortion are typically related to planning, design, and implementation problems in the VoIP network. The root cause of these issues can vary from physical impairments to codec issues to problems with echo cancellation. The VoIP provider needs to take proper steps in planning and designing the network to avoid these issues. Other issues such as garbled voice and related problems are primarily caused by network problems such as delay, jitter, and packet loss. We discussed the key network management standards, including TMN, FCAPS, ITIL, and eTOM, that define a framework for comprehensive network management. The strategic nature of a communications network in general and VoIP in particular demands a comprehensive network management strategy that covers the application and the underlying transport network.

The next chapter discusses a comprehensive network management strategy for VoIP networks and describes how this approach helps service providers/enterprises in performance management and optimization of the VoIP network. The performance management and optimization strategies are discussed in detail in Chapters 7 and 8.

Reference

1. Biswas, Suparno. *From FCAPS to ITIL: An Optimized Migration*. February 21, 2005.

2. International Telecommunications Union website for information about ITU-T. http://www.itu.int/ITU-T.

3. Parker, Jeff. FCAPS, TMN & ITIL—Three Key Ingredients to Effective IT Management. May 6, 2005.

4. Davidson, Jonathan, James Peters, Manoj Bhatia, Satish Kalidindi, and Sudipto Mukherjee. *Voice over IP Fundamentals*, Second Edition. Indianapolis, IN. Cisco Press, 2006.

5. Siddiqui, Talal. "Managing Voice Quality in Converged IP Networks," BRKVVT-2301 presentation. Cisco Systems Inc., July 2007. The TeleManagement Forum. http://www.tmforum.org.

6. NetQoS. "Performance First." http://www.netqos.com, 2009.

7. Werbach, Kevin. "Using VoIP to Compete." *Harvard Business Review*, Vol. 83, No. 9, September 2005.

## Chapter 2

A Metrics-Based Approach for Managing the VoIP Network

This chapter describes a methodology for managing the Voice over IP (VoIP) network through a key performance metrics-based approach. It builds on the overview given in Chapter 1, "Voice over IP (VoIP) and Network Management," and describes how this metrics-based methodology can be used to go beyond the day-to-day network element management of a Fault Configuration Accounting Performance Security (FCAPS) model. The methodology helps isolate key trends in the network for proactive and preemptive network management. We use the terms *key metrics* and *key performance indicators (KPI)* interchangeably throughout the book.

After a VoIP network gets deployed, it needs to be managed and optimized. This phase is also called Day 2. In this phase, the typical focus is on alarm and trouble ticket management. As mentioned earlier, the KPI-based approach enables the network operator to better manage the network because of the availability of trending data. The approach described in this chapter is deployed at the Manager of Manager or the Operations Support Systems (OSS) layer.

Systemic issues are hard to isolate and track across the VoIP network. The localization of fault also becomes a challenge because the VoIP network is composed of a large number of technologies and vendor products integrated together, and the overall loss in downtime as a result of rectifying these issues is significant. These issues are presented in this chapter along with an approach that can help to identify them. Specific metrics from various VoIP infrastructure segments are used to derive this methodology.

A VoIP network is composed of the following key functional groupings: an Internet infrastructure component that includes all call-related signaling and handling, a customer infrastructure, a public switched telephone network (PSTN) component, and a Session Border Controller (SBC) termination of the VoIP network through Session Initiation Protocol (SIP) trunking. The typical voicemail and announcement are also covered. We cover customer or VoIP last-mile infrastructure use cases in Chapter 3, "VoIP Deployment Models in Service Provider Networks," and Chapter 4, "Internet Telephony."

The KPI-based methodology is applied to managing these functional components. This chapter presents a service-level agreement (SLA) management strategy as a use case for this methodology.

VoIP Networks Require a Layered Management Approach

A VoIP network is a large and complicated solution that encompasses many integrated technologies. It presents a problem for VoIP infrastructure managers because each technology brings its own network management challenges. As a result, the network management focus gets divided into specific areas within the network, depending on where the problem is on a particular day. A set of dashboards must be built on the KPIs that track the VoIP functional segments. This enables the network operator to develop a layered approach of KPIs across the various vendor and solution boundaries. The layered KPI-based functional grouping helps isolate the problems similar to peeling back the layers of an onion. The VoIP functionality in essence can be tracked through service flows.

A service flow is composed of a set of key metrics that can be tracked for identifying the flow. An example of the service flow is a voicemail functional segment. The voicemail segment includes all the signaling and media-related metrics that determine the voicemail functionality. We cover the service flow–related concepts at length in Chapter 7, "Performance Analysis and Fault Isolation." The basic idea with this approach is to allow the network operator to develop a layered methodology to troubleshoot, perform proactive monitoring, and do capacity planning across the various vendor and functional solution boundaries. In essence, they start at the topmost layers of KPIs, which represent an aggregated view of a particular VoIP segment. They further have the capability to drill down to a specific segment component KPI layer.

An organized layered approach needs to be developed to effectively manage the VoIP network. This can be achieved by focusing on the performance management aspect of the network, and then utilizing the KPI to isolate the systemic, localized, and subjective issues. The network operator thus can develop a holistic management view of the network and work toward reducing the downtime and impact to the network.

Figure 2-1 depicts the layered approach at a high level. This is a simplified representation of a typical management infrastructure that a service provider or even the enterprise provider ends up deploying in its network. The system view can also be considered a manager of manager layer that basically represents a topmost aggregated view. The system view can cover various servicing market areas or, for an enterprise environment, various functional data centers or campuses.

The systemwide view, jointly with the Network Management System (NMS), Element Management System (EMS), and Network Element (NE) management, gives an operator the ability to take a comprehensive approach in handling the entire VoIP network operations. The system dashboards represent KPI data collected from all over the VoIP network. Figure 2-1 also depicts the NEs managed by EMS and NMS. The bulk of the provisioning software sits at the NMS layer, and its scope is across the EMS layer.

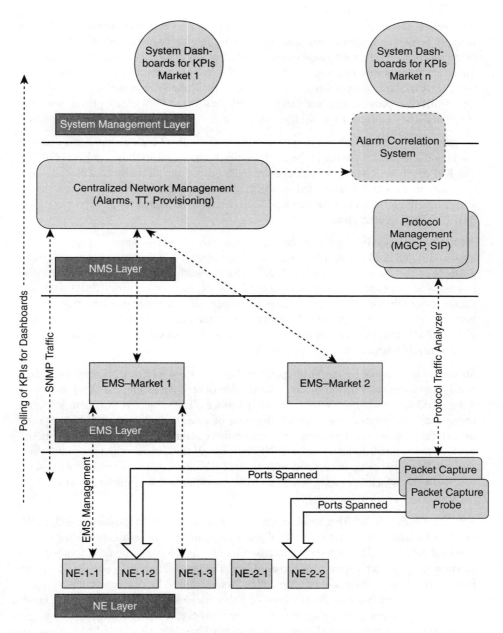

Figure 2-1   *Layered Approach to Network Management*

The NMS layer is also primarily used in handling fault notification (alarms) and trouble ticket management across VoIP segments or markets. EMSs are mostly used for specific VoIP functional components that can be composed of numerous NEs. Protocol analyzers are also a key here and are usually deployed at the NMS layer, where they can collect information and help in analyzing data collected from packet probes deployed in the network. In some cases, alarm correlation systems are also deployed at the NMS layer; they help in identifying the root cause of the problem by continuously analyzing the alarms. These are complicated rule engine-based systems. In VoIP networks, Call Management Switches (CMS) play a centralized role in the network. These systems perform the entire dial plan, subscriber provisioning, and other class IV, class V, and PBX-related provisioning and switching. They fall under the NE layer and are a key instrument in generating the KPIs.

We now further define the top-down layered approach. It is based on managing and tracking the network from the scope of the entire VoIP system, as illustrated in Figure 2-1. The KPI dashboards at the system level are fed metrics from all the system components, and these metrics are consolidated in VoIP segment functional views. In essence, the network element metrics are collapsed into VoIP network functional building blocks that provide an aggregated view.

An issue can trickle up the functional layer stack, but is easily isolated to a specific or localized segment of the network because of the layered KPI dashboard approach. An example of this issue can be an individual malfunctioned Media Terminal Adapter (MTA), causing a Media Gateway Control Protocol (MGCP) restart in progress (RSIP) storm, thus affecting the overall call processing. If it's more of a systemic problem, that also becomes apparent through the KPI dashboards. In the case of an isolated issue, the specialized EMS, along with other diagnostic tools, such as packet capture devices, can be used to troubleshoot and rectify the problem.

Subjective issues seen by the VoIP operator also play a key role, and on numerous occasions, they are usually driven by end-user trouble tickets. An example of a subjective issue is bad call quality or even loss of voice services. The subjective part here is related to the user complaints about voice loss because of some basic connectivity error on the user end device or even a bad end device (a phone) causing bad voice quality. The network operator needs to have the capability to track all these aspects and root out false reports. The network operators can achieve this by keeping real-time track of KPIs across their network, which in essence would provide a real-time view of the health of the network.

Systemic issues, on the other hand, can be related to networkwide problems, such as the impact of a large DNS outage. The localized issues can be a single bad Multimedia Terminal Adapter (MTA), which means that basically the customer home gateway causes an entire voice switch performance to go down. The layered approach enables an operator to track the entire system and notice systemic issues caused by deep anomalies spread across the network, such as the DNS problem that trickles across a VoIP network. Finally, the subjective issues are more quality-driven and can originate from user complaints. Simply observing and comparing against the normal baseline metrics and noticing no difference shows whether the subjective issue is indeed a problem. We discuss these topics in the following sections.

Tracking Systemic Performance Issues

VoIP network systemic issues not only represent the problems affecting the entire network but can also mask a specific or localized issue. Thus, tracking these network problems through a KPI-based approach at the system level not only helps isolate the systemic issues but also helps to rapidly isolate the issues to specific segments. The systemic issues can be voicemail not functioning properly, which can be tied to equipment failures, key DNS servers being down, or even oversubscription of SIP trunks to the voicemail servers. Tracking of SIP errors for SIP trunks as KPIs is a starting point and can isolate the issue to the specific functional or service grouping. Chapter 7 covers this topic in depth.

A visual dashboard representation of the KPIs from the VoIP network segments facilitates the tracking and trending of the network. The KPIs from the network are the building blocks for tracking the VoIP services. As mentioned earlier, the voicemail service, a key VoIP component, can be tracked by SIP counters tied to its SIP transport trunks, further tracking the voicemail success and failure KPIs from the Call Management System (CMS), creating a voicemail transport–to–voicemail serviceability mapping view. The increase in voicemail errors, as seen in the Call Digital Records (CDR) from the CMS, can be cross-checked against the SIP errors to isolate the issue. In some cases, the voicemail configuration errors can be the root cause, but following this approach of KPIs leads to the issue in an organized manner and leads to quick voicemail service issue resolution. Chapter 7 and Chapter 8, "Trend Analysis and Optimization," cover examples through dashboards related to these concepts.

The tracking and trending of the network through these KPIs on a daily basis captures Busy Hour traffic behavior. This allows you to establish a baseline for peak hours and nonpeak hours. The Busy Hour KPI behavior tracked also includes metrics like Busy Hour Call Attempts (BHCA). The collected KPI data also yields accurate capacity planning information. The tracking, if done on a real-time basis, isolates any anomalies in the network. Basically, deviations from the baseline traffic pattern highlight these anomalies. Further, high and low water marks can be placed on these visual dashboards to help the network operators easily track abnormal traffic patterns.

The key in trending and isolation of issues for a particular voice service is to visually stitch together a set of KPIs representative of that service, develop baseline behavior of the metrics, and then identify acceptable high and low water marks. Following this methodology in a network operations center (NOC) allows the network operator to effectively isolate the system issues.

The graphical dashboard in Figure 2-2 provides an example of the voice trunk usage counters in a VoIP network that are part of the KPI set. The vertical axis depicts the Signaling System 7 (SS7) trunk counter utilization as a percentage, whereas the horizontal axis represents the trunk usage stats.

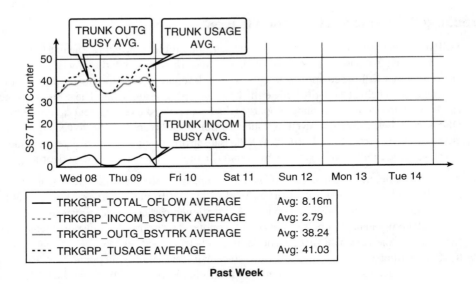

Figure 2-2   *Voice Trunk Usage Stats*

The trunk dashboard reflects KPIs for PSTN trunk capacity. It represents aggregate inbound and outbound voice trunk traffic. The KPIs tracked are as follows:

- **TRKGRP_OUTG_BUSY_AVG:** The number of trunk circuits in the reporting trunk group marked as busy with originating calls taken every 100 seconds during the interval divided by the total number of trunks. The dashboard represents them as TRKGRP_OUTG_BSYTRK_AVERAGE.

- **TRKGRP_AVERAGE_USAGE_AVG:** The incoming usage and the outgoing usage counters for the reporting trunk group divided by the number of trunks in the reporting trunk group. The dashboard represents them as TRKGRP_TUSAGE_AVERAGE.

- **TRKGRP_INCOM_BUSY_TRK_AVG:** The number of trunk circuits in the reporting trunk group marked as busy with terminating calls taken every 100 seconds during the interval divided by the number of trunks in the reporting trunk group. The dashboard represents them as TRKGRP_INCOM_BSYTRK_AVERAGE.

- **TRKGRP_TOTAL_OVERFLOW_AVG:** The number of outbound trunk call attempt failures because of all trunks in the reporting trunk group being in a busy state divided by the number of trunks in the reporting trunk group. The dashboard represents them as TRKGRP_TOTAL_OFLOW_AVERAGE.

You can observe a pattern in Figure 2-2. Most of the trunk counters representing the trunk seizures are for outgoing trunks. The number of incoming trunk seizures are low and the overflow of trunks is also low. A potential capacity issue can be deduced as the total trunk usage and overflow of trunk usage goes up. In this example, there is no overflow of trunks.

> **Note**   The total PSTN SS7 trunk utilization of the network for a particular switch is tracked at any given time, as seen in Figure 2-2. A set of these figures when combined as a dashboard enables the network operator to track the SS7 trunk behavior as a service. In Chapters 7 and 8, we cover this and similar KPI dashboards in more depth. A combined dashboard similar to Figure 2-2 and other figures, which can include the SS7 network elements, the CMS, and the SS7 signaling protocol-related counters, represents the layered KPI methodology. The combined dashboards enable the network operator to quickly traverse the dashboards and see the impact of the problem.

Localized Performance Issues

A local issue is related to a specific element or functional component of the VoIP network, while a systemic issue has a wider footprint and needs to be fixed across it. Tracking performance issues at a system level through KPIs and dashboards allows the isolation of anomalies and trends of a particular segment in the VoIP network. Following the approach of the KPIs from the earlier trunk dashboard example, we can identify and further drill down to specific localized issues. This localization can be related to a particular set of trunks leased from a certain service provider (SP), or in the case of an enterprise network, they might reflect a set of leased SIP trunks. This drilling down from the system level allows the operator to identify the potentially bad terminating trunking line cards as the root of the problem. The operator can also observe per-trunk behavior for oversaturation and add more capacity. NMSs, like the EMS, can be used further to isolate the PSTN components that need to be overhauled to increase capacity. Figure 2-2 gives an aggregate view of the trunk usage; it can be used as a starting point for this analysis, and nested dashboards with detailed KPI data can be leveraged to track per-trunk behavior and provide trunk-to-line card mapping.

Subjective Performance Issues

Occasionally, the problems reported in the VoIP network are subjective. Subjective problems are defined as problems that cause impact to voice quality. Subjective problems are driven by customer-reported issues or, in small cases, in-house testing by the VoIP provider. The customer-reported problems include mostly voice quality issues. These problems are subjective in nature, because they often result from user or pilot errors such as configuration issues at the customer premises equipment (CPE).

If all the key VoIP metrics are being tracked, the network operator has a normal functioning baseline in place. The reported issues can be cross-validated against the current metrics dashboards to detect any anomalies.

The VoIP operations need to have processes in place to perform this cross-validation and walk the customer through standard home checks to root out a lot of these common problems. The processes entail walking the customer through common equipment problems and performing some VoIP-related tests for that particular subscriber's profile.

The VoIP provider handles one of the main challenges of voice quality by performing Mean Opinion Score (MOS) tests; this allows the provider to gauge the subjective (audible) voice quality of the network. The MOS is a range from 1 to 5, with 5 being the best audible quality. The MOS is also computed from the network on a periodic basis through monitoring applications, which is a metrics-based analysis computed by tracking traffic from packet captures from the VoIP network. As a best practice, it needs to be correlated with live call testing. The live call testing is what the VoIP operator should introduce as a regular practice and tie it the MOS obtained from the applications to have conformance and the collected baselined KPIs. These three aspects combined together in standard checks allow independent sources of VoIP health validation. The provider also must perform a regression test plan upon turn-up of new VoIP switches and services. The regular call tests to keep on top of voice quality and the regional or campus VoIP feature verification also need to be performed to provide an additional set of validations to the baseline data.

The customer experience is what drives the revenues or the productivity of an organization. The customers might be an SP subscriber or an enterprise VoIP network's user. It is the handling of subjective issues that becomes the top priority for VoIP providers.

In subjective issues, the focus can be rooting out false alarms, but at the same time, an increase in the number of customer-reported problems can also indicate a network-wide anomaly. This situation needs to be isolated, as mentioned in earlier sections. Chapter 7 expands on the topic more.

In short, subjective performance issues can be user errors but also can be deeper rooted and need to be isolated through VoIP network KPI trending, monitoring, and analysis.

Downtime and Impact

Adapting an approach to track the systemic, localized, and subjective issues helps reduce predictable and unpredictable downtime and impacts the customer service positively. The core approach outlined in this book is to identify and track the KPIs. The issues identified need to be acted upon in a timely and efficient manner to achieve a highly available network.

We now look at a scenario where a service provider's customers experienced intermittent voicemail service performance issues in particular markets. After spending a couple of days investigating the problem, it turned out that the VoIP switch SIP trunks to the voicemail servers were getting oversubscribed. The SP provisioned more trunks, and the problem was resolved. So the impact was high and the customer sentiment was negative, because a key feature was not accessible to him. If the SP had tracked the SIP trunk-related KPIs for all markets, it could have predicted the capacity issue and possibly avoided the impact to customers.

To summarize, issues can be localized quickly to be acted upon. The actions range from increasing the capacity, shifting of load from overutilized resources to the underutilized ones, and creating baselines for all VoIP functional groups so that they can drive these changes and at the same time help with issue isolation.

Note   Reduced downtime can be achieved by clearly identifying the service flow segments in the VoIP network, followed by identifying the KPIs needed to track the service flows, and lastly by deploying a management system to collectively track all service segments.

Proactive Monitoring Concept

The proactive monitoring concept is based on catching the problems before they happen at the high level. In practice, it is an extension of reactive fault management that focuses on quickly identifying the fault and its source, repairing it, finding the root cause, and conducting impact analysis. The proactive monitoring concept takes this approach to the next level, where network operators continuously strive for identifying problems as they happen in the network before they impact the end user by keeping a close watch on the network and call processing entities' behavior by way of trend analysis of performance indicators. This is a continuous process, and its practice varies from network types to size.

Therefore, it is imperative that the NMS offer a good degree of flexibility and provide a portfolio of element and network management tools that would address and solve the most compelling problems that the SP or an enterprise would face in the course of a operating an IP network carrying voice and other media traffic.

An NMS designed to proactively monitor the network determines what data should be collected and how frequently the collection should be. It also sets proper thresholds for exception conditions. It controls data aggregation by correlating this data in the proper context and performs statistical analysis to support various reports. It also sets proper thresholds for exception conditions. Moreover, it contains a predefined set of reports and the ability to allow the user to customize the reports. These reports allow users to monitor the performance, troubleshoot performance problems, perform service-level management, and obtain trending information for capacity planning and network optimization. The following definitions give a high-level overview of some of the key operations functions:

■ **Fault data management:** Fault management is concerned with collecting unsolicited network events from the network, typically alarms and Simple Network Management Protocol (SNMP) traps, and presenting this information to users and other applications. Functionality includes maintaining alarm lists, deduplicating alarms, filtering alarms according to different criteria, correlating alarms from multiple devices and network level, maintaining historical alarm information, and so on. A more comprehensive fault management scheme includes an integrated trouble ticketing system to track repair information and provide feedback to the SLA management entity about network availability and downtime. The SLA management system is additionally responsible for managing root cause and impact analysis. This can be a valuable resource for training NOC staff.

- **Configuration management:** A network can undergo scheduled configuration updates, including moves and additions. Fault repair requires reactive changes in the network to rectify a problem either as a temporary workaround or often a permanent solution. This system leverages common Application Programming Interfaces (API) such as Simple Object Access Protocol (SOAP), eXtensible Markup Language (XML), or SNMP to configure the network elements. It might also be capable of managing configurations using the traditional command-line interface to cover even the basic configuration tasks using Secure Shell (SSH) or Telnet capabilities.

- **Accounting management:** This includes a collection of resource usage by the subscribers in the process of using the various network services such as voice services or data services. The accounting information collected on a per-call basis includes such things as duration, uplink and downlink packet and octet counts, and the quality of connection parameters such as dropped and lost packets and media jitter. On the data side, the traditional usage parameters needed can be segregated by the transport protocols and protocol port numbers and data services accessed.

- **Performance management:** This includes a collection of performance data from the network through various mechanisms (polling, file upload, and so on) and conversion to a common syntactic format, processing of performance data for aggregation and statistical analysis such as trend analysis, calling behavior based on call detail and diagnostic records, and calculation of statistical means, standard deviations, and so on. This provides a network baseline and helps in establishing thresholds. Threshold crossings can be converted into alarms or vents that can be propagated to a central management console for immediate attention before the problem is spread on a wider scale.

Figure 2-3 describes a framework of the flow of information and interaction among the various elements of the overall network management system. This figure shows events coming up from managed devices; they can be XML-based events generated by XML agents on these devices. XML agents are being introduced more and more, because these agents facilitate a flexible and easily extensible XML-based communication channel.

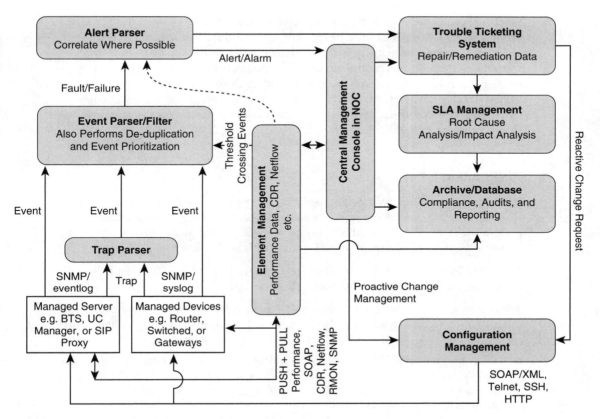

Figure 2-3 *Information Flow in a Fault Management System*

This section introduced concepts of proactive monitoring and tied it with the interactive flow shown in Figure 2-1. Basically the traditional way of managing a network through FCAPS is described, and we discussed how it's used in a VoIP network. We continue our discussion of KPIs, describing the metrics from various VoIP functional groups to stitch together a VoIP service view.

KPIs

We have emphasized the significance of tracking the KPIs and described how it relates to effectively managing a VoIP network. The goal in this section is to provide an overview of metrics available from various VoIP network functional areas and to discuss how they can be stitched together to provide a uniform view of the network. The following sections go over a set of VoIP-related KPIs that can help achieve this goal in a typical large VoIP service provider environment. VoIP signaling KPIs help track all signaling protocol-related metrics, whereas the VoIP media KPIs help track general media quality. The signaling KPIs allow functional determination of all building blocks for a VoIP network. The functional component can be the PSTN interfacing leg (OFFNET), or it could be the IP network (ONNET) leg. It can also be the VoIP infrastructure leg (like cable), an SIP leg, and so on. The key concept here is that all these legs can be tracked by metrics corresponding to the protocols controlling these legs. Similarly the metrics related to delay, jitter, and packet loss can track the functional health of the media aspects of the VoIP networks. Tracking them together provides coverage for the entire VoIP network. Chapter 6, "Managing VoIP Networks," and Chapter 7 cover both the metrics for the enterprise and service provider networks at length.

VoIP-Signaling KPIs

We present a use case for VoIP-signaling KPIs. The use case example is just to set the stage for the reader so that the reader understands that it's not only the call quality or performance that can be tracked but also the scope is to cover all the VoIP network functionality this way. We elaborate on the signaling KPIs in Chapters 6 and 7.

The following VoIP-signaling KPIs are collected from a Cisco BTS 10200. The BTS VoIP switch has telco Class V switch equivalent functionality.

The following KPIs are representative of a VoIP service provider environment over a cable infrastructure. These or a subset of these can act as key indicators for any VoIP network:

- **ISDN user part (SS7/PSTN) signaling protocol-related information:** ISDN user part (ISUP) is part of the SS7 signaling used for call setup, a key indicator for profiling the Internet-to-PSTN call interface. These counters are reported by Internet Transfer Points (ITP) and the CMS.

- **MGCP signaling protocol-related information:** MGCP is used to communicate between the CMS, the Media Gateways (MGW), and the CPEs. We can profile key performance issues in the VoIP network by reviewing average MGCP attempts and failure counters. These counters are reported by MGWs, MTAs, and the CMS.

- **SIP-related information:** The SIP signaling protocol is used for communication in several VoIP network segments. This includes SIP trunks between two CMSs, SIP trunks for interfacing with voicemail equipment, and SIP communication with user agents. SIP total inbound and outbound traffic counters for each of these segments act as key indicators for profiling VoIP network issues.

- **Call-processing specific information:** Key call-processing-related information can be collected through CDRs at the CMS, along with the call-processing performance counters. The CDRs help profile the call traffic call termination cause code, while the performance counters capture the call origination, termination, and success data.

- **Trunk Group (TG) usage information:** The TG usage information can help profile the trunk utilization by trunk type: SIP, PSTN Bearer, Channel Associated Signaling (CAS), and ISDN. The key metrics for trunk profiling include the inbound, outbound, and total utilization along with out-of-service and trunk overflow data.

The metrics corresponding to the protocols with other metrics when tracked on a real-time basis allow the operator to be on top of ingress and egress service patterns of the network. The trunk example shown in Figure 2-2 is a good representation of this concept. We next discuss the VoIP media KPIs that glue together VoIP metrics concepts.

VoIP Media KPIs

VoIP media KPIs directly affect the audible quality of VoIP traffic. They are dictated by the following key metrics:

- **End-to-end delay:** The time taken from point A to point B in the network. Delay can be measured in either one-way or round-trip delay. VoIP typically tolerates delays up to 150 ms one way before the quality of the call is unacceptable.

- **Jitter:** The variation in delay over time from point A to point B. If the delay of transmissions varies too widely in a VoIP call, the call quality is greatly degraded. VoIP networks compensate for this by having jitter buffers. Jitter should not exceed 30 ms for acceptable voice quality.

- **Packet loss:** The drop of packets in the data path, which severely degrades the quality of the voice call. Packet loss should not exceed 1 percent for good-quality voice.

Table 2-1 shows KPI threshold numbers for delay, jitter, and packet loss. Variances in any one of these metrics can affect voice quality. In most VoIP deployments, the VoIP CPE (this could be an MTA) reports these numbers at the end of each call. This information can be collected from the CDRs or from other mechanisms like the Cisco IP SLA, which can also be engaged to determine a network-wide snapshot of these metrics. We also cover this topic in Chapter 7.

Table 2-1 *General Media KPI Threshold Numbers*

	Good	Acceptable	Poor
Delay (Round Trip)	0–150 ms	150–300 ms	>300 ms
Jitter	0–20 ms	20–30 ms	>30 ms
Loss	0–0.5%	0.5–1.5%	>1.5%

The VoIP provider should perform periodic MOS tests to maintain voice quality standards. MOSs of 4 and higher are considered to represent good voice quality. These tests are subjective but provide a validation of quality to attain positive customer sentiment.

We have put together a methodology with a layered approach of metrics. To recap, the layered approach is basically using a signaling KPI set to drill down into the particular voice segments. At the same time, the media-related KPIs allow a complementing metrics set to validate the voice quality. We further gave examples of VoIP network signaling and media KPIs and explained that these components should be the building blocks of the service-tracking methodology.

Next, we discuss how to tie the approach with various functional segments of the network through the concept of service flows.

VoIP Network Segments and VoIP Service Flows

We discussed the concept of service flows and VoIP functional segments. We continue to build on this concept here. The following sections highlight key VoIP network segments; each segment has its respective signaling and media traffic components. The segments represent different functional groups of the VoIP network, and they consist of the signaling and media KPIs.

The tracking of these KPIs in pseudo-real-time fashion defines the service flows. The service flows are in essence the customer-affecting (internal or external) segments providing real-time VoIP functionality. These KPIs can be collected from the CMS; other possible collection points are the NEs.

The sections that follow define the key segments. We describe the segments in detail in Chapter 7 and cover the VoIP network view, which depicts these segments through the KPIs.

The segment use case presented here represents a cable infrastructure service provider. Equivalent metrics from other products can also be mapped to the use-case metric set.

The key takeaway is the segments and the stitched-together KPI concept behind them to form a traceable service. The following sections cover some of the key service segments and the corresponding KPIs. The service segments are defined as components that need to be tracked to achieve a VoIP service.

Voicemail Segment

Voicemail services are a key component of the VoIP network. The voicemail segment consists of the signaling and media KPIs. This segment covers the case of VoIP service providers consisting of SIP communication for both the media and signaling.

SIP transmit (Tx) and receive (Rx) counters form the basis for the KPIs for this segment. Thus, the voicemail segment can be managed by tracking these Tx and Rx KPIs. As an example, tracking the Rx and Tx SIP 500 error counters allows visibility to key errors in the voicemail service. Along with SIP error KPIs, tracking the total outgoing and incoming SIP messages and total outgoing and incoming SIP successful messages easily allows the tracking of the voicemail flows and in turn the entire segment.

Announcement Segment

The announcement segment is another key component of the VoIP network. This segment, and in turn the flows, can be tracked by the MGCP KPIs. The announcement segment basically consists of the CMS, which instructs the announcement devices through MGCP to play announcement Real-time Transport Protocol (RTP) to ingress media gateways (MTA or MGX). MGX is the Cisco voice trunking gateway. The service flow is managed by tracking MGCP Tx and Rx counters.

Voice Termination Point Segment

The voice termination point is defined as call origination and termination endpoints. The call control KPIs that communicate to these endpoints allow the determination of the IP and the PSTN segments.

The media or voice termination point can be on the IP side, which implies that both VoIP call legs (the origination and termination sides) are on the IP network, or it can be an ONNET to OFFNET call, where either the origination and termination leg can be to a PSTN. The following section expands on the KPIs in these segments.

Voice ONNET Call Leg Segment

In most cases, the ONNET (IP infrastructure) segment can be tracked by MGCP and SIP-related KPIs. Proprietary protocols need to be examined to identify the key KPIs, but for general purpose segment tracking, the ingress and egress performance counters for the respective protocols can be used. These KPIs are also reported by the MTAs and captured in the CDRs. They can also be extracted from the CMS logs, because both the CMS and the MTA-reported information is logged in the CMS logs.

Basically, the KPI set is not only reflected in the performance counters that are available on a periodic basis, but the end device (MTA) also reports statistics at the termination of the call. These counters are logged in the CMS syslog for both MGCP and SIP endpoints.

Voice OFFNET or PSTN-Bound Segment

There are three components to a PSTN segment: a media resource seizure component, an SS7 call setup component, and a component that carries the voice media a voice bearer trunk. We cover the first two in this section. The next section examines the third component.

The PSTN call segment can be tracked by MGCP-related KPIs to the MGW. This segment basically tracks the trunk resource seizer for a call. MGX, the Cisco voice trunking gateway, has MGCP termination cards that can be polled for the Tx and Rx MGCP counters for tracking the connection flow going OFFNET to the PSTN leg. The PSTN-bound call segment consists of tracking SS7-related signaling KPIs. For VoIP networks, these segments are tracked by ISUP counters obtained from the CMS and the SS7-related signaling counters from devices like the Cisco Internet Transfer Point (ITP). It is a product for transporting SS7 traffic over IP (SS7oIP) networks.

PSTN Bearer Traffic Segment

The bearer traffic segment consists of all the voice traffic riding over dedicated Inter Machine Trunk (IMT) lines toward the OFFNET (PSTN) network. The traffic flows can be tracked by ingress and egress trunk usage counters, along with any Rx- and Tx-related error counters. You saw an example of these counters in Figure 2-2.

The segments and the corresponding KPIs, when tracked on a periodic basis, allow the VoIP operator to develop a functional operational strategy for troubleshooting, capacity planning, and SLA.

We have given an overview of the layered approach of tracking the KPIs. We described some of the problems that occur in a large VoIP environment. We then continued on to describe the various KPIs and how they can be used to track various functional segments. These segments represent the services riding through the network. The bundling together represents an overall VoIP service. Each of these services has a direct impact to the customer in most cases. We showed that the KPIs need to be stitched together to compose a VoIP service or segment view. This was done to track the error KPIs along with the corresponding successful connection-related metrics.

Next, we discuss an SLA use case and describe how it can be tied to this layered approach of tracking KPIs.

Service-Level Agreement (SLA) Management

We present an SLA use case for an SP. We cover a service provider's environment and some of its key externally leased services. The same concepts are also applicable to enterprise networks. We touch on a 5 9s impact on network uptime. The 5 9s terminology is used in the industry to track high availability. We then describe a few VoIP service segments and discuss how the uptime SLA can be directly applied to them. The service components are for SIP-terminating session border controllers (SBC), followed by the PSTN-facing trunk and corresponding SS7 link network. We also include a section on vendor accountability through SLA. Finally, we cap off the discussion with how we track the SLA with an overview of a tool strategy. SLA management in VoIP networks also involves tracking the following items:

- **Signaling quality:** Measures call-signaling and setup performance.

- **Delivery quality:** Measures transport performance, including metrics such as long latencies, lost packets, jitter, and call signal strength. All of these are factors in the overall call satisfaction.

- **Call quality:** Measures a user's overall call experience.

The voice quality is a key for the VoIP operator, and identifying false reports becomes challenging. Call quality metrics, such as the MOS, enable VoIP network operators to perform this identification. A more trackable approach is through trended KPIs, baselined for a good voice call quality period. Cross-validating any customer-reported complaints against this baseline roots out false reports.

All these factors can be included in SLAs to drive accountability for a VoIP operator's business segments, external SPs, and equipment vendors. You achieve this by tracking the KPIs on a periodic basis and tying them back to network issues and reported, valid voice quality problems. Then, you correlate them to externally leased services and vendor equipment.

Note   When tracked, the KPIs can help to develop throughput and quality accountability, thus deriving the SLAs. The service segments are tied at the surface to a particular KPI set, but in reality there is Layer 1–Layer7 impact. The following sections describe segments or service flows that can be tracked at the protocol Layer 4 (SIP, ISUP, and so on), but they have critical path dependency to lower layers, and at the same time, the call-processing Layer 7 has a dependency on them.

Tracking the service segments defined so far is critical. As you can see from the 5 9s data in Table 2-2, downtimes can have big impact on a yearly basis. In a VoIP market, where the competition is high and quality of service can have a huge impact, achieving 5 9s should be the goal. Anything less would affect the customer satisfaction and revenue.

Table 2-2   *5 9s Impact*

Availability, %	Downtime per Year	Downtime per Month	Downtime per Week
90	36.5 days	72 hr.	16.8 hr.
95	18.25 days	36 hr.	8.4 hr.
98	7.30 days	14.4 hr.	3.36 hr.
99	3.65 days	7.20 hr.	1.68 hr.
99.50	1.83 days	3.60 hr.	50.4 min.
99.80	17.52 hr.	86.23 min.	20.16 min.
99.90	8.76 hr.	43.2 min.	10.1 min.
99.95	4.38 hr.	21.56 min.	5.04 min.
99.99	52.6 min.	4.32 min.	1.01 min.
99.999	5.26 min.	25.9 sec.	6.05 sec.
99.9999	31.5 sec.	2.59 sec.	0.605 sec.

Obviously, VoIP network downtime needs to be reduced. Reduced downtime can be achieved by clearly identifying the service flow segments in the VoIP network, followed by identifying the KPIs needed to track the service flows, and lastly by deploying a management system to collectively track all service segments. The management system should provide proactive resolution of issues and help in performing accurate capacity planning. The actual VoIP services availability data, when continuously tracked against the 5 9s benchmark, allows the operator to make the network more predictable and stable.

We discuss a few leased services in the following sections and identify the potential KPIs that can be used to track SLAs around them. These SLAs can be the measuring stick that the service provider uses to keep on top of the availability on the leased services.

SBC Trunk Uptime

SBCs are being deployed more often in VoIP networks. In a lot of these deployments, they sit at the edge, interconnecting the different service providers through SIP trunking. They are also used as NAT devices, where they hide the internal networks. The focus in this book is toward SBC SIP trunking and how to track this segment of the network through KPIs. Managing the SBC trunk uptime and utilization is critical to the VoIP service provider, because most of their PSTN traffic and even long-distance traffic can be terminated at these devices. If another SP is terminating these trunks, it needs to be accountable for the agreed-upon traffic SLAs. The accountability can be achieved by tracking the SIP-related KPIs as reported by the SBCs.

PSTN/IMT Trunk Uptime

PSTN IMT trunk is the mechanism through which the VoIP provider sends bearer traffic over to the OFFNET PSTN network. PSTN IMT trunk utilization can be measured through the IMT trunk KPIs collected from the CMS. Similar to SBC setup, the IMT trunks are mostly leased from other service providers. The CMS sees all the trunk utilization information. The trunk KPIs can be utilized to track the uptime and thus provide a source of measurable accountability and SLAs.

Signaling SS7 Link Uptime

The SS7 signaling component of the PSTN controlling the call setup and termination is also leased through the external SPs. This can be done through an A-Link or D-Link or other link configurations. These links need to be tracked for their utilization and, in turn, uptime. The SLAs can be developed around ISUP-related KPIs obtained from CMS and the ITP to keep that map to the SS7 links.

So far, you saw how some of the leased services can be tracked. We now touch on vendor accountability, because the VoIP network is a complex integration of multivendor solutions, tracking or rooting out the unstable components in a measurable way so that more data is provided to the vendor to either improve the quality or be replaced.

Vendor Accountability

A VoIP network is a complex environment with many vendor devices and solutions integrated to provide an overall service flow. All these vendors and their solutions need to be accountable in a constructive and methodological way. Tracking the different VoIP segments through the service flows and their respective KPIs enables the network operator to develop metrics for each segment. Thus, the operator can hold the specific vendors accountable through these metrics for their specific product quality and in turn SLAs.

The vendors advertise their own product SLAs. Typically, they are driven from Mean Time Between Failure (MTBF), which is the average time a network or component works without failure. Also, the Mean Time to Repair (MTTR) and average time required to recover (or restore) from a network outage or repair a failed component provide a mechanism to measure availability.

Availability is expressed as a percentage and is measured by a ratio of two numbers: MTTR and MTBF. It is calculated using the following equation:

$$\text{Availability} = [\text{MTBF} / (\text{MTBF} + \text{MTTR})] * 100$$

The previous formula is also commonly mapped as follows:

$$\text{Availability} = [\text{Uptime} / (\text{Uptime} + \text{Downtime})] * 100$$

The KPI tracking for the various service segments enables the operator to identify downtimes and degradation in service for a specific set of equipment. This can be done simply using a naming convention on the various service-tracking KPIs to associate them to the vendors. This mechanism allows a service to be tied to the equipment responsible for hosting it.

This downtime tracking on a real-time basis can be tied to an advertised MTBF from the vendor. In the industry, the MTBF numbers are used for vendor product SLA accountability.

Tools Utilized

To effectively manage the VoIP network, some amount of specialization is needed. Specialization is necessary to create a pseudo-real-time dashboard for the service KPIs that are not readily available through the normal EMS and NMS products. Tools are needed to extract KPI information from CMS, SBC, MGW, ITP, and other devices. This data then needs to be parsed and correlated to create the dashboards and periodic SLA reports for all the VoIP service segments. We describe tools in Chapter 7.

The basic idea is to identify key metrics for tracking various service flows or segments. Furthermore, the metrics need to be transformed into meaningful information to derive the SLAs. This requires specialized tools to be implemented to determine what the meaningful information is. An approach for achieving this is through development of customized scripts; these scripts would then generate the meaningful information for building service dashboards. They would also generate events for trouble ticketing systems.

The service dashboards and the trouble ticketing systems are two key aspects that can help the network operator act on resolving issues and tracking them effectively.

Summary

The layered VoIP network management approach is critical for managing the VoIP network in a systematic way. The layering is from the perspective of drilling down from the system view to more granular views. The top-down metrics view combined with the bottom-up metric collection approach constitutes the layering methodology. A layering approach enables the VoIP operator to understand not only specific localized issues but also systemwide issues.

This chapter also presented a proactive monitoring approach based on a typical FCAPS model. To effectively run a VoIP network, you need more than just FCAPS. You need to have specialized dashboards, alerts, and reports to track all service flows and to keep on top of capacity and SLAs challenges. These mechanisms should be integrated into the existing network management frameworks for the trouble ticketing, fault management, and SLA. The KPI, service segment, and service flow concepts are fundamental and set the stage for upcoming Chapters 6, 7, and 8. Lastly, we show a use case through SLA for various key service segments and describe how KPIs can be used to derive these SLAs.

Reference

1. Wikipedia, the free encyclopedia. "High availability."

## Chapter 3

VoIP Deployment Models in Service Provider Networks

This chapter gives you an understanding of how Voice over IP (VoIP) is deployed in service provider (SP) networks. This chapter focuses on describing a use case in which the VoIP infrastructure and the transport and the access are managed by an SP. Chapter 4, "Internet Telephony," focuses on VoIP networks in which only the VoIP infrastructure is managed. Different network components and their functions are described to illustrate how various call functions are implemented to provide voice services to residential and business customers. Figure 3-1 depicts a block architecture of the SP scenarios discussed in this chapter. Here, the service provider also owns the last-mile network access. Later chapters cover scenarios where the SP does not own the access network.

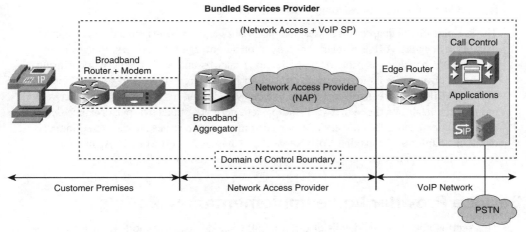

Figure 3-1 *Service Provider Architecture Overview*

This chapter provides a high-level view of the connectivity between different components in a VoIP SP. You learn about the common VoIP networks and the corresponding components. The intention of this chapter is not to provide design guidelines or technology-specific reference material, which is outside the scope of this book, but to offer a collection of metrics from across the various VoIP architectures. As a general note, the acronym *KPI* (key performance indicators) is used throughout the book to refer to key protocol counters or metrics.

This chapter covers various VoIP applications in the SP market; residential application is geared toward providing primary- or secondary-line voice services to SP's residential customers. These customers include existing high-speed data subscribers and new subscribers who are looking at either replacing their current circuit-switched telephone line or adding additional phone lines to their household. This gives SPs a chance to provide bundled services to their customers.

Another application covered in this chapter is Small/Medium Business (SMB) application, which is geared toward business customers. SPs can provide high-speed data and digital voice services to their business customers using their IP infrastructure. For the SMB, using IP infrastructure can be a more cost-effective way of getting voice services as compared to a traditional circuit-switched phone line from the telephone company.

Other applications discussed in this chapter include IP trunks, which are used for traffic offload and public switched telephone network (PSTN) bypass, and Sessions Border Controllers (SBC), which are used for offloading VoIP traffic to the PSTN, network hiding, and voice transcoding.

The latter part of the chapter highlights some of the security-related issues in SP voice networks. These issues include denial of service (DoS) attacks, theft of service, and other issues that are common in existing IP networks today.

The last part of the chapter discusses common issues and problems related to voice in SP networks. Because VoIP is primarily deployed on a converged IP network, it faces many of the same challenges as other data applications, such as failures in the network, routing protocol convergence issues, oversubscription of network resources, and so on. However, because VoIP is more sensitive to things like delay and jitter, it's important to proactively monitor the health of the SP network and prevent network outages or performance degradation that can cause loss of service to its customers. These issues are discussed in more detail in Chapter 6, "Managing VoIP Networks," Chapter 7, "Performance Analysis and Fault Isolation," and Chapter 8, "Trend Analysis and Optimization."

Service Provider Voice Implementation Models

This section goes into the details of different SP voice deployment models. Various network components and their functions are discussed with illustrations. There are two different VoIP implementation models in SP networks:

- **Centralized Switching Model:** In this model, the call-processing functions are controlled by a central entity such as a Softswitch (Call Agent or Call Management Switch [CMS]), which passes call control information to different network elements, sets up and tears down calls, and keeps data records for the calls as Call Detail Records (CDR). The endpoints do not need to have intelligence in regard to initiating or terminating calls; they receive the information from the Softswitch and carry out the necessary call functions.

- **Distributed Switching Model:** In this model, the call-processing functions are distributed to different network elements. A single entity does not control the various call functions. In this model, the endpoints have call intelligence and can initiate and tear down calls without a centralized entity controlling them. The current VoIP SPs are hesitant to go this route, because it makes the end VoIP clients fatter or richer in features and they do not need to subscribe to the SP's premium services. IP Multimedia Subsystem (IMS) is the route that SPs are looking into where presence servers are used to track the end clients.

This chapter primarily focuses on the centralized switching model because most of the current SP deployments are based on this model. The other common distributed switching model is introduced briefly, but it is discussed in more detail in Chapter 4, which also covers some of the current Peer-to-Peer Distributed switching models. The next section covers how the centralized and distributed switching models are deployed in different SP networks.

Residential Applications: Voice over Broadband

In a voice over broadband deployment model, the SP uses the IP infrastructure to provide residential IP telephony services to its customers. An example of such an implementation model is the PacketCable architecture defined by Cable Television Laboratories (CableLabs) PacketCable specifications. The PacketCable specifications define a framework of how VoIP can be implemented over the Data Over Cable Service Interface Specification (DOCSIS)/IP infrastructure. Figure 3-2 provides a high-level overview of the PacketCable architecture. The system uses IP technology and QoS to provide high-quality transport for the VoIP network.

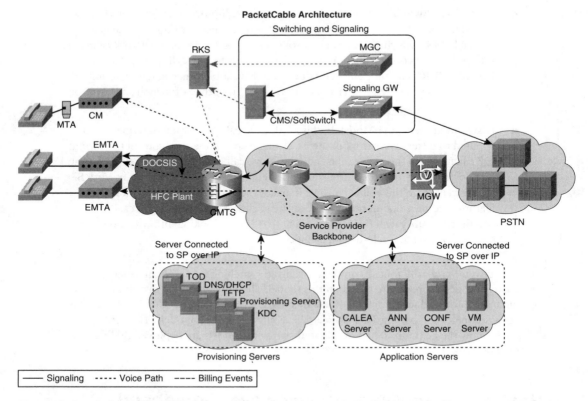

Figure 3-2   *PacketCable Architecture Overview*

The following are some of the key elements of the PacketCable network:

- **Call Management Server (CMS):** The CMS is responsible for providing call control and signaling for the endpoints using Media Gateway Control Protocol/Network-Based Call Signaling (MGCP/NCS) protocol. In a centralized switched model, all the intelligence resides on the CMS, which is responsible for instructing other network elements on their functions.

  The CMS is composed of several logical components, such as Gate Controller (GC), Media Gateway Controller (MGC), Signaling Gateway (SG), and Announcement Controller (ANC). The GC is responsible for quality of service (QoS) authorization and control. The MGC provides call control and signaling for PSTN Media Gateways. The SG communicates call signaling to the PSTN using protocols such as Signaling System 7 (SS7). The ANC interfaces with the Announcement Player (ANP) to play network announcements.

- **Cable Modem Termination System (CMTS):** The CMTS sits at the edge of the network and connects the endpoints to the SP infrastructure such as provisioning servers, CMS, Media Gateway (MGW), and so on over the DOCSIS Hybrid Fiber

Coax (HFC) network. It also allocates resources for voice calls when instructed by the CMS and upon receiving requests from the endpoint.

■ **Media Terminal Adapter (MTA)/Embedded-MTA (EMTA):** MTA connects the subscriber equipment, such as a host PC or analog phone, to the SP network over the DOCSIS (HFC) network. It establishes a physical connection with the CMTS and forwards traffic between the SP network and the subscriber equipment. It contains a network interface, radio frequency (RF) interface, CoderDecoder (CODEC), and all signaling and encapsulation functions required for VoIP transport, class features signaling, and QoS signaling.

■ **Media Gateways (MGW):** The MGW provides bearer connectivity to the PSTN and is used for off-net calls (when an SP customer calls someone connected to the PSTN, basically an IP-to-PSTN network call).

■ **Provisioning Servers:** Figure 3-2 includes a setup of servers; they perform provisioning and billing functionalities. These servers include the Dynamic Host Configuration Protocol (DHCP) server for assigning IP addresses and other network parameters to the endpoints, Domain Name Servers (DNS) for name resolution, Trivial File Transfer Protocol (TFTP) for downloading configuration files to MTAs, and optionally other servers such as syslog server and Ticket Granting Server (TGS), which are used in the PacketCable network.

■ **Application Servers:** These servers include voicemail (VM) servers for providing voice mailbox service to subscribers, conferencing servers for audioconferencing service, announcement servers for playing network announcement messages, and Communications Assistance for Law Enforcement Act (CALEA) servers for subscriber wiretapping for law enforcement agencies.

■ **Record Keeping Server (RKS):** These are used for billing purposes. They store call detail record information through PacketCable Event Messaging.

Residential gateways in the form of MTA embedded in a cable modem are also known as Embedded Multimedia Terminal Adapters (EMTA). VoIP access is provided at the customer premises. By plugging a standard analog telephone into the MTA device, a user can make phone calls to another Multiple System Operator (MSO) customer directly across the IP network or to anyone outside the SP or MSO network through an MGW.

CMSs and MGCs provide centralized call-control processing by passing control information and setting up connections between residential MTAs. After these connections are established, voice passes directly between gateway endpoints in the form of RTP packet streams, as shown in Figure 3-3. Most connections with the PSTN are through voice bearer trunks with a Media Gateway providing the bearer connections and a Signaling Gateway (SG) providing the signaling connection into the SS7 network. Multi-Frequency/Channel Associated Signaling (MF/CAS) trunks are provided for some specialized requirements, such as Operator Services.

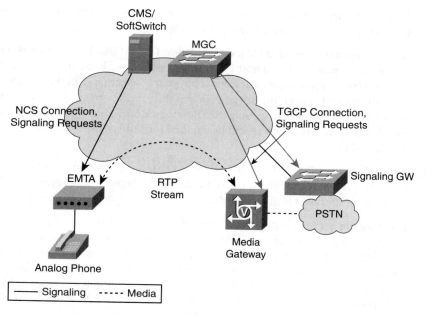

Figure 3-3 *PacketCable Signaling Architecture*

The PacketCable Network-Based Call Signaling (NCS) protocol is used to communicate with the MTA endpoints. The PacketCable Trunking Gateway Call Signaling Protocol (TGCP) is used to communicate with Media Gateways. NCS and TGCP are profiles of the Multimedia Gateway Control Protocol (MGCP), which belongs to the xGCP suite of protocols. These protocols allow a central call control mechanism to control customer premises equipment (CPE) devices for voice services.

Small/Medium Business Applications (Voice over T1/E1/PRI/CAS)

The SP uses small/medium business applications to provide IP-based telephony to SMB customers, typically when only a low number of devices are needed at the customer premises (usually this number is less than 50).

SPs often use an Integrated Access Device (IAD) to provide voice and data service to the customer premises. The IADs are connected to a Local Exchange Carrier (LEC) leased Primary Rate Interface (PRI)/Channel Associated Signaling (CAS) line, which is aggregated through a bigger transport connection like DS3 to the VoIP SP. The CMS residing at the SP is used to provision the IADs.

MGCP is the signaling protocol used to communicate to the IADs, and both signaling and the data ride the same leased line. Figure 3-4 reflects such a topology. The SP also routes PSTN/SS7-bound calls originating at the customer premises through the CMS that's managing the IAD/MGCP link. The trunking gateway and the SS7 gateway (which can be a Cisco Internet Transfer Point [ITP]) handle the PSTN-bound bearer and signaling traffic, respectively. The CMS acts as a central switching point and is thus an ideal place for collecting key performance indicators (KPI) because it acts as central switching component. The MGCP-based communication counters for announcement servers, trunking gateway and IADs, along with SIP-based communication counters for voicemail server and PSTN-related SS7 signaling (SIGTRAN) protocol counters make up the KPIs for the Small Business model.

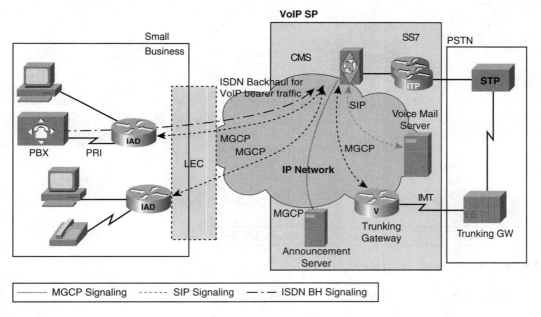

Figure 3-4 *Small/Medium Business Deployment Architecture*

Integrated Access Device (IAD) is also one of the key elements for the small-business network. It can be a collection point for MGCP-related metrics. An IAD is a device used to multiplex and demultiplex traffic in the customer's premises. The IAD is used primarily to route traffic and signaling over to a single T1 line or to an ISDN PRI trunk. This is also called a voice gateway that utilizes E1 lines in the case of non-U.S. markets.

IP Trunks

The SP market is converging to IP, but the PSTN is still the prevalent infrastructure and will be utilized for a while. However, more SPs use the IP network where possible. They achieve this by placing a trunking VoIP switch, which provides Class IV or Tandem switch functionality, to most commonly address two needs. First, it serves as a long-haul SIP trunking switch to carry traffic between SPs of different regions. Second, it acts as a PSTN bypass and an inter-SP trunk interconnect to offload long-distance traffic. The architecture presented in Figure 3-5 touches on various technologies, but the discussion in this section is focused on signaling protocol–related metrics.

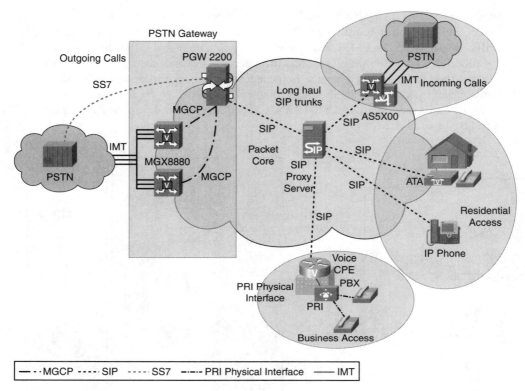

Figure 3-5 *IP Trunk Deployment Architecture*

Some of the key components of the IP trunk architecture are as follows:

- **The Cisco PSTN Gateway (PGW):** The Cisco PGW 2200 is a carrier-class call agent that performs the signaling and call-control tasks (such as digit analysis, routing, circuit selection, and more) between the PSTN and the IP infrastructure. PGW is also called a trunking switch, and it performs Class IV–type functionality.

- **Analog Telephone Adapter (ATA):** The Cisco ATA 186 is a handset-to-Ethernet adapter that turns traditional telephone devices into IP devices, which enables the analog phones to be connected to an IP network. Customers can take advantage of the many new IP telephony applications by connecting their analog devices to Cisco ATAs.

- **SIP Proxy Server:** The Cisco SIP Proxy Server is a call control engine that enables SPs to build scalable, reliable VoIP networks today. Based on the SIP, the Cisco SIP Proxy provides a full array of call-routing capabilities to maximize network performance in both small- and large-packet voice networks.

- **Cisco Access Servers:** The Cisco access gateway provides universal port data, voice, and fax services on any port at any time. It is used as a common gateway for terminating IP trunks that carry VoIP and other types of traffic. It can be a collection point for signaling and media metrics. Cisco MGX 8850 and AS5400 are examples of the access gateways and are depicted in Figure 3-5.

IP phones also communicate through SIP trunk and SIP proxy servers. Figure 3-5 highlights a deployment of a Cisco PGW VoIP trunking switch in a residential broadband network. It touches aspects of PSTN and IP architecture connectivity. In this particular architecture, the PGW sits at edge of the IP network and deals with offloading the VoIP traffic to the PSTN. The traffic is routed through SIP proxies onto the trunking gateway. Figure 3-5 represents a mixture of networks with various integration boundaries. This shows the SIP connectivity from various sources: the business access, PSTN incoming and outgoing calls, and residential access. All the services provided by these networks need to be tracked. They can be tracked by signaling and media metrics and can help in sizing and service-level assurance (SLA) for the integration points. The SIP, MGCP, and SS7 protocol-related metrics are some of the key metrics that need to be tracked in the IP trunk deployment architecture.

The other major use of IP trunks is across international boundaries, where H.323 networks are prevalent, as seen in Figure 3-6. Figure 3-6 also shows trunk connectivity between the two PGW gateways that are respectively part of large, complex networks. Note the various integration points and the diverse protocol networks. To manage the VoIP service, the capacity and the SLA across these network integration points also become complex. The signaling and other key metrics can help in tracking, trending, and isolating service issues and better plan for capacity and manage SLA.

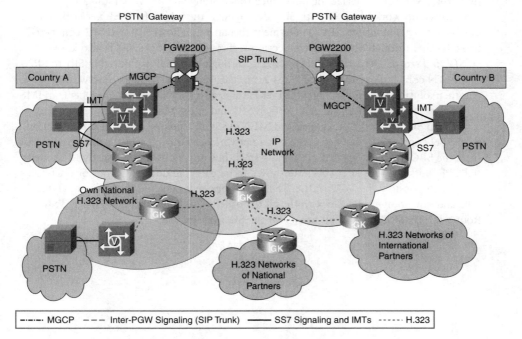

Figure 3-6 *IP Trunk Deployment Across International Boundaries*

In cases in which it is economical to route the traffic over to IP, providers offload the long-distance traffic to another provider rather than using the PSTN. This offloading is provided by a switch that is performing Class IV or toll-switch functionality. You can see in Figure 3-6 that country B connects to country A through an SIP trunk. That way, it can reach H3.323 networks. The PGW keeps track of all CDRs and is extensively used to apply policies for routing traffic through it to optimize cost. In general, services provided by this critical switch need to be tracked. The call and protocol metrics provided by the switch are crucial for running the VoIP network in an efficient way. Thus, the network management capabilities that facilitate this collection and dashboarding become the key to running the VoIP network.

Figure 3-6 shows SIP, MGCP, and H.323 protocols being used for signaling communication. The corresponding traffic counters represent the KPIs needed to effectively monitor the network. The counter collection points are the respective switching, aggregation, and endpoints. Chapter 7 covers these KPIs in detail.

Session Border Controller (SBC) Models

SBCs are also heavily used by VoIP SPs for a variety of reasons. This section discusses some of the common-use cases to continue the discussion of VoIP networks and the corresponding components.

In VoIP networks, the most prevalent and common use for SBCs is offloading VoIP traffic to the PSTN. Other usage includes voice transcoding and network hiding. Figure 3-7 describes a network of an SP with SBC deployed at the edge of the SP network. Figure 3-7 also shows various SBC deployment scenarios. It shows SP1 and SP2 connected to each other through SBCs; this connection is through SIP trunks and is also known as an *SIP tied trunk*. Here, the call flows occur through these SIP trunks from SP1 to SP2. The PSTN connectivity is handled by SP2, representing a PSTN offload network. Another solution for a Call Control Server Farm is also shown in the figure; here, an IMS server is communicating through an SBC to SP1 and at the same time through a set of servers (an SIP Proxy, a Gatekeeper, and MGX/AS5400 trunking gateways) representing the PSTN connectivity network. The other network scenarios reflect the SBC connection from the SBC to an enterprise1 network, a small-business network, and lastly to residential CPE units. All these networks are shown to be carrying different types of traffic along with possibly the VoIP traffic. The key component depicted in the figure is the SBC and how it interconnects with the all the other VoIP networks. The SBC and SIP Router Proxy (SRP) are the tap points for the SIP metrics.

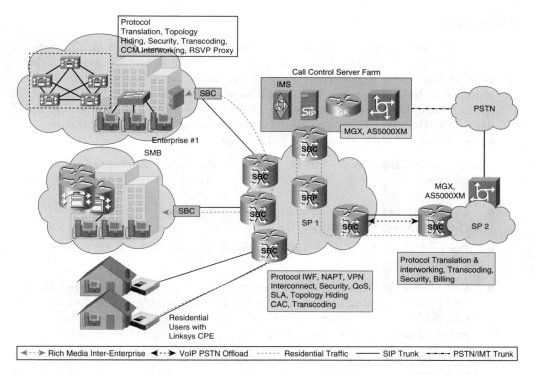

Figure 3-7   *Session Border Controller Deployment Architecture*

Key Components Used in SBC Models

Some of the key components used in SBC deployment models are as follows:

- **SBC:** SBCs are used at the edge of the network; they manage and control the SIP session streams traversing the borders of the networks they sit at. The SBCs provide the functionality for hiding customer networks by NATing. The SIP session streams represent SIP signaling and media communication.

- **SIP Router Proxy (SRP):** SIP Proxy and SRP perform the same functionality. SRPs basically help route requests to the user's current location, authenticate and authorize users for services, implement provider call-routing policies, and provide features to users.

PSTN Offload

VoIP SPs are offloading the handling of PSTN traffic through PSTN SPs. Typically, these are referred to as the traditional telcos. The telco and the VoIP SP have SBC devices at the edges of their networks, and through SIP trunks between these SBCs, the VoIP traffic is offloaded to the PSTN network. The VoIP SP maintains the quality of the VoIP offload service through SLAs. Telcos being the hosts of the VoIP offload also report the SLAs on this service. The VoIP offload allows the VoIP SP to focus and improve the IP-centric traffic and not worry about managing a PSTN-based network.

The needs or challenges for the VoIP SP that drive this design are numerous and include the following:

- Turning up, maintaining, and growing the VoIP network and interfacing with the PSTN involves a lot of overhead.

- The VoIP SP has to plan in advance the trunk capacity needed to service its subscriber base.

- VoIP SPs not only have to deal with the bearer traffic aspect of the service but also the PSTN/SS7 signaling network. Both of these call bearer and signaling aspects have their own infrastructure, and thus are an overhead that needs to be monitored and maintained. The monitoring through select KPIs allows effective operations.

- The PSTN trunk turnup procedure requires countless hours of interaction with the telco; thus Opex cost is high.

- Maintenance is another challenge, especially if operations work is being performed on network augmentation and Emergency/911 circuits. The downtime for this kind of service has many repercussions, some of which can even lead to lawsuits, if the E911 services are impacted. Effective monitoring of the circuits through KPIs allows quick resolution of the outages. The PSTN offload allows the VoIP SP to basically hand over the liability of managing the 911 trunks to the PSTN provider.

- The VoIP SP has to constantly monitor the capacity and continue to profile the traffic to keep up with the subscriber growth.

In addition to the aforementioned needs or challenges, additional aspects include monitoring and managing the complex network of SIP trunks, which are used to interconnect the VoIP switches or the SBCs. Another key challenge is to keep on top of SIP trunk utilization and call performance metrics. As you will see in Chapters 7 and 8, monitoring and correlating these metrics yield an effective VoIP network management system. The SIP traffic counters constitute the KPIs needed to effectively monitor the SIP network. The main theme here is to provide a background of the various operational overheads. Chapters 7 and 8 look at how the metrics are used to track both the hosted PSTN network and the VoIP offload network, which leads to tracking of key metrics.

Network Hiding

Previously, we mentioned that SBCs are deployed at the edge of the VoIP SP network. This section provides more context for this discussion. The SBC enables VoIP signaling and media to be received from and directed to a device behind a firewall and Network Address Translator (NAT) at the border of an adjacent network. The SBC achieves this by rewriting the IP addresses and ports in the call-signaling headers and the Session Description Protocol (SDP) blocks attached to these messages. SDP is basically used for multimedia session setup. This functionality is offered by all SBCs. This NAT functionality enables the VoIP SP provider to hide the network. Hiding allows the SP to not expose its internal network to the outside work, because NAT translates the internal network to another external-facing network. The VoIP SP's SBC basically gets a tied SIP trunk to the SBC of the PSTN provider and does NAT for the back-end internal network.

Being able to look into the traffic enables the SBC to perform a wide range of functionality, including antispam, QoS, and billing. These features also help the VoIP SP to potentially improve the VoIP QoS, better track the connection billing records, and detect any security violations like spam attacks. To summarize, the SBC-based network is effectively monitored thorough KPIs comprised of SIP traffic counters. The collection, correlation, and reporting of these counters are important items that a VoIP SP should perform.

Voice Security in Service Provider Networks

This section discusses some of the security issues related to deploying IP-based telephony services in SP networks. It covers the security challenges seen at the network element and discusses what can be done to address them. Also, signaling and media encryption is discussed to address some of the security issues.

Securing VoIP Network Elements

To prevent DoS attacks and theft of service, the SP can implement security features on network elements that can minimize the chances of an outsider gaining access to valuable network resources and/or free service.

These security measures can include deploying stateful firewalls in the network that allow only authorized traffic to enter the SP network. Configuring access control lists (ACL) on the edge routers can help prevent unwanted traffic in the network. The endpoints that are connected to the edge of the SP network should be authenticated using the Authentication, Authorization, and Accounting (AAA) protocol. Unauthorized users should be denied access to network resources by either black-holing their traffic or assigning them a low bandwidth class of service that would not allow them to send or receive a significant amount of traffic. The key concept behind black-holing is to stop the propagation of this kind of traffic; on the other hand, a low class of traffic has residue in the network.

Securing Call Signaling and the Media

Because call signaling is used for setting up new calls, tearing down calls, and modifying the state of existing calls, it is important that these signaling messages are secured. In the case of a centralized switching model, such as PacketCable, this is accomplished by having a security association between the endpoint and a trusted network device. A security association is a set of provisioned security elements (for example, security keys) on both the endpoint and the trusted network device. By having a set of security associations, the trusted device can authenticate the endpoint when interacting with it. The interaction that takes place between the endpoint and the trusted device can be encrypted. IP Security (IPsec) is one of the mechanisms used to achieve this with the preprovisioned (preshared) keys. This ensures that all traffic between the two devices is from known sources and encrypted.

Similarly, to protect customer privacy, conversations must be kept private. To do this, the media streams generated by customer conversations must be encrypted. The endpoints in the conversation can negotiate a set of ciphersuites (type of authentication and encryption to be used) and then encrypt all their traffic using the negotiated method. Some of the ciphersuites negotiated by the endpoints can include Hash-based Message Authentication Code Message-Digest Algorithm 5 (HMAC-MD5) and Hash-based Message Authentication Code Secure Hash Algorithm (HMAC-SHA) authentication algorithms, and Data Encryption Standard (DES), Triple DES (3DES), and Advanced Encryption Standard (AES) encryption algorithms.

Common Issues and Problems When Deploying IP-Based Telephony Services

This section discusses some of the common issues encountered by providers when deploying IP-based telephony services over their IP infrastructure. The issues are ongoing and need to be maintained for the life of the VoIP service.

Convergence-Related Issues

As discussed earlier in the chapter, VoIP is primarily deployed on converged IP networks. Recall that a converged network is defined as a network capable of transmitting all types of traffic including data, voice, video, and images. Most existing SP IP networks have been designed to carry primarily data traffic and are geared toward data applications such as email, web traffic, and so on. The VoIP traffic is sensitive to time, the packets need to be delivered within a specific time period, and the network needs to facilitate this through various mechanisms. Deploying VoIP in such networks introduces new challenges for the SP's operations staff that needs to carefully monitor the health of their network and work closely with other groups in the company to provide VoIP continuity across the network. For example, in a DOCSIS/IP network, different groups are responsible for managing and maintaining the HFC/RF network, and another group is responsible for the IP network. Although these groups have totally different job responsibilities and technical background, they both need to work closely to provide quality service to the cable providers' customers.

Issues in Media Affecting Quality

Unlike circuit-switched voice networks that have dedicated paths and fixed bandwidth for every call, VoIP networks can share the same resources for data and voice traffic. In some cases, link overutilization or poor media characteristics (noise on RF links) can result in dropped or delayed packets.

VoIP traffic or packetized voice traffic needs to be sent at fixed intervals at the transmitting end so that the receiving end can predictably receive these packets and decode them. Because of serialization delay at the transmitting end, network delay, and jitter, these packets can arrive at the receiving end at varying intervals.

VoIP endpoints use dejitter buffers to compensate for variance in delay during media packet transmission through Real-time Transmission Protocol (RTP). If the dejitter buffers overflow because of excessive delay, this can impact voice quality so that it might sound like a robotized voice. Excessive packet loss can cause issues such as choppy voice quality.

Other voice quality issues can be caused by things such as codec mismatch, where each endpoint uses a different codec (for example, G.711 versus G.729).

Issues in Signaling Affecting the Services and Features

Voice signaling protocols such as MGCP, NCS, and SIP carry important information about how voice calls need to be set up, how resources need to be allocated, and how QoS needs to be provided to the voice traffic.

Network congestion and resource oversubscription can adversely affect voice-signaling protocols that can affect the services and features these protocols support. Voice signaling–related issues can also be caused because of improper network design and misconfigurations.

If voice signaling gets impaired, it can have a range of effects from delayed call setup to failed call setup, from loss of dial tone to one-way voice, and so on.

These issues are often caused by interoperability issues between equipment from different VoIP vendors. Even though they claim to be compliant with protocol standards, they can still have varying implementations of protocol stacks in their products.

IP Routing–Related Issues

SPs might deploy various routing protocols such as Open Shortest Path First (OSPF), Intermediate System–to–Intermediate System (ISIS), Border Gateway Protocol (BGP), and so on for providing IP connectivity across their infrastructure. These routing protocols carry network information that is used for calculating the most efficient path for carrying customer traffic through the SP network.

The failure of these routing protocols can result in a loss of IP connectivity or degraded service for the SP's customers. Such failures can severely impact VoIP traffic. If a link or node in the SP network fails, causing the routing protocol to reconverge or recalculate its routes, the voice traffic might be sent over a low-bandwidth link that can cause voice degradation. For this reason, the SP needs to carefully tweak routing protocol timers to make sure that the network can converge in a timely manner, minimizing the impact to VoIP traffic.

High Availability and Convergence for Business Continuity

A lot of non-Tier1 and non-Tier2 SPs might not implement redundancy in their network when deploying data-only applications. This becomes a critical issue when VoIP is deployed in such networks. A failed router or switch in the SP core network can cause loss of service to data and voice customers.

Therefore, it is critical for the SP to implement redundancy in the network so that a loss of a link or node does not result in loss of service to its users. Implementing redundancy in the network might involve deploying hardware and software with high-availability features. Redundancy is implemented at a device level where the hardware has active and standby components. If the active component fails, the standby can take over without causing a network outage. Redundancy is also implemented at a link level where multiple links provide connectivity to other network resources, so if a link fails, the other links can carry all the traffic. In some cases, the SP might also choose to deploy redundancy in the form of additional hardware that can take over if certain devices in the network fail. Redundancy can also be implemented in software, such as in routing protocol implementations, which can provide alternate routes through the SP network if the primary/best path fails.

The focus here is not the type of redundancy, or the various specific challenges, but the impact of the failures. Thus, effectively tracking key metrics can help in sustaining the VoIP service. These metrics can range from protocol to Layer 2/3 and h/w uptime metrics.

Summary

This chapter described some of the common SP voice deployment models that are deployed by major SPs in the United States and around the world. These networks are evolving, and a mind-set needs to be created for identifying a key performance indicator (KPI) across these networks. Thinking at the protocol layers and identifying network elements that can be collection points for these protocols are the key. After these KPIs are identified, monitoring them through a series of dashboards allows the SP to effectively manage its services.

References

1. Riddel, Jeff. *PacketCable Implementation*. Indianapolis, IN: Cisco Press, 2007.

2. Davis, Brian. *PacketCable Primer*. Cisco Systems, Inc.

Chapter 4

Internet Telephony

This chapter gives you an understanding of how Voice over IP (VoIP) is deployed over a publicly shared infrastructure such as the Internet. In these deployment models, the company providing VoIP services to end customers might not own the entire infrastructure that is used for deploying this service. It might use infrastructure owned by other entities to provide VoIP as an overlay service by deploying some of its own network components required for offering the VoIP service. Examples of such providers include Vonage (http://www.vonage.com), Skype (http://www.skype.com), and so on.

Service providers (SP) that offer VoIP services using the public Internet infrastructure are referred to as VoIP SPs. Because the VoIP SP might be an entirely different entity from the Network Access Provider (NAP), which is an entity that provides last-mile access to subscribers, it uses the NAP infrastructure to provide VoIP services to end customers. The NAP, on the other hand, offers not only the last-mile connection to the end user but can also offer connectivity to the public Internet.

Figure 4.1 shows a simplified view of how a VoIP SP uses the Internet and the NAP infrastructure to provide services to end users. The NAP provides Internet connectivity to end users, whereas the VoIP SP provides VoIP services. The VoIP SP's network components Session Initiation Protocol (SIP) proxies (gatekeepers, gateways, and so on) might be geographically dispersed over the Internet to provide better coverage to users in different locations (different cities and countries) and to provide scalability and redundancy in case the VoIP SP network resources get congested or experience a failure.

The challenge with this deployment approach is that the VoIP SP has control only over its own network components; it does not control the NAP infrastructure or other devices on the Internet. This makes it challenging for the VoIP SP to provide quality of service (QoS) to VoIP traffic and offer high-quality VoIP services to end customers. This deployment model is different from the models discussed in Chapter 3, "VoIP Deployment Models in Service Provider Networks," where VoIP services are provided by the broadband SP that also provides Internet connectivity to end users. Because the broadband SP, which is also the VoIP SP in this case, owns the last-mile connection to end users and uses its own infrastructure to provide VoIP services, it has a better capability to provide QoS to VoIP traffic and offer high-quality VoIP services to end users.

Figure 4-1 illustrates the domain of control boundary for the VoIP SP.

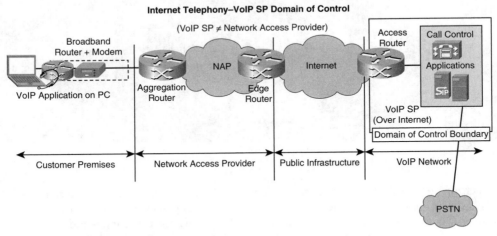

Figure 4-1 *Internet Telephony—VoIP SP Domain of Control*

This chapter also describes the various components deployed in the VoIP SP's network and their functions to illustrate how call functions are implemented to provide voice services to residential and business customers in this deployment model.

Internet Telephony Deployment Model

As discussed in Chapter 3, there are two different VoIP implementation models in SP networks:

- Centralized Switching Model
- Distributed Switching Model

Because the Centralized Switching Model was already covered in Chapter 3, this chapter covers the details of the Distributed Switching Model and explains how this model is used by some VoIP SPs to offer services to end customers by deploying VoIP as an overlay service over existing high-speed data networks such as cable, digital subscriber line (DSL), Fiber to the Home (FTTH), and so on. Because the VoIP SP might not own the end-to-end infrastructure needed to provide VoIP services to end customers, it faces several challenges when trying to provide QoS and service-level agreements (SLA) for VoIP services to its customers (especially business customers). These challenges and issues are discussed in more detail in the section "Common Issues and Problems with Internet Telephony," later in this chapter.

Even though the VoIP SP might not own the last-mile connection to the end users and the network infrastructure on the public Internet, it still needs to deploy its own network gear to provide VoIP services. Figure 4-2 illustrates a common deployment model for offering VoIP services over the public Internet and some of voice-related network components.

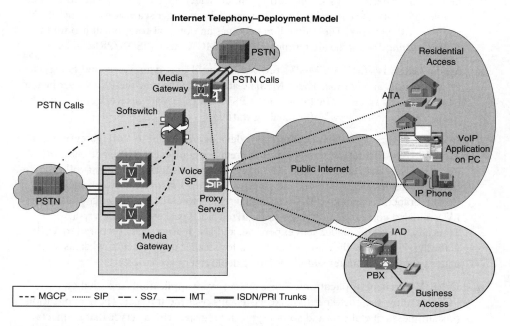

Figure 4-2   *Internet Telephony Deployment Model*

As illustrated in Figure 4-2, both business and residential customers connect to the VoIP SP network over the Internet. The VoIP endpoints (Analog Telephone Adapter [ATA], IP phones, VoIP applications on PCs, and so on) register with the SIP proxy server in the VoIP SP network before they can make VoIP calls. On-net calls (calls that originate and terminate within the VoIP SP network) are forwarded to different endpoints connected to the VoIP SP's network, whereas off-net calls (calls that originate within the VoIP SP's network but are destined to the public switched telephone network [PSTN]) are forwarded to the Media Gateways (MGW) to be handed off to the PSTN. There are two different connections to the PSTN illustrated in Figure 4-2. The MGW on the top connects to the PSTN through PRI or ISDN trunks and carry signaling and voice bearer traffic. The MGW on the left, which is connected to the Softswitch, connects to the PSTN through Inter-Machine Trunks (IMT) and carries only voice bearer traffic because the signaling is carried over the Signaling System 7 (SS7) links.

Some of these network components and their functionality are discussed in the following section. Note that some of the network components illustrated in Figure 4-2 overlap with network components discussed in Chapter 3, but their application is different in the deployment models discussed in this chapter.

Internet Telephony Network Elements

Some of the key components of the Internet Telephony model are as follows:

- **Softswitch:** The Softswitch is a call agent that performs the signaling and call-control tasks (such as digit analysis, call routing, circuit selection, and so on). The Softswitch enables SPs with the capability to route voice calls at larger scale between the PSTN and the VoIP network. The Softswitch becomes an optional component if both the VoIP signaling and media are terminated at the MGWs using ISDN/PRI trunks.

- **Media Gateways (MGW):** MGWs provide universal port data, voice, and fax services on any port at any time. These Media Gateways connect to the PSTN over bearer trunks to carry voice traffic between the PSTN and the VoIP network and connect to the call agent to carry VoIP signaling traffic.

- **Integrated Access Devices (IAD):** These devices are similar to VoIP gateways and carry VoIP traffic between the customer premises equipment (CPE) and the VoIP network. They interface with the SIP proxy server to negotiate different call functions (call setup, call teardown, and so on) and in the customer network such as a Private Branch Exchange (PBX) over digital trunks to carry analog voice traffic. The IAD can also convert analog voice to digitized VoIP packets and vice versa. Typically, IADs are smaller, less expensive, and less feature rich compared to VoIP gateways. Therefore, they are better suited for smaller deployments, whereas VoIP gateways might be better suited for larger deployment scenarios.

- **Other VoIP client applications:** These are software applications installed on user PCs that allow the customer to make VoIP calls over the Internet. These applications (instant messaging clients and so on) typically register with a server that connects multiple users over the Internet. After the users are connected, they can make voice calls between their PCs using these VoIP applications.

- **Private Branch Exchange (PBX):** A PBX (non-IP PBX is discussed in this deployment model) is a privately owned telephone switching system for handling multiple analog telephone lines without having to pay the phone company to lease each phone line separately. The PBX typically connects to the IAD, gateway, or the PSTN using digital trunks.

In this deployment model, residential customers typically use an ATA, an IP phone, or a VoIP application on their PC to make voice calls. Because residential customers typically subscribe to one or two phone lines, this functionality can easily be supported using small/cheap endpoints. On the other hand, business customers typically require more phone lines (depending on the size of the business, number of employees, and so on). In this case, the business customers could deploy an IAD that can support 24–48 phone lines (depending on the number of ports). The IAD can connect to the VoIP SP network over the public Internet on its uplink and provide VoIP services to the customers connected to its FXS ports. The IAD can also connect to an existing PBX in the customer network using digital trunks and connect to the VoIP SP network on its uplink to provide VoIP services to these customers. This is marked as Business Access in Figure 4-2.

Internet Telephony Applications

There are several options available to end users for subscribing to VoIP services over the Internet. These options vary anywhere from free VoIP services using PC-based software applications to paid services using an ATA or similar device. The paid services are similar to the traditional voice services provided by a telephone company in the sense that they might require the user to sign a contract with the VoIP SP and the billing is done on a monthly basis. The charges for these types of paid services are based on the package selected by end customers. For instance, some VoIP SPs might offer a flat-fee-based service that gives the end customer unlimited calling and other features such as a mailbox, three-way calling, call forwarding, and so on. Other packages are usage-based, where the customer is charged based on the number of minutes consumed and based on the features selected.

PC-Based Software Voice Applications

PC-based VoIP applications are commonly deployed in the residential customer space. The end customer does not need to purchase any special hardware to subscribe to the voice services. All the user needs is a PC-based software application that can make outgoing and receive incoming voice calls and a high-speed connection to the Internet (for good voice quality). The users make use of the built-in or externally connected microphone and speakers of the PC for the voice conversation.

The PC-based software application can be an instant messaging client (such as Yahoo! Messenger, MSN Messenger, Skype, and so on) that can be downloaded from the Internet free of charge. The user does not need to sign a contract with a VoIP SP or the ISP to receive voice services. The user does need to register with the provider of the instant messaging service and create an online account.

After the user creates an online account and signs on to the instant messaging service, she can communicate with other users registered and signed on with the same provider or a different provider of instant messaging service. If a user wants to make a VoIP call to another user, both parties need to be signed on to the instant messaging service, and both parties need to know each other's instant messaging ID. The ID is created when the users register for the service and is unique to each user.

If the user wishes to communicate with other users, he can create a buddy list (list of people the user frequently communicates with) or a group that includes a list of users identified by their instant messaging ID or their email address.

This solution is a good example of the distributed switching model because the end devices (applications) can directly initiate and terminate voice calls without being controlled by a centralized server or a Softswitch. One of the great benefits of these PC-to-PC-based voice calls is that the users can be physically located anywhere in the world and still be able to communicate with each other for free as long as they have an Internet connection and are signed on to the instant messaging service.

This can pose a challenge to the traditional voice service providers, such as the telephone companies, that generate revenue by providing voice services to these same customers. In some countries/regions, government entities or telephone companies might even block VoIP communication, especially cross-border or international voice communications using the PC-based applications mentioned previously.

In some cases, the provider of the instant messaging service can offer voice services for a minimal charge if the user wants to communicate with other users connected to the PSTN. Users can also get a directory number from the provider so that they can receive incoming calls from other users connected to the PSTN.

ATA-Based Voice Applications

If the customer wants to subscribe to other voice services such as a voice mailbox, three-way calling, call forwarding, caller ID, and so on, it might be more suitable to get voice services using an ATA. In this case, the customer needs to sign up with a VoIP SP that cannot only provide the hardware needed for this voice service but also the features listed previously. The voice service is typically a flat-fee service that offers the customer unlimited calling to certain destinations and international calling options for a per-minute service charge.

For this service to operate, the customer needs to wire the ATA device to his home IP network using an RJ-45 Ethernet cable so that the ATA can receive an IP address, connect to the Internet, and register with the VoIP SP's SIP server. (Note: SIP is the most commonly used protocol; therefore, it is referenced here for illustration purposes.) The ATA device also needs to be connected to an analog telephone that can be used for placing outbound calls and receiving incoming calls.

The ATA is typically configured with the VoIP SP's SIP server address so that it can connect to the server and register with it. The SIP server can facilitate the call setup and call teardown functions, and provide other voice services mentioned previously. The ATA device has a heartbeat mechanism, where it sends a registration message to the SIP server every few seconds to let the server know that it is still present and to verify that the server responds to these messages.

The advantages of this type of solution include the following. The customer typically does not need to sign a long-term contract with the VoIP SP. Customers can easily connect the preconfigured ATA device to their analog phone and their home IP network, and start using the voice service right away. The voice service is typically much cheaper than the service offered by the traditional telephone company. Additionally, the service is somewhat portable in the sense that the customers can take the ATA device with them anywhere they travel and still be able to make voice calls as long as they have connectivity to the Internet and the site's security system does not block the VoIP SP system.

Traffic Profiling

In Chapter 3, we discuss SP-managed VoIP deployment models where there are more powerful tools available for monitoring VoIP traffic. In this chapter, we discuss traffic profiling because it might be the only method available for the NAP to monitor VoIP traffic transported through its network.

For example, if the NAP offers voice services to its customers, it can treat its voice traffic at a higher priority compared to voice traffic from another VoIP SP by applying a different QoS policy for these two types of voice traffic. (One type of voice traffic is its own, whereas the other type is voice traffic from another VoIP SP.) Applying different QoS policies to voice traffic is critical; in the case of network congestion, the NAP can drop voice traffic from another VoIP SP in favor of its own voice traffic. This can help provide better service to customers who subscribe to voice services offered by the NAP as compared to customers who receive voice service from another VoIP SP.

The NAP can also block certain applications based on source/destination IP address (Layer 3) and User Datagram Protocol (UDP)/Transmission Control Protocol (TCP) port numbers (Layer 4 information). This can cause problems for the VoIP SP unless it has an agreement with the NAP to transport its VoIP customer traffic over the NAP infrastructure. This also poses some other challenges for the VoIP SP in trying to provide high-quality voice service to end customers, as discussed in the section "Common Issues and Problems with Internet Telephony," later in this chapter. There are other options for traffic control, including deep packet inspection and control of applications and content at Layer 7.

Potential Bottlenecks

As discussed in Chapter 3, bottlenecks can be avoided in a managed SP VoIP deployment model. In the Internet Telephony deployment models discussed in this chapter, the VoIP SP might not have the capability to avoid potential bottlenecks, because it might not own the last-mile connection to the end users.

Apart from the challenges faced by the VoIP SP because of its interaction with the NAP and other entities on the public Internet, the VoIP SP needs to worry about some of its own challenges and potential bottlenecks in its network. One of the possible areas for bottlenecks might be the SIP proxy server deployed by the VoIP SP. The server can easily be overloaded if a large number of end devices try to register with the server at one time. Therefore the VoIP SP needs to implement redundancy on the SIP proxy server and other devices in the voice network.

Another potential bottleneck can be the end devices themselves. If the end customer has a low-bandwidth last-mile connection, it can impact the quality of the VoIP call if the user is downloading a large data file off the Internet at the same time while making a VoIP phone call. If the end customer is using PC-based VoIP software, there can be a bottleneck because of processor resources on the PC. The last-mile connection bandwidth and end device/application-related issues are discussed in more detail in the section "Common Issues and Problems with Internet Telephony," later in this chapter.

Wholesale VoIP Solution

An IP-based voice network makes it possible to offer more network-based call transport services with advanced features at considerably lower costs by using the same network while maximizing the available bandwidth. SPs leverage the scalability of packet-based (generally IP) networks to deliver these services to residential, commercial, and even large enterprise customers while interconnecting with the PSTN. This interconnectivity increases the intrinsic value and network entropy. To extend the reach of the VoIP network, SPs have to build a platform that is standards based by leveraging International Telecommunications Union—Telecommunications Sector (ITU-T)- or Internet Engineering Task Force (IETF)-based protocols for compatibility with other systems and SPs. An IP network can be extended for global reach by providing an SS7 interface for interconnectivity to the PSTN. This increases the SP's reachability to other networks. The benefits can be transferred to its customer base, such as business-to-business communication, global Virtual Private Networks (VPN), and cheaper toll charges. SS7 is the signaling protocol defined by ITU-T (and ANSI in the United States) for the worldwide public switched telephone network, including mobile services such as GSM and Code Division Multiple Access (CDMA). The SS7 protocol has variants for most of the countries in the world. To connect to the PSTNs in multiple countries for global reach, a wholesale VoIP solution must address interoperability among various SS7 variants. Figure 4-3 depicts the wholesale VoIP solution, showing the key elements and their interconnectivity.

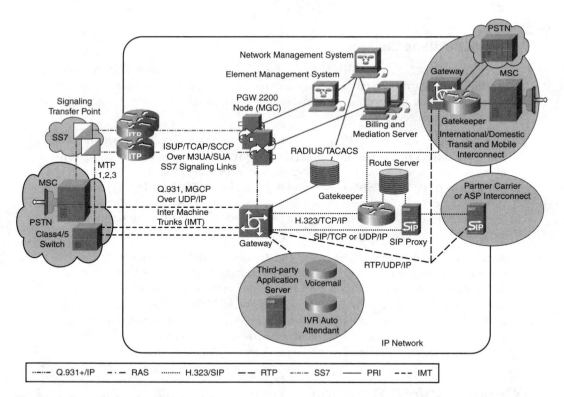

Figure 4-3 *Wholesale VoIP Model*

Key Network Elements

Some of the key components of the Wholesale VoIP Solution are discussed in the following sections.

Media Gateway Controller (MGC)

An MGC or a Softswitch is a central component of the wholesale voice network. PGW2200 (PGW) is a Cisco implementation of MGC, as shown in Figure 4-3. PGW is responsible for providing centralized call control including advanced call routing, numbering analysis, and features. These capabilities are based on its ability to interface with the PSTN and the IP network.

PGW2200 has ISDN and SS7 interfaces that interpret most of the variants of SS7 protocols based on ANSI (mainly North America) and ITU-T (international) standards. This capability enables PGW2200 to interface with the PSTN. PGW2200 terminates the ISDN User Part (ISUP), Telephone User Part (TUP), Transaction Capabilities Application Part (TCAP), and Signaling Connection Control Part (SCCP) that provide the functions defined for the transport, session, and applications layers in the OSI model for the SS7 protocol. The Message Transfer Part (MTP) of the SS7 protocol is terminated on an IP Transfer Point (ITP).

ISUP defines the protocol used in the call setup, feature implementation, and maintenance and releasing of the trunk circuits. The voice-traffic PGW2200 translates the ISUP messages into ISDN Q.931 or Media Gateway Control Protocol (MGCP) to send the instructions for call setup, feature activation, and call teardown to the Media Gateways, which are not directly capable of SS7 terminations for control signaling. SCCP and TCAP are used to provide advanced routing functionalities, such as global title translation (GTT) for resolving toll-free (for example, 800/888) numbers, calling card numbers, and mobile subscriber identification numbers. TCAP carries noncircuit-related data in the SS7 network. PGW uses TCAP to provide advanced features (for example, equipment identification, roaming, and SMS) by carrying Mobile Application Part (MAP) messages in mobile telephony networks, such as GSM. This provides the PGW with a capability to interconnect Mobile Switching Centers (MSC) in the mobile network with the VoIP network. MSCs perform a similar function in mobile phone networks as the Softswitch does in a VoIP network, including handling calls, SMS, handover, and call billing recordkeeping.

This Media Gateway controller uses H.323, MGCP, and SIP protocols to communicate with the Media Gateways and other devices in the IP-based packet network. This Softswitch can be considered a bridge between circuit-switched and packet-switched networks, providing greater interoperability options between the two networks and hence increasing the intrinsic value of both networks to wholesale voice network subscribers. The concept network's intrinsic value is based on the number of subscriptions and interconnectivity to other systems for greater reachability. It generates and stores call detail records. PGW provides interfaces for network management applications for operational purposes including provisioning and monitoring.

IP Transfer Point (ITP)

ITP performs physical SS7 link termination and terminates Message Transfer Part (MTP) Layers 1 through 3. MTP Layer 1 has information about the physical, electrical, and functional characteristics of the digital signaling link that are typically E1 (2048 kbps; x32 64-kbps channels), DS-1 (1544 kbps; x24 64-kbps channels), V.35 (64 kbps), DS-0 (64 kbps), and DS-0A (56 kbps). MTP Layer 2 is responsible for point-to-point transmission integrity by implementing flow control, message sequence validation, and error checking. MTP Layer 3 provides message routing in an SS7 network using point codes.

The Internet Engineering Task Force (IETF) defined a set of protocols as part of RFC 2719 known as SIGTRAN, which is derived from SIGnalling TRANsport, to facilitate the transport of SS7 messages over IP networks. SIGTRAN defines the Steam Control Transmission Protocol (SCTP), which operates directly on top of IP at the same level as the TCP SCTP's basic service, which is a connection-oriented reliable transfer of messages between peer SCTP users. SCTP facilitates the adaptation of SS7 MTP layers, including the MTP3 User Adaptation (M3UA) layer. M3UA supports the transport of any SS7 MTP3-user signaling, including ISUP and SCCP messages to ITP. Similarly, the SCCP User Adaptation (SUA) layer defines a protocol for the transport of any SS7 SCCP-user signaling message such as TCAP, or MAP over IP using SIGTRAN services.

SS7 links are typically terminated in specific telecommunication facilities. ITP can be placed in these facilities to terminate these links. MTP3 and above layers containing call setup information, including ISUP and TUP, can be transported to PGW2200 for call processing over an IP link. These higher SS7 layers are transported back from ITP to the PGW2200 Softswitch over the IP network using M3UA/SUA with SCTP as a reliable transport mechanism. ITP can be placed in a distributed fashion to provide greater SS7 connectivity options to centralized PGW2200 Softswitches. Alternatively, ITPs can be colocated with the PGW2200 (MGC) hosts, as shown in Figure 4-3. This enables greater flexibility to deploy PGW2200 or any other MGC in any data center and leverage common ITP for multiple MGCs.

Route Server

Route Server enhances routing capabilities of the wholesale voice solution by implementing complex routing logic based on business rules. Route Server can be an adjunct device for SIP proxies, H.323 gatekeepers, or the Media Gateway Controller, as shown in Figure 4-3. The Cisco implementation of Route Server is based on Carrier Sensitive Route Server (CSR), which uses Gatekeeper Transaction Message Protocol (GKTMP) to interface with the gatekeepers. The CSR has a database that contains business rules based on the least-cost path, time of day, percent allocation, and QoS. This database is highly optimized for per-call routing performance and supporting dynamic updates. CSR has a well-defined data scheme to facilitate imports and back-office system integration for dynamic updates of tariff information from all the carriers that can be accessed by the wholesale VoIP solution and call admission control/QoS parameters. This enables the system to place the call using the most efficient route from the most cost-effective carrier. Route servers from other vendors can include all, some, or more features such as user-preferred routes chosen by dialing steering codes.

Gatekeepers

Gatekeepers are H.323-based devices that provide call admission control and central dial plan resolution for all registered entities, including IP or ISDN trunks from Softswitches and IP-based PBXs, Media Gateways, and a variety of voice- and video-capable endpoints (IP Phone Voice CPE and videoconferencing equipment) in a zone managed by the same provider. Gatekeepers are used to provide scalability to large H.323-based networks. Gatekeepers can be networked together in a layering topology or clustering configuration for greater coverage and scalability. Directory gatekeepers are used to provide call routing between large wholesale VoIP networks. These VoIP network components facilitate cost-effective routing, resource management, and greater reliability while scaling the network for increased media traffic processing capacity. In that sense, gatekeepers are basically voice traffic cops, as shown in Figure 4-3.

In an SIP environment, SIP proxies provide an equivalent function including call admission control and route resolution. They can also be leveraged to connect with another partner carrier or SP using SIP instead of SS7, as shown in Figure 4-3. An SIP proxy can be used for peering with Application Service Providers (ASP) to use their services, including external directory lookup, name-to-number resolution, and roaming services.

Application Servers

Application servers can add a rich set of features to a wholesale voice network. Application server vendors include Broadsoft, Sylantro, and Vocaldata. In addition to PGW2200, Media Gateways, SIP proxies, and H.323 gatekeepers, application servers can offer the following:

- Differentiated services (for example, enhanced web-controlled services including voicemail and call-forwarding features)

- Usability of services (for example, subscriber GUIs instead of star codes such as *69 for callback)

- System Administrator control (for example, moves, adds, and changes of IP endpoints, users, and trunks done by the customer and not the service provider)

- Single-link connectivity to the desktop (no separate voice and data connectivity for subscriber)

Element Management Systems (EMS)

Element Management Systems (EMS) and Network Management Systems (NMS) provide fault and performance monitoring, provisioning, and processing billing records for network components, including Softswitches, Media Gateways, gatekeepers, proxies, and infrastructure elements such as LAN switches and routers. Cisco MGC Node Manager

(CMNM), Billing and Mediation Server (BAMS), Cisco Access Registrar (CAR), and Cisco Unified Operations Manager (CUOM) are examples of an EMS.

An EMS provides an application programming interface (API) for NMS interaction and large data repositories for ease of management and storing critical data for operational purposes. NMS consolidates data points from all the individual Element Management Systems to correlate fault monitoring and to coordinate flow-through service provisioning, accounting, and performance. It provides a comprehensive view of security through network surveillance (Fault Configuration Accounting Performance Security [FCAPS]) for the overall network. This information flow is explained in Figure 2-3 in Chapter 2, "A Metrics-Based Approach for Managing the VoIP Network." NMS products include Netcool, Cisco Information Center (CIC), RADIUS or TACACS server for managing network access, and HP Openview. These systems are managed and monitored by Network Operations Center (NOC) staff on a continuous basis to provide service assurance.

Figure 4-3 describes the Wholesale VoIP Model; its various applications are discussed in the next section.

Wholesale Voice Applications

A wholesale voice network provides an infrastructure on which to build various applications and services that leverage the IP routing framework and reachability through their network. The sections that follow describe some of these applications.

Prepaid and Postpaid Calling Solutions

Wholesale voice networks based on PGW2200 can provide prepaid and postpaid card calling services. This solution incorporates a RADIUS server and an Interactive Voice Response (IVR) server. A Cisco IOS–based Media Gateway can also run IVR service through the Voice XML (VXML) service feature, which is capable of audio file playout to the caller. VXML capability leverages eXtensible Markup Language (XML) to deliver voice prompts with greater flexibility over IP networks. Alternatively, an external IVR server can be used that might use other protocols such a Computer Telephony Interface (CTI) or VXML. Figure 4-4 depicts the interconnectivity of the solution components.

A caller can dial an access number of a trunk terminating on a Media Gateway that is controlled by the PGW. This call is established with the aid of PGW2200, which interprets the SS7 call setup messages from the PSTN caller that are contained in the ISUP layer. PGW220 then translates these messages into ISDN Q.931 or MGCP and relays them to the gateway, which terminates the trunk that is identified in the ISUP message by Circuit Identifier Code, or CIC. This Media Gateway is referred to as an originating gateway in the wholesale voice network shown in Figure 4-4. The originating gateway invokes the IVR prompts using VXML service, as configured in its dial plan through dial peers, after successful termination. Dial peers on the Media Gateway contain the dial plan information, including the dialing pattern to match to a destination, which could be another gateway or a trunk, codec selection, and other call characteristics such as caller ID or references to VXML prompts. The caller is prompted to enter the card number and PIN information for validation through a RADIUS server. This information is collected by the gateway using DTMF and sent through the Authentication, Authorization, and Accounting (AAA) protocol to a RADIUS server for processing.

Figure 4-4 illustrates the Prepaid Service Model and its various components.

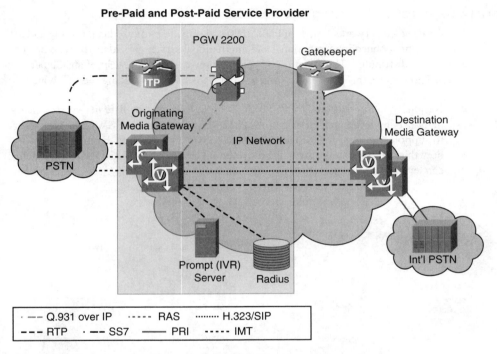

Figure 4-4   *Prepaid and Postpaid Service Model*

After the caller is authenticated by the RADIUS server, the originating gateway plays another IVR prompt (either natively or leveraging an external IVR prompt server) instructing the caller to enter his desired destination number. This information is collected through DTMF and is cached by the gateway while it sends the digits to the RADIUS server. The caller is authorized to make a call to the destination of his choosing, or not, based on his credit information stored in the RADIUS database. The originating gateway, after getting a validation notification from the RADIUS server using the AAA protocol, sends the digits corresponding to the destination number using H.323 or SIP to extend the call through the IP network. This call might be routed through an SIP proxy in the case of an SIP or a gatekeeper if H.323 is used. These devices contain the call-routing information to locate the appropriate destination gateway in the SP's network.

The gatekeeper determines the network bandwidth availability to the destination gateway. After permission is granted by the gatekeeper using the H.323 RAS protocol to the originating gateway, it extends the call using either the SIP or H.323 protocol to the destination gateway while starting the call accounting with the RADIUS server for billing purposes.

The destination gateway further extends the call through its trunks to the PSTN based on the called party information that it received from the originating gateway. The duration of the call is controlled by either the calling or called party. It is commonly superseded by the credit allowed (and maintained) by the RADIUS server.

Network Transit and Trunking Applications

Wholesale voice networks based on PGW2200 can provide optimized call routing and a variety of interconnections to national and international service providers that can be geographically distributed. This function is typically provided with the help of an adjunct route server. This enables this platform to offer a variety of services, including the following:

■ National and international transit (routed through an intermediate network or clearinghouse) by leveraging either SIP- or H.323-based routing or PGW2200's capability to support a variety of ANSI and ITU-T SS7 ISUP/TUP variants. Figure 4-5 shows how this can be achieved by using either an SIP proxy or an H.323 gateway, which can leverage a route server for the most optimum routing.

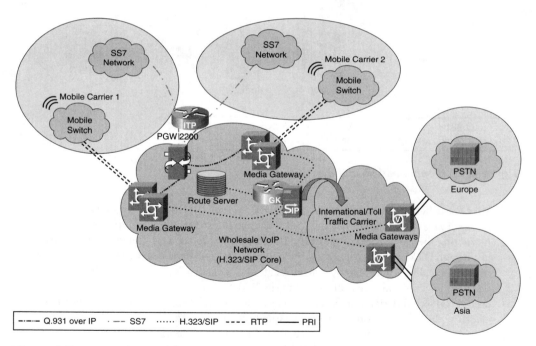

Figure 4-5 *Network Transit and Trunking Applications*

■ Mobile Switching Center (MSC) trunking service for mobile carriers can be offered through this platform because it facilitates the implementation of a VoIP transport solution to reduce leased line costs. This solution can provide MSC-optimized call-routing options to access the wireline local, national, and international PSTN with signaling transparency between switches over the IP infrastructure. Figure 4-5 illustrates how a wholesale VoIP SP can provide interconnectivity and transit to multiple mobile SPs and access to the international PSTN using SIP or H.323 in the IP network instead of SS7 in the core.

Managed Services for Enterprises

SPs can manage voice services on behalf of their enterprise customers rather than just providing trunk access. These managed services—including dial plan maintenance; phone or user moves; adds, changes, deletions (MACD); and advanced features such as devices mobility, single number reach, and voicemail—can be offered through this IP-based platform to business customers. The benefits include reduced operational expenses through a converged infrastructure (leveraging multiservice VPNs), administration staff reduction, on-net calling by way of consolidated and managed dial plans, and lower toll charges by virtue of optimized call routing for the enterprise customers. This creates an additional revenue stream for the SP with an expanded services portfolio for enterprise customers or even residential consumers. Multilocation businesses work more seamlessly using abbreviated dialing within their global corporate network leveraging a central dial plan in PGW2200. Application servers can provide value and service to small and medium businesses. The end users for these business customers can be empowered through web-based user interfaces to control feature subscription and customization. These features can include call forwarding, unified messaging involving voicemail delivery through email or transcribing voicemail to text for delivery through SMS, and single-number reach features, where an incoming call can be sent to a list of numbers assigned to the called party's profile.

Applications and Benefits for Service Providers

A wholesale VoIP network enables SPs to integrate with their IP infrastructure and roll out multiservice networks and other revenue-generating services. SPs can also benefit by reducing their expenses by consolidating access trunks and IP trunks for access to the PSTN. Because SS7-based interconnects with Inter-Machine Trunks (IMT) incur reduced tariff charges compared to ISDN/PRI-based facilities, this solution can reduce recurring charges for the service providers while providing them with greater scalability and reachability options. Scalability is achieved by adding more IMTs on demand that are controlled by the existing SS7 signaling link and the associated dial plan. SS7 also provides reachability options using local number portability (LNP), international access through support of different SS7 variants, and database lookup capability by accessing a Service Control Point (SCP) in an SS7 network.

Common Issues and Problems with Internet Telephony

This section discusses some of the challenges and common problems encountered by a VoIP SP when providing VoIP services over the Internet to residential and business customers. Some of these scenarios are described in Figure 4-6.

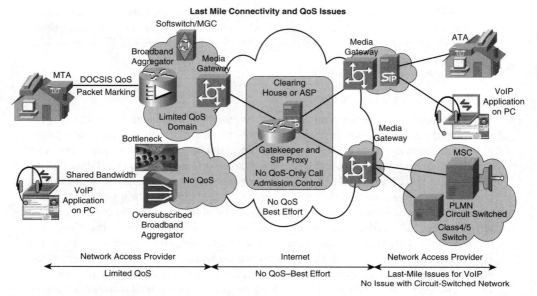

Figure 4-6   *Last-Mile Connectivity and QoS Issues*

The VoIP SP network components, including Softswitches, Media Gateways, and other call-processing devices (SIP proxies, gatekeepers, and so on), are deployed over the Internet. In some cases, these components provide registration services to VoIP end-points, for example, in cases of instant messaging clients, where they need to register with the SP's server before they can make VoIP calls.

Last-Mile Connection Bandwidth

If the VoIP SP does not own the last-mile connection to the end users, it becomes diffi-cult to provide QoS to voice traffic. If the last-mile connection is congested because of traffic overload, customer traffic including voice might get dropped, causing degraded service to end users. Because voice traffic is time-sensitive and is impacted more than data traffic by packet loss, jitter, and latency in the network, the VoIP SP needs to make sure that voice traffic is always prioritized over best-effort data traffic. Some broadband ISPs bundle voice services with their basic data service offerings. This enables them to provide basic QoS (for example, using RSVP/DOCSIS as discussed in Chapter 3) for voice traffic prioritization over data traffic; however, this provides QoS only between the subscribers of the same ISP. The QoS domain is limited only to the provider-controlled network.

The VoIP SP needs to *try* to make sure that voice traffic is prioritized, but it can't always do so successfully when it doesn't control the end-to-end path of the voice traffic, which is always the case in the Internet Telephony service model discussed in this chapter.

End devices, such as the ATA and VoIP client applications, can be configured to provide a higher priority to voice traffic to provide the best possible voice quality to the users, but this might not always guarantee processor resources (for software-based VoIP applications) or network resources to voice traffic. One of the techniques used on end devices is to mark voice traffic with a higher class of service as compared to data traffic. Typically, voice-signaling traffic is marked with an IP precedence value of 3, whereas voice bearer traffic is marked with an IP precedence value of 5. This might help the voice traffic get better treatment when network resources are congested, but this approach can work only if the IP precedence values are preserved end to end and intermediate devices are configured to provide QoS based on the IP precedence values. This approach would not be practical in the Internet Telephony deployment models discussed in this chapter. It might work for the deployment models discussed in Chapter 3.

In some cases, the SP providing the last-mile connection might reset the IP precedence value of the VoIP SP's VoIP traffic to 0, causing this traffic to be treated just like best-effort data traffic. This makes it even more challenging for the VoIP SP to provide QoS to its VoIP traffic.

Another factor that can impact the quality of VoIP calls is a different connection bandwidth in the last mile. For example, if User A makes a VoIP call to User B, and User A has a higher-connection bandwidth (> 1 Mbps) than User B (< 64 kbps), the voice traffic might still be impacted because of congestion, which causes dropped packets. This is common when using PC-based VoIP software to make voice calls to other PCs and other VoIP end devices.

If the VoIP SP owns the last-mile connection (the VoIP SP is also the ISP), it makes it possible to provide the appropriate QoS to VoIP traffic because the VoIP SP has more control over network resources. Therefore, it can preserve the IP precedence values of on-net voice traffic end to end and provide preferential treatment to VoIP traffic as compared to other data traffic in the network. This approach might not work when the VoIP SP's customers call other users connected to the PSTN or users who subscribe to a voice service from a different provider.

End Device/Application-Related Issues

A properly configured IP network with end-to-end implementation QoS parameters provides predictable timing behavior of the media packet flows using Real-time Transport Protocol (RTP) as they egress the network boundary. Assured and timely delivery of media packets with reasonable delay and minimum jitter are tremendous benefits of QoS. However, they can be lost when the endpoints in the VoIP network are equipped with conventional operating systems such as Windows OS, Linux, Mac OS, and so on. This is typically the case with personal computers and handheld devices such as smartphones or PDAs. VoIP client application developers only have access to rudimentary tools for processor resource management, such as process priority adjustment. Most media client applications get resource allocation on a best-effort basis. The resource allocation and reservation for these applications are typically made at the launch of the application and remain in effect during its entire runtime. Because these applications offer more than voice and video capability, such as instant messaging and file sharing, the resource requirements of all subtasks are unknown in advance. This makes it extremely difficult for the application to invoke a CPU management routine for resource reservation, even if it is allowed by the underlying operation system. This poses a problem for these media client applications to digitize the voice and video signal using DSPs (either a sound card or software-based DSP implementation), affix appropriate headers, and put media packets on the network with manageable delay and minimum jitter.

However, there are media engines (software or hardware resources with media-processing capabilities including transcoding) especially made for soft clients for PCs, smartphones, and small CPEs. Their goal is to optimally manage system resources for the media-capable soft client so that it can put media packets on the network with minimal delay and without jitter. Such media engines are commonly used by Skype, AOL, Apple's iPhone, and other soft client development vendors. Issues such as version compatibility, operating system firewalls, and other competing applications can still pose a challenge for processor resource management for soft client applications.

No Customer Service-Level Agreements (SLA)

One of the challenges faced by the VoIP SP that provides VoIP services over the public Internet is that it might not be able to provide any service-level agreements (SLA) to its end customers. This is a more critical issue when providing services to business customers as compared to residential customers.

For business customers, losing their VoIP service because of a network outage or experiencing degraded service because of network congestion can translate into loss of revenue and lost business opportunities. If they do not have an SLA in place with their VoIP SP, they need to absorb the cost to recover their lost business opportunities, which affects their bottom line. On the other hand, for residential users, losing their VoIP service might not be as critical, especially if they do not rely on the VoIP service as their primary line for voice.

From the VoIP SP's standpoint, loss of VoIP service or degraded service can definitely cause higher call volume into their call centers and directly affect customer satisfaction. It might also make it more challenging for the VoIP SP to sell its services to business customers who might be guaranteed an SLA from the traditional voice providers such as the telco.

Issues with Emergency Calls (E911)

The original 911 technology standard was developed to transmit a caller's critical location data to the closest public safety answering point (PSAP). This is done by a database lookup against the calling party number contained in the automatic number identification field (ANI) with the Automatic Location Information (ALI) database that is based on a master street address guide (MSAG) in the PSTN. ALI and MSAG are maintained by the SP. This technology was improved by E911 (Enhanced 911) as a result of the Federal Communications Commission's (FCC) directive in the United States with the advent of devices that don't have a fixed location, such as cell phones; hence, it is difficult to track their location. The main challenges came from wireless phones. A mobile phone user can be anywhere in the cell coverage area that now typically encompasses the entire globe. Similarly, a VoIP subscriber can potentially use the assigned phone number associated with a location many miles away from his actual physical presence. Because of the global nature of Internet, the user can be in a different country while using his domestic directory number. Whereas this is a convenience, it poses a challenge to provide 911 services.

This was a manageable challenge for wireless E911. In Phase 1 of the E911 plan imposed by the FCC, the location of the wireless 911 caller was determined by the associated cell tower within one mile. In Phase 2, the location accuracy was increased to 100 meters by adopting a triangulation technique that involved signals of three or more nearby cell towers.

However, VoIP E911 remains problematic because of the nature of the Internet. An FCC regulation adopted in 2005 only mandates VoIP service providers to send the 911 call with the reference of the registered physical location. However, because the soft clients can be anywhere where there is Internet connectivity, it is not always an effective approach. SPs can choose to have their subscribers sign waivers at the time of subscription to make them aware of this limitation and to mitigate possible legal ramifications.

Another challenge with providing E911 service with VoIP is that this technology requires interconnectivity that depends upon routers, access modems, and PCs, and a power outage can cause the VoIP E911 system to fail. Some VoIP endpoints might have a battery backup system to provide power in the case of a power outage, but the majority of them might not, especially if the end user buys an inexpensive off-the-shelf product (readily available on the Internet and in local electronics stores). If the VoIP SP network does not have battery backup, a power outage can result in a VoIP service outage, causing E911 to fail. This is not an issue with traditional voice services provided by telephone companies where the user can still get a dial tone and place E911 calls even if there is a power outage.

Security Issues

Because RTP packets containing voice are essentially data packets, an access to a network segment carrying VoIP traffic using a packet sniffer enables a hacker to capture the packets and play them out using a variety of codecs. Techniques for improving VoIP security, such as encryption, physical security, and VLAN segmentation of the network, are possible in enterprise networks, but practically impossible on the Internet.

Similarly, sending spurious traffic to VoIP services or endpoints can launch a denial of service attack to disrupt the normal service. This is especially an easy technique where an SIP is used for call setup.

Theft of service is also a threat to VoIP SPs that use Internet to connect their subscribers through remote Media Gateways.

Generally, encryption coupled with compression for better performance in the Internet is prescribed for mitigating most of the security risks. However, even compression and encryption do not provide a guarantee against VoIP call tapping. The algorithms involving preservation of the length of voice patterns when used with variable bit rate (VBR) compression techniques can reveal a significant amount of information about the conversation, enabling hackers to reconstruct it with fairly good accuracy.

Based on research conducted by Charles V. Wright, Lucas Ballard, Scott E. Coull, Fabian Monrose, and Gerald M. Masson of Johns Hopkins University, which is documented in their research paper "Spot Me if You Can: Uncovering Spoken Phrases in Encrypted VoIP Conversations," "On average, our method achieves recall of 50 percent and precision of 51 percent for a wide variety of phonetically rich phrases spoken by a diverse collection of speakers."

The VBR compression method uses the higher compression rate for simpler sounds and a lower compression for more complex sounds. Because these bit patterns are also preserved in length-preserving encryption, it is possible to identify the spoken sounds. The researchers considered the dictionary size, word variation, and impact of noise when developing this bit rate–based speech recognition technique. Padding techniques can make the voice pattern length more constant, which makes it harder to detect the speech pattern. However, the voice encryption standards and protocols do not yet specify standards for padding.

Summary

Providing VoIP services over the public Internet can be a challenging proposition for the VoIP SP. Not only does the VoIP SP face challenges in trying to provide QoS to the VoIP traffic end to end, but it also faces issues that might not be directly under its control, such as problems caused because of end user equipment such as PC-based VoIP software applications, cheap-quality handsets, and so on.

It is definitely more challenging to sell VoIP services to business customers as compared to residential customers because of the reasons mentioned in this chapter. However, in the end, if the VoIP SP can provide VoIP services at a much lower cost as compared to the traditional voice providers such as the telco, the cost savings might outweigh the QoS issues. It all depends on what the end user is looking for (a reliable primary-line voice service versus an additional secondary phone line) and whether their goal is to subscribe to a low-cost VoIP service versus a guaranteed SLA-based service.

References

1. Wright, Charles V.; Ballard, Lucas; Coull, Scott; Monrose, Fabian; and Masson, Gerald. "Spot Me if You Can: Uncovering Spoken Phrases in Encrypted VoIP Conversations," IEEE Symposium on Security and Privacy in 2008; May 18-22, 2008, ISSN: 1081-6011, ISBN: 978-0-7695-3168-7.

2. Broersma, Matthew. "Compresson lets attackers tap VOIP calls," *Network World*. June 17, 2008.

3. Domjan, Hans and Gross, Thomas R. "Extending a Best-Effort Operating System to Provide QoS Processor Management"; *Quality of Service* — IWQoS 2001, June 2001, ISBN: 978-3-540-42217-4.

4. Edwards, John. "The Essential Guide to VoIP and E911," *VoIP News*. April 4, 2008.

## Chapter 5

VoIP Deployment Models in Enterprise Networks

This chapter explains the main components that form the Cisco Unified Communications (UC) solution as deployed on a converged IP network in enterprise networks. The intent of this chapter is to describe the most commonly deployed scenarios for Voice over Internet Protocol (VoIP) in enterprise networks to build a foundation for voice-specific network management and optimization topics in the later chapters.

Various deployment models are described that are used in a typical enterprise network including the fundamental models: central call processing and distributed call processing. Large campus deployment schemes are also discussed. This chapter examines the differences in hosted and managed services around UC solutions. It also presents a brief overview of IP Contact Centers (IPCC), which are essentially an extended functionality of a UC solution. The idea is to give the user background information to understand the management strategy and optimization recommendations without going into the design details needed to build these networks from the ground up.

Some of the security concerns that arise when media is converged with data applications are also discussed here. Where a UC solution provides benefits to an enterprise by enhancing productivity through a rich set of features and functionalities and reduced cost of ownership, it also raises some challenges for network administrators. This chapter discusses issues involving the effect on the media quality, possible signaling messages flow impairments, and IP routing issues affecting the overall UC solutions. For business continuity, UC solutions are generally deployed with redundant components, multiple call paths, and provisions in aiding predictable failover and fail-back of services while preserving the media sessions.

This chapter introduces the serviceability elements, namely, billing records such as Call Detail Records (CDR), debug, and system logs in addition to the common network-monitoring tools such as Simple Network Management Protocol (SNMP). A detailed discussion of comprehensive monitoring strategy appears in Chapter 6, "Managing VoIP Networks," and Chapter 7, "Performance Analysis and Fault Isolation."

Unified Communication Solution Components in Enterprise Networks

This section describes the key components of Unified Communications solutions in enterprise networks, including Unified Communications Manager, gateways, gatekeepers, session border controllers, messaging applications, and other rich media applications.

Unified Communications Manager/CallManager

Cisco Unified Communications Manager (CUCM or UCM), formerly known as Cisco CallManager (CCM), is the central component of Enterprise Unified Communications networks. UCM is a software application that runs on a fault-tolerant Media Convergence Server (MCS). Unified Communications Manager (CUCM) performs the following main functions:

- Keeping track of all the endpoints by registering them, providing firmware images, maintaining their configurations, and processing call setup requests. Endpoints include various types of IP phones, IP adapters for analog phones, and fax machines.

- Interacting and often controlling voice gateways, gatekeepers, and trunks to other call-processing systems to provide access to external systems while managing call admission control.

- Managing dial plan functions including routing paths for dialed destination, implementation of Class of Restriction/Service (CoR/CoS), tracking and maintaining bandwidth consumption for different locations such as Call Admission Control (CAC), and providing alternative call routing paths when needed.

- Providing features such as Conferencing (Ad Hoc and Meet ME, Call Barge, and Call Join), media transcoding, Call Park, mobility, Extension Mobility, personal assistant features (IPMA), address book, Speed Dials, Abbreviated Dialing, and Music on Hold (MoH), to name a few.

- Provide connectivity to legacy PBX- and TDM-based components for the VoIP networks that are in transition.

- Providing serviceability elements including application programming interface (API) for provisioning such as eXtensible Markup Language and Simple Object Access Protocol (XML/SOAP), billing records for all the calls, system logs, SNMP-based alarms and alerts, and debug logs for fault detection. CUCM also acts like a management proxy for its endpoints to indicate their availability, registration status, and location by providing information on their status using system logs and alerts.

- Providing a web interface on phones that enables new phone services to be added, such as stock ticker quotes, timecard applications, weather information, and so on.

Note   This book refers to Cisco Unified Communication Manager as Unified CM or CUCM. UC Manager is also synonymous with Cisco CallManager or CCM. Normally, UCM or CUCM is used when referring to versions v5.x to v7.x. Versions 4.x and below are normally referred to as CCM. This book uses CUCM.

To provide greater scalability and redundancy at the same time, UCM servers are grouped to form a cluster. A cluster can have multiple call-processing servers in addition to a publisher, Trivial File Transfer Protocol (TFTP) server to manage device configurations, and Music on Hold (MoH) server. The number of subscribers is based on the number of endpoints registered with the CUCM cluster. The maximum number of subscribers supported in the cluster is related to the specifications of the database used by the specific CUCM software release. A publisher in the CUCM cluster contains a read/write database that stores the CUCM configuration information, although the subscribers in the cluster that are primarily responsible for processing calls and providing features might provide the ability to change the database if the publisher is offline.

This clustering scheme is supported over a WAN to provide geographical redundancy. However, splitting clusters over WAN links must meet criteria involving network latency and bandwidth as defined in Unified Communications Solution Reference Network Design (SRND) guides as published on Cisco.com. See http://www.cisco.com/en/US/netsol/ns818/networking_solutions_program_home.html.

Voice Gateways

An autonomous VoIP network can be defined as a network that has a common call admission scheme, a common dial plan with the same dialing rules, feature implementation, and routing scheme. A UC network in an enterprise network is an autonomous network or a collection of autonomous networks. For a single autonomous network, it needs to be connected to the rest of the world. An enterprise with multiple autonomous VoIP networks leverages a service provider's network to connect these networks. For example, different enterprise campuses located in geographically dispersed regions utilize service provider network resources such as Multi-protocol Label Switching (MPLS) networks, Virtual Private Networks (VPN), or any other form of IP transport to link with each other.

Voice gateways are used in UC networks to interconnect enterprise Voice over IP (VoIP) networks to various other traditional telephony systems such as PBX, a key system that is a small-scale PBX, and public land and mobile network (PLMN). Voice gateways also connect the various part of the enterprise such as remote branches, international locations, or mobile units.

Cisco voice gateways support a wide range of standards-based signaling protocols such as H.323, Media Gateway Control Protocol (MGCP), and Session Initiation Protocol (SIP) as well as CAS and CCS (ISDN BRI & PRI and Q.Sig) for digital connections. These gateways can also be used to support analog devices such as fax machines, modems, and point of sales (PoS) devices. The analog interfaces include "ear & mouth" or "recEive and transMit" (E&M), Foreign Exchange Office (FXO), and Foreign Exchange Station (FXS).

Cisco voice gateways range from entry-level to carrier-class sizes. They can be feature rich and integrated with routers or Catalyst switches. The interconnect capability of these gateways is supported through a variety of interfaces ranging from STS, T3, E3, E1/T1, and FXS/FXO on the telephony side. IP connectivity is provided through Ethernet, serial, ATM, or POS interfaces.

A voice gateway's geographical placement primarily determines the telephony interface, VoIP control protocol, and its size. For example, a centrally located gateway serving a large number of users is a large-capacity gateway with digital trunks. If it provides inter-connectivity, it might even have IP trunks. A local voice gateway in a branch location is a smaller gateway and can have fewer analog FXO lines based on the limited need of the users in that location. An FXS port on the gateway might provide connectivity to fax machines or point of sales modems.

A Cisco gateway most likely serves as a digital signal processor (DSP) farm to facilitate features such as media transcoding and conferencing in addition to PCM conversion (Codec selection G.711, G.729, and so on) of voice signals between IP telephony and the public switched telephone network (PSTN). A voice gateway running Cisco IOS Software can also provide fallback support for IP phones that are connected behind it in case these endpoints lose connectivity with the UC cluster. This feature is known as Survivable Remote Site Telephony (SRST) in the Cisco implementation of Unified Communications networks. The IP phones registered to the local voice enable a router that functions as a Call Processing Agent or Call Agent until the connectivity to the UC cluster is restored.

Gatekeepers

A gatekeeper is an H.323 device in a VoIP network that facilitates address translation and network access control (like a traffic cop) for H.323 endpoints, including gateways and IP trunks to call-processing entities like a CUCM cluster. IP trunks are essentially IP connections between the devices that are capable of carrying voice traffic and the associated signaling traffic similar to a TDM trunk such as Primary Rate Interface (PRI). Gatekeepers also provide other services such as bandwidth management, accounting, and centralized dial plan resolution. Their presence in a VoIP network is optional provided that these critical functions are offloaded to other call control entities.

Gatekeepers employ a concept of a "zone." A *zone* is the logical collection of H.323 nodes such as gateways, terminals, and MCUs registered with the gatekeeper. There can be only one active gatekeeper per zone. According to H.323 standards, a gatekeeper has the following mandatory functions:

- **Address translation:** Translates H.323 IDs (such as gateway-A@domainX.com) and E.164 numbers (standard telephone numbers) to endpoint IP addresses.

- **Call Admission Control (CAC):** Controls endpoint registration into the H.323 network by allowing (or rejecting) registration requests. Facilitates CAC to permit a certain number of calls as statically defined by the administrator and/or dynamically calculated by the CAC algorithms.

- **Bandwidth control:** This function is about managing endpoint bandwidth requirements. Gatekeeper uses H.225 RAS messages that are used to provide this function by managing bandwidth based on codec utilization and interzone bandwidth.

- **Zone management:** Provides zone management for all registered endpoints in the zone including their registration process and alternate destinations.

Optional gatekeeper functions as defined in the H.323 standard include the following:

- **Call authorization:** Restricts access to certain terminals or gateways and/or has time-of-day policies restrict access.

- **Call management:** Maintains active call information and uses it to indicate busy endpoints or redirect calls.

- **Bandwidth management:** Rejects admission when the required bandwidth is not available. This function leverages H.225 RAS methods. H.225 RAS is part of the H.323 protocol family that provides call signaling for endpoint registration, processes, admission requests, and the maintenance of bandwidth changes.

- **Call control signaling:** Routes call-signaling messages between H.323 endpoints with the use of the Gatekeeper-Routed Call Signaling (GKRCS) and Gatekeeper Transaction Message Protocol (GKTMP) for advanced features leveraging intelligent call control entities.

Gatekeeper deployments are generally done in a redundant fashion using alternative gatekeepers or a Cisco-proprietary gatekeeper clustering scheme for greater resiliency.

Session Border Controller

A Session Border Controller (SBC) exists in the VoIP networks to control signaling and sometimes media streams as well. An SBC often defines a point of demarcation between two VoIP networks owned and/or controlled by different entities. Typical examples include an SBC between two different enterprise networks, between an enterprise network and a service provider network, or between two different service provider networks. SBC provides the following functions in an enterprise VoIP network:

- **Security:** SBC can provide security by hiding an enterprise's private network space from the peering entity.

- **Protocol translation:** A Cisco SBC might convert one protocol to a different protocol. For example, it can aid in joining an H.323 network with an SIP network. This functionality can be expanded to normalize billing and call detail records, in which a call setup involving different signaling protocols appears as a single call.

- **Media normalization:** A Cisco SBC can help in transporting often complex dual-tone multifrequency (DTMF) tones and various fax protocols, and normalize codecs used between two networks by using its transcoding feature aided by on-board DSPs.

- **QoS and bandwidth management:** A Cisco SBC can aid in QoS marking of media frames or Real-time Transport Protocol (RTP) packets using TOS or DSCP. This will aid the SBC itself and possibly other network devices in the call path to enforce bandwidth using RSVP and codec filtering.

SBCs are now frequently deployed in enterprise networks to take advantage of the new IP-based interconnectivity options (as a time-division multiplexing [TDM] alternative) offered by various service providers. As enterprises realize the benefits of collaboration between companies for productivity gains, SBC can also act as a back-to-back user agent (B2BUA) between enterprises to facilitate secured sessions across network boundaries. This enables on-net dialing, access to the company directory, and other advanced features such as presence status and videoconferencing.

SBCs are generally deployed in redundant schemes leveraging Domain Name Service (DNS) or Server Load Balancing (SLB) techniques, or using the Hot Standby Router Protocol (HSRP).

Note   Cisco has used various terms for SBC, including IP-to-IP Gateway and Cisco Unified Border Element (CUBE). This book refers to SBC as CUBE when discussed as the Cisco IOS–based SBC. The products mentioned here support SBC or CUBE functionality at the time of this writing.

Messaging Application

Voicemail is a critical communication method in an enterprise. In a Cisco Unified Communications framework, Cisco Unity delivers a feature-rich unified messaging solution for Microsoft Exchange or Lotus Domino environments. In Unified Messaging (UM), email, voice, and fax messages are sent to one inbox. The subscriber of a particular mailbox has an option to retrieve voice messages through a telephone user interface (TUI) or by opening as an item in his/her inbox through Microsoft Outlook for Exchange deployments or through Lotus Notes for Lotus Domino deployments. This framework also supports voicemail in a standalone format that does not require integration with Microsoft Exchange. This option also allows the user to access his voicemail messages through a GUI or a web interface accessible from a PC or a smartphone.

Cisco Unity can be integrated with CUCM using the Cisco-proprietary Skinnny Call Control Protocol (SCCP) or can be integrated with other standards-based voicemail systems using the Simplified Message Desk Interface (SMDI). Cisco Unity can also integrate with proprietary voicemail systems such as Lucent, Avaya, or Nortel using the Cisco Digital PBX Adapter DPA7630 and DPA7610.

Cisco Unity also provides a set of tools to enhance the serviceability of this solution. This toolset includes bulk provisioning tools, tools to generate usage reports, configuration wizards, and integration wizards.

Rich Media Applications

A wide range of applications are offered by Cisco under the Unified Communications framework for enterprises. These applications provide more features and capabilities to the overall communication network to enhance collaboration and productivity. Some of the applications are discussed here briefly.

Cisco Unified MeetingPlace and WebEx

The Cisco Unified MeetingPlace application provides a platform for collaboration in enterprise environments by facilitating voice, video, and web conferencing. It can be integrated with CUCM and Microsoft Exchange in a Unified Communications network. MeetingPlace enables a user to schedule conferences through a web interface or through the Microsoft Outlook calendaring application. For a scheduled conference session, participants call in to a central number, after which they are placed into the conference call after going through an optional authentication process. Alternatively, the MeetingPlace application can be configured to dial the participants of the scheduled conference call. It also enables ad hoc conferencing sessions that can be launched on demand.

Cisco WebEx is a hosted model for collaboration that provides a subscription-based option for voice, video, and web conferencing. In this Software as a Service (SaaS) model where the provider, in this case Cisco WebEx, hosts the entire infrastructure for a collaboration platform, enterprise and commercial customers can subscribe to a variety of collaboration services and are billed based on usage. The end users of this platform use a secure web connection over the Internet to connect to a collaboration space for sharing documents, presentations, and their desktops. The WebEx platform allows the end user to be connected using voice and video for the same session by providing a very feature-rich and coherent collaboration platform that can scale based on demand. WebEx applications can be integrated with other applications such as Microsoft Office, allowing the users to instantly launch collaboration sessions, schedule meetings, and share documents.

Cisco Unified Presence

Cisco Unified Presence enhances the value of an enterprise's VoIP network by advertising the users' status and willingness to communicate and collaborate with fellow users on the same network. Cisco Unified Presence consists of two main components:

- Cisco Unified Presence Server
- Cisco Unified Personal Communicator (client)

Cisco Unified Presence Server integrates with CUCM to retrieve the presence status of the subscribers. It also integrates with other applications, including Cisco Unity and

MeetingPlace, to provide a comprehensive unified communication platform. The presence server can also integrate with third-party presence servers such as Jabber, Microsoft Office Communications Server (OCS), and IBM Lotus Sametime to extend the presence capability to these desktop applications in the enterprise network. It can also aid in forming federations to extend the presence capability between two distinct forests in a Microsoft Active Directory environment.

Cisco Unified Personal Communicator (CUPC) is a client application running on Microsoft or Macintosh desktops. It can also be used as a soft phone with video capability that requires it to be registered with the UCM. This integration also enables CUPC to control the desktop IP phone and check voice messages. CUPC subscribers can use this client to subscribe to the presence status of their subscribers of interest, watch their presence status, initiate text chat with them, promote the chat session to voice or video session, or collaborate using a feature-rich MeetingPlace application from within the client.

CUPC authentication can be done either through a local end user database on UC manager or through a Lightweight Directory Access Protocol (LDAP)–capable server such as Microsoft Active Directory.

Cisco Emergency Responder

A Unified Communications network provides flexibility to move the IP phones from one location to another without the involvement of an administrator, and at the same time, it creates a challenge to track the current location of IP phones. This is a concern in the United States and Canada, where users dialing 911 to make emergency calls to get services of ambulance, fire department, or police must provide their location. The main purpose is to route the call to the appropriate public safety answering point (PSAP) and secondarily to dispatch the appropriate emergency service personnel to the right geographical location. The location is determined by querying an Automatic Location Information (ALI) database that maps the phone numbers of the 911 caller to the caller's physical location. This is also referred to as Enhanced 911 or E911.

In larger organizations that maintain their own phone systems, the onus of responsibility lies on them to maintain the ALI database information and provide updates to the ALI database maintained by the local exchange carrier (LEC) from time to time.

Cisco Emergency Responder (CER) is designed to address the E911 requirement for multiline telephone systems (MLTS) deployed in North America. The term *MLTS* covers systems such as PBX and key systems that provide multiple extensions that are typically routed through common pilot numbers. These multiple lines or extensions can be spread over a large geographical area. CER dynamically tracks the location of the phone based on the switch port to which the phone is attached using an Emergency Location Identification Number (ELIN). The ELIN contains the LEC's ALI information. When someone calls 911, the ELIN is passed as the caller ID, and when it reaches the 911 operator, the ALI information is queried based on the ELIN that it receives. This allows the 911 operator to dispatch the emergency personnel to the exact location without relying on the caller to provide this information.

CER periodically polls the switches that connect the IP phones through SNMP and updates the location database that is locally stored. The ports on the switches are statically mapped to a physical location. This polling establishes a link between the phone and the physical location by switch port association. In the case of Wi-Fi endpoints, the location is tied to a wireless access point.

Cisco Unified Contact Center

Cisco Unified Contact Center Enterprise (UCCE) or Cisco Unified Contact Center Express (UCCX) provides enterprises with a contact management or customer relationship management (CRM) solution with Automated Call Distribution (ACD), Interactive Voice Response (IVR), and Computer Telephony Interface (CTI) functionality. The ACD queues and distributes based on administrator-defined rules. For example, the incoming calls can be routed based on the caller-entered information through IVR and then matched with the skill set (defined in its database) of the appropriate contact center representative. If the representative, also known as an agent, is not readily available, the call is queued. Upon an agent's availability, the call is connected to the appropriate agent. The agent gets the caller-entered information through the CTI that can also help to open a screen on the agent's desktop with the caller's account information.

The connectivity between the contact center routing engine (on either UCCE or UCCX) and CUCM is maintained by Java Telephony API (JTAPI) and the CUCM Computer Telephony Integration Manager (CTIM). CTIM is used to manage multiple JTAPI connections on separate subscribers in a CUCM cluster.

The choice of either Cisco Unified Contact Center Enterprise or Express depends on the organization's call volume and size of the deployment. UCCE is for large deployments or those with specialized needs. UCCX is for small to medium deployments.

Cisco Unified Application Environment

The Cisco Unified Application Environment offers developers a platform to build feature-rich converged voice, video, and data applications. These applications can be integrated with any other core application used in the enterprise such as Oracle, SAP, medical-imaging applications, inventory tracking systems, hospitality applications, and gaming or entertainment applications. This tremendously increases the value of the Unified Communications platform by integrating it with an organization's core business to drive productivity and enhance the user experience.

Common Enterprise Deployment Models

IP telephony networks not only link legacy PBXs, but they can also replace them with an IP-based central call-processing entity such as a UC Manager. UC Manager enables enterprises to leverage their existing data network for delivering media traffic to all the locations that are connected it. Unified Communications solutions on a converged IP network can be deployed using one of the following models.

Centralized Call Processing

This model is suitable for large enterprise campuses that have a large group of employees located at a central location and a smaller number of employees in remote branches that have WAN connectivity with the central site. This model can also be applied to a business that is comprised of a large number of branches, such as a bank with a national or even international presence. In this scenario, the central site can be a data center with WAN aggregation points with only the essential IT staff for operational support. Essentially, it is most suitable for hub-and-spoke topologies. Figure 5-1 shows the centralized call-processing model with dispersed branches.

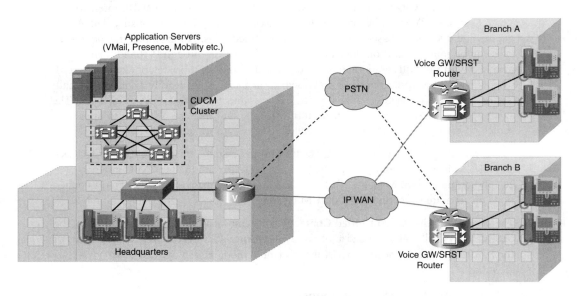

Figure 5-1 *Centralized Call-Processing Deployment Model*

The centrally located UC Manager cluster provides dial tone, all the features to the central site, and all the remote branches over the IP WAN. Calls between all the endpoints controlled by the UC Manager are routed over the converged network. If the WAN link coming into the remote site does not have enough bandwidth, the local branch gateway can provide direct connectivity to the PSTN. In this scenario, the call signaling, which consumes less bandwidth, is still provided from the central UCM cluster over the WAN link.

Therefore, the quality of service (QoS) is enabled throughout the converged network starting from the endpoint, which is typically an IP phone, and then is extended through an access switching layer, distribution layer, core layer, and then the WAN access devices or routers to ensure smooth delivery of call-signaling and media packets. The calls destined to a destination that is not a part of this unified communications network (off-net calls) egress through voice gateways to access the public land and mobile network (PLMN).

Note   The term *public switched telephone network (PSTN)* is used for traditional telephony networks. The *public land and mobile network (PLMN)* covers the global network that IP networks might interface with for universal connectivity. This book uses PLMN and PSTN interchangeably, but both terms refer to the global network maintained by the mobile and landline service providers around the world.

The central site has a large-capacity gateway or multiple gateways. These gateways in the central site almost always terminate digital trunks. Alternatively, an IP trunk, if offered by the service provider, can be terminated at the central site. IP trunks are typically terminated on a Session Border Controller (SBC). The number or size of the trunks is determined by performing traffic engineering analysis during the planning and design phases of deployment. The remote branches also have gateways that might be appropriately sized according to the number of employees in each branch. The branch gateways terminate either analog or digital trunks based on the size of the branch. Typically, larger branches use digital trunks such as T1 or E1. Smaller branches may only need a few analog trunks to satisfy their capacity requirements.

The gateways in remote branches also provide failover support to the endpoints, including IP phones, fax machines, and other devices. This is a Cisco-proprietary feature known as Survivable Remote Site Telephony (SRST) in case of WAN outage. The goal of SRST is to make the user experience transparent in case of WAN outage. This allows the users in a remote branch to continue the use of the telephony system without interruption, although some of the advanced call features might not be available in SRST mode, such as presence status of the user, that requires periodic updates from the central entity like CUCM.

The users might lose some advanced features such as presence status, but they can continue to make and receive calls on their IP phones. Because the internal call traffic flow over the WAN links with finite capacity, this bandwidth is managed by the CUCM cluster in the central location. The CUCM recognizes each branch as an individual location in its database and tracks its availability. When a call is made between any two branches or between a branch and the central site, the consumed bandwidth is tracked by the UC Manager. This bandwidth management capability is known as CAC. If CUCM determines that there is not enough bandwidth at a central location, it uses an alternate route through the PSTN. This capability is known as Automated Alternate Routing (AAR) in the Cisco UC context. AAR makes the entire experience of using an IP phone to reach an on-net destination transparent to the user in the case of WAN congestion. It is recommended to use wideband codecs such as G.711 or G.722 within the location locally because of the higher bandwidth availability.

In most scenarios, the WAN bandwidth is limited and a precious commodity because these circuits are leased from service providers and hence incur recurring expense. Although media traffic is prioritized over data traffic in the converged network because QoS is enabled, it should not affect the data applications. Therefore, it is recommended to use lower-bandwidth codecs such as G.729. The trade-off is justified because the Mean

Opinion Scores (MOS) for G.729 are close to G.711 scores but provide significant bandwidth savings on these WAN links.

The centralized call-processing model manages the dial plan centrally using MGCP with a limited subset in branch gateways that are also SRST routers. If the design employs the H.323 protocol, a significant portion of the dial plan configuration resides on the gateways. The local H.323 dial plan on the gateways enables them to continue to route calls in failover mode.

This model also allows consolidation of valuable PSTN trunk resources in one central location, where one large gateway or a stack of gateways terminate the PSTN trunks. Because the other applications, such as Cisco Unity for voicemail, Unified Contact Center Express, Cisco Emergency Responder, and Cisco Unified Presence servers, are centrally located, it is a cost-effective deployment model that also streamlines the administration of the unified communications network in particular and the overall converged network in general.

Distributed Call Processing

In a distributed call-processing model, the call-processing resources such as UC Manager, Cisco Unity, IVR, or other applications are located at each geographically separate site. The size of the site determines the number and size of resources. For example, Figure 5-2 shows that larger remote sites have CUCM clusters.

Figure 5-3 shows that smaller sites can deploy Cisco Unified Communication Manager Express (CUCME).

The dial plan and the bandwidth management tasks are also distributed in this model. Each CUCM cluster and/or CUCME maintains its own dial plan and tracks the bandwidth required.

Gatekeeper is commonly used to consolidate and streamline the bandwidth management and overall dial plan administration. All the CUCM clusters and/or CUCME nodes can connect through the central gatekeeper or the gatekeeper cluster for greater scalability. This is accomplished by the CUCM cluster by registering an intercluster trunk with the gatekeeper. Similarly, IOS-based CUCME registers itself as a gateway with the gatekeeper. The gatekeeper maintains the bandwidth between all the sites by associating them with a zone. When the gatekeeper determines that the bandwidth allocated between the two zones (sites) is completely depleted, it sends an admission reject (ARJ) message back to the CUCM cluster or the CUCME. This triggers the AAR function in the CUCM, whereas the IOS-based CUCM can select an alternate dial peer to use the PSTN/PLMN link that is local to that site.

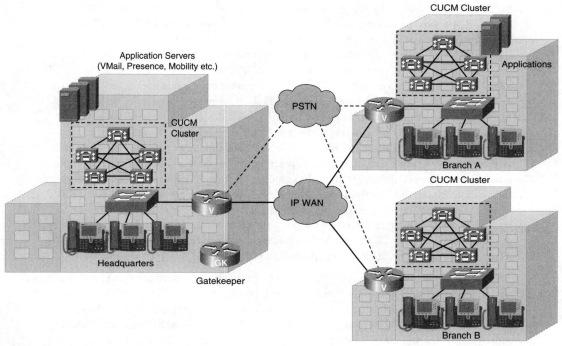

Figure 5-2 *Distributed Call-Processing Deployment Model with CUCM Clusters*

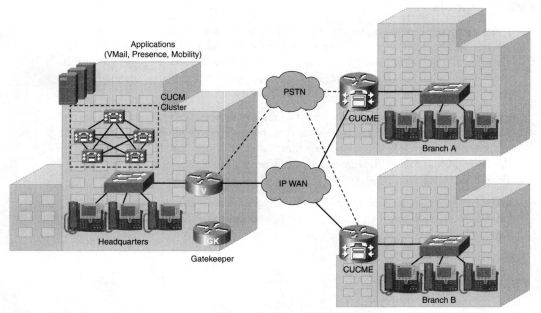

Figure 5-3 *Distributed Call-Processing Deployment Model with CUCME Routers*

Bandwidth management by the gatekeeper is topology agnostic, which means it will not be aware of topology changes and their effect on actual bandwidth availability. This creates a challenge in MPLS networks or when an alternative network path is utilized as a backup. We recommend using the Resource Reservation Protocol (RSVP) in these scenarios. RSVP is a topology-aware signaling protocol. With RSVP as a CAC scheme, an RSVP-enabled application such as a CUCM attempts to make a resource reservation for a specific amount of bandwidth for a media session between two devices by querying the output interfaces of all the intermediate RSVP-enabled devices in the network path. The amount of bandwidth depends on the session (voice or video) and the codec used (for example, G.711, G.729, or H.261 for video). RSVP uses the underlying routing protocols to traverse the network path to make this reservation; hence it is topology aware. All the intermediate network devices in the call path might not be RSVP aware. In this scenario, they simply ignore the RSVP reservation request. This poses a potential risk, where RSVP unaware/incompatible devices run out of bandwidth and still ignore the RSVP query (as opposed to denying the request). This is not a significant issue if these devices don't pose as the network bottleneck. However, their bandwidth utilization should be measured using alternate schemes such as an explicit (and static) configuration of session limit.

This model reduces the reliance on the IP WAN links, which reduces the failure domain to a single site. This deployment model can be resource intensive if other applications such as voicemail, presence, and mobility in addition to call control using CUCM are present at every site. This also increases the administration tasks. Therefore, this model is most suitable for an enterprise with multiple campus locations of significant size.

Hybrid Models

Large campuses can be composed of one single large building, several buildings that are geographically close, or a split campus where several buildings may be located in different parts of a metroplex area. These large-scale deployments often employ central and distributed call-processing methods because of geographical, topological, and business requirement variations and the associated complexity.

CUCM, voicemail, IVR application servers, presence servers, voice gateways, and mobility servers should be colocated in a single large building, for example, a high-rise building, as shown in Figure 5-4. All the endpoints, including IP phones and fax machines, are on the same LAN as the call/feature processing servers. This is also referred to as a stand-alone deployment model.

If the campus has several buildings, the call-processing servers, especially a CUCM cluster, can be split between multiple buildings for greater resiliency. This is shown in Figure 5-5.

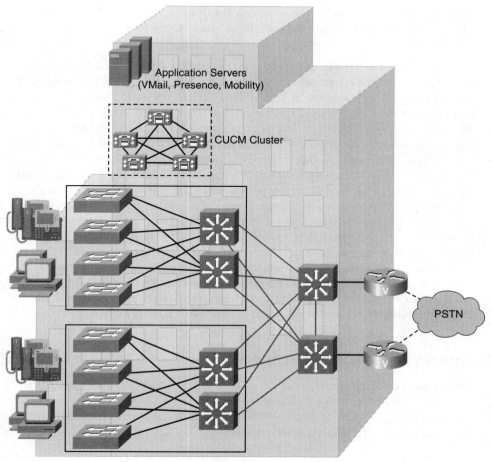

Figure 5-4 *Standalone Campus Deployment Model*

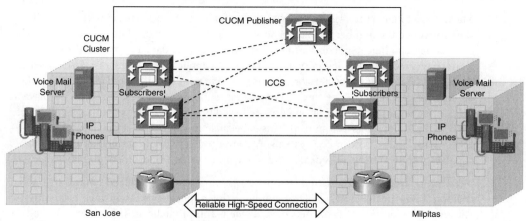

Figure 5-5 *Split-Cluster Design for Distributed Campus or Data Centers*

This split-cluster model can also be used for a distributed call-processing design, in which the CUCM cluster is split between the two data centers for geographical redundancy.

If all the servers in a CUCM cluster are not in the same LAN, an additional 900 KB of bandwidth is needed for intracluster communication and signaling. Typically, all campus buildings are connected through high-speed links that are at least 100 Mbps. Enterprises generally own these links because they are on their premises. If the buildings of a large enterprise are geographically split in a metroplex area, they might be connected through high-speed metro links if offered by the local provider. The CUCM cluster servers can be split between the two geographically dispersed locations with the highest employee concentrations. It might be necessary to place gateways in different building if they fall under different LATAs. Often IP trunk service providers can consolidate the dial plan for a large-enterprise customer, which eliminates the need to split the voice gateways. The dial plan consolidation on the service provider's (SP's) end ensures that the directory number assigned to different LATAs can be routed through the same trunk that is terminated on a central gateway in the enterprise network. Alternatively, placement of voice gateways in different sites with local PSTN connectivity ensures that in the case of metro link outage, the off-net voice calls are preserved and the on-net calls can complete over the PSTN. With a split CUCM cluster, all the features are preserved except for the ones that need to rely on the publisher of the cluster, such as the extension mobility feature.

Common Issues and Problems

VoIP deployment in enterprises results in some new issues, such as security and exacerbation of the existing communication network problems such as voice quality and signaling impairments. This section discusses some of the most common issues that the network administrator has to monitor and mitigate.

Convergence-Related Issues

Management in an enterprise is always looking for cost savings. This motivates the IT staff to reduce the number of parallel networks that serve various business-critical applications that can be categorized into two broad categories: data and media. IP-based networks are typically designed initially for data-only applications. Media is carried over a dedicated communications network that is set up in parallel with the data system, including its own cabling infrastructure. Of course, greenfield enterprises, which do not have a preexisting legacy communication network based purely on TDM technology, might set up converged networks as a startup. This is because it is not concerned about interoperability with legacy PBX or key systems. A converged network is defined as a network capable of transmitting all types of traffic including data, voice, video, and images.

Network convergence is justified by operational cost savings, reduced complexity, and the possibility of delivering richer communication applications to enhance productivity. Although the journey toward this convergence is full of challenges, it requires a mind-set change in the IT organizations that typically have separate and often disjointed data and telecom groups that now need to collaborate for the sake of convergence. The challenges are overcoming reluctance on both sides, learning new technology, and having sensitivity toward each other's concerns to make the change happen successfully. For example, data network administrators can be reluctant to open the network and deploy quality of service and other required changes to make the network ready for voice deployment. Similarly, the telecom administrators have always regarded "dial tone" as a mission-critical application for the enterprise. This might make them reluctant to transition voice systems over data networks that are often considered as best-effort and unreliable networks for critical and real-time media traffic.

Issues Affecting Media Quality

Circuit-switched networks have a dedicated path and fixed bandwidth for every single call. The analog voice signal is digitally coded using pulse code modulation (PCM) for transmission in the circuit-switched network. In a PSTN network, several PCM streams can be aggregated into a larger digital stream for transmission over a single physical link. This aggregation technique is known as time-division multiplexing (TDM).

In contrast to circuit-switched networks, packet-switched networks might not always send the media packets through the same path, and the available bandwidth is elastic. The PCM-encoded voice is converted into media packets using codecs on the receiving end (typically an IP phone) and then decoded back to the PCM on the sending end (a voice gateway interfacing the PSTN). This process introduces serialization and packetization delay. These media packets are prone to drop, delay, and jitter because of the nature of a packet-switched network.

The stability of an IP network is even more critical for communication-related applications than for data applications. Real-time media traffic must be transmitted as a steady stream with minimum and constant delay without any drop. Converged VoIP networks employ dejitter buffers to compensate for variance in delay during media packet transmission through the Real-time Transport Protocol (RTP), which in turn creates additional delay as discussed in Chapter 1, "Voice over IP (VoIP) and Network Management." If there is occasional single-packet loss or the delay is beyond the bounds of the dejitter buffer playout period, the voice sounds like a synthetic (or robotic) voice. Consecutive packet loss during transmission makes the voice choppy.

Codec impairments or mismatch can make the voice sound like an underwater voice. A codec impairment, which causes the signal to attenuate, makes the voice soft, whereas too much gain can make the voice sound fuzzy. Codec mismatch can also have one-way audio or not allow a call to complete.

Hybrid circuits in any voice network are the cause of voice echo. This is because of the impedance mismatch in the hybrid (analog and digital) circuits between 2-wire (analog) to 4-wire (digital) conversion. There is additional delay in packet-switched (converged) networks because of packetization of voice, transmission through varying network paths involving queuing, and dejitter buffers. This causes the inherent echo (caused by impedance mismatch in the hybrid circuit) to become exacerbated to a degree that makes it perceivable to human ears because there is a significant gap between actual conversation and the reflected (echoed) signal of the same conversation.

Quality of service (QoS) implementation is imperative in converged IP networks to minimize these issues. However, bandwidth elasticity and underprovisioning of the priority queues dedicated for voice traffic may result in tail drops, which occur when the dejitter buffer proves to be insufficient and overflows. CAC techniques are deployed to prevent this possible situation. Gatekeepers can be used to implement CAC, as discussed in the section "Centralized Call Processing," earlier in this chapter. Gatekeepers are network topology agnostic. Hence they may not be effective, and as a result, may not be helpful in preserving voice quality. Alternatively, RSVP can be deployed to provide CAC that is network topology aware. However, if all the nodes in the network path are not RSVP aware, it may render this scheme ineffective, especially if they represent the network bottleneck.

Voice-Signaling Protocol Impairments

Voice-signaling protocols, such as H.323, MGCP, SCCP, and SIP, are all prone to impairments because of improper network design, misconfiguration, or instability. This affects the services and features these protocols support.

These impairments can range from delayed call setup to failed call setup attempts. The worst-case scenario includes where the caller might not get either correct or timely feedback when entering digits for reaching the intended called party or using features because of DTMF transport issues by the signaling protocols.

Different vendors of VoIP communication gears, even though they claim to be compliant with protocol standards, might have varying implementations of a protocol stack in their products. This is a cause for incompatibility between different call-processing agents in the call path, resulting in a call setup failure. For example, SIP interoperability is a challenge because the extent of standards-based implementation, including features and options, might be different between vendors. Also, vendors often have varying interpretations of the SIP standards.

Voice Security in Enterprise Converged Networks

VoIP inherits all the security-related issues that exist in the IP network and are applicable to data applications. This includes denial of service attacks, spamming, eavesdropping, and connection hijacking leading to call fraud. These problems are exacerbated if a more vulnerable access medium, such as wireless LAN, is involved in the call path.

Voice is a unique application for the IP networks because for every session, the endpoint opens a separate socket for the media traffic using RTP, in which the call is established using the signaling protocol on a different socket. RTP chooses a random port number for every session. This creates a security challenge because the range is quite wide (from 16384 to 32767). This is in contrast to Internet services, where ports are statically assigned and are carefully allowed by the firewalls. This RTP behavior is also a challenge for Network Address Translators (NAT) because the call setup protocols assume that the endpoint advertising the socket (IP address and port number) is routable from the remote host and unmodified by the network. This is contradictory to the NAT principle employed by the firewalls to define a boundary between the Internet-routable address space on the outside and the enterprise network on the inside.

Although open-standard protocols such as SIP are advantageous for greater interoperability to increase the intrinsic value of the network, it makes them prone to vulnerabilities such as DoS attacks. Also, the greater extensibility of SIP opens it to greater potential attack scenarios in general. These scenarios and the mitigation techniques are discussed in depth in Chapter 8, "Trend Analysis and Optimization."

Summary

This chapter discussed various components that formed a Unified Communications network for an enterprise. The chapter provided a high-level overview of commonly used deployment models in different enterprise networks. Some common issues for a converged network were also highlighted in the end. This chapter focused on setting a context of discussion for the preceding chapters in Part III of this book.

Although the benefits of converging unified communications with the enterprise's IP network are unquestionable, the risk and issues must be managed. Chapters 6 and onward discuss in detail the available tools and describe how to maintain these networks while constantly optimizing their performance to derive maximum benefit of the convergence of Unified Communications with the IP network.

References

1. Kaza, Ramesh and Asadullah, Salman. *Cisco IP Telephony: Planning, Design, Implementation, Operation, and Optimization.* Indianapolis, IN: Cisco Press, 2005.

2. Cisco Unified Communications Solution Reference Network Design 6.0, http://www.cisco.com/go/srnd.

3. Dantu, Ghosal, and Schulzrinne. "Securing Voice over IP," IEEE Network, September/October 2006.

## Chapter 6

Managing VoIP Networks

To ensure voice quality and to optimize media delivery over the IP, it is crucial to properly plan, design, implement, and manage the underlying network. This chapter discusses the best practices for planning media deployment over IP networks. It starts with how to assess the readiness of the network, traffic engineering, high availability, and managing the IP network and its integrated components that process voice and other media transmissions. These concepts are essential because managing poorly planned and designed networks is a vain effort. Managing a Voice over Internet Protocol (VoIP) network starts from a discovery process that maps out the entire network by identifying all the network elements, their roles, and the key performance indicators (KPI) to track and establish a baseline and effectively monitor it on regular basis.

This chapter covers the most common monitoring mechanisms available to network administrators and their corresponding scopes and effectiveness in managing VoIP. All networks are bound to experience outages. Managing outages and correlating network problems with end-user problems is critical. This chapter also explores how incident management systems, including trouble ticketing systems, are integrated with network management. In summary, this chapter establishes the baseline for managing VoIP networks.

Requirements for Enabling Voice in IP Networks

The task of managing VoIP networks begins even before the implementation phase when the decision is made to deploy a VoIP network. Because traditionally voice communications have been regarded as a mission-critical service that supports public safety access points for emergency services and critical business support over time-division multiplexing (TDM) networks, the same level of service is expected from VoIP networks because they are tasked with providing the primary method of voice communications.

Network Readiness Assessment

To ensure a consistent level of service, the underlying IP network must

■ Have the bandwidth and performance to handle converged services including media and data

■ Meet the demand of high-availability voice services by providing resiliency to mitigate the effect of network outages

■ Be modular, hierarchical, and consistent to promote consistency and manageability

This predeployment IP network infrastructure assessment should address each of these areas. The infrastructure assessment is accomplished by gathering the needed information from network engineering staff and direct scanning and performance data gathering from the network devices to evaluate the current or planned network implementation, including hardware, software, network design, security baseline, network links, and power/environment.

The following sections examine areas of the infrastructure assessment that must be evaluated.

Network Design

This section discusses the following aspects of the network design that must be evaluated during network readiness assessment for VoIP deployment:

■ **Hierarchy and modularity:** Network hierarchy is perhaps the single most important aspect of network design resiliency. A hierarchical network is easier to understand and easier to support because consistent, expected data flows for all applications occur on the network over similar access, distribution, and backbone layers. This significantly reduces the complexity of network management, increases the understanding and supportability of the network, and often results in decreased traffic flow problems and troubleshooting requirements, and improved IP routing convergence times. Network hierarchy also improves the scalability of the network by enabling it to grow without major network changes. Hierarchy also promotes address summarization, which is important in larger IP routing environments.

Network modularity can be defined as a consistent building block for each hierarchical layer of the network and should include like devices, such as configurations and identical software versions. By using a consistent "model" for each layer of the network, supportability can be improved as it becomes much easier to properly test modules, create troubleshooting procedures, document network components, train support staff, and quickly replace broken components. Each defined hierarchical layer should have a basic solution that is used repeatedly throughout the network. If enhancements or special requirements are added to specific modules, special attention should be paid to testing, documentation, and supportability.

■ **IP routing:** IP routing is a design issue for all larger IP network environments. The primary issue is that the routing protocol converges quickly following various failure scenarios. In addition, the routing scales differently in the particular network environment. The readiness assessment process should include IP routing protocol selection, configuration, IP summarization, and IP routing protocol safeguards to prevent IP overhead and undesirable routing loops. In general, Cisco recommends Open Shortest Path First (OSPF) or Enhanced IGRP (EIGRP) for improved convergence with added variable-length subnet mask (VLSM) support. In environments that anticipate an excess of 100 routes in a routing table on core and distribution layer devices, IP summarization is recommended into the core, generally configured at the distribution layer. For WAN environments, additional summarization might be needed at the edge to reduce overhead on slower WAN links. Routing protocol safeguards are also recommended to reduce overhead and prevent unexpected routing behavior, such as routing across user or server virtual local-area networks (VLAN). In larger WAN environments, it is also recommended that only routes from a particular area be routed into the core and that a particular site should not be a potential reroute point for core or other major traffic.

- **IP addressing:** The IP addressing scheme's evaluation should investigate specifically how the allocation of current IP address space affects the allocation of IP addressing for IP phones and other communications devices. In most cases, an organization does not have available within its current allocation an IP address space within the existing user VLANs or subnets for an IP phone rollout. The following strategies exist for the allocation of space:

  - Increasing subnet size.

  - Providing additional VLANs and subnets for phones using either additional access ports or 802.1Q trunking on user ports.

  - Using secondary interfaces to allocate additional address space where 802.1Q trunking is not supported.

  - Network Address Translation (NAT) should be used with caution. The VoIP network architect should explore the impact on voice signaling protocols when packets from endpoints with translated IP addresses traverse firewalls and proxy servers.

- **Hot Standby Router Protocol (HSRP):** HSRP is a Cisco-specific software feature implementation as described in RFC 2281 that permits redundant IP default gateways on server and client subnets. HSRP can be configured for router prioritization to identify the primary and backup device, preempt the capability to return the gateway to a higher-priority router, and provide "backbone track" support to track the availability of a backbone or WAN interface on the router. On user or server subnets that require default gateway support, HSRP provides increased resiliency by providing a redundant IP default gateway.

- **Quality of service (QoS):** Voice quality on a network is ensured by the use of QoS features. These features must be enabled and available end to end in a network to provide high-quality voice services on the converged network.

Voice quality is affected by two major factors: lost packets and packets with varying delay (jitter). Packet loss causes voice clipping and skips. The industry-standard codec algorithms used in Cisco digital signal processors (DSP) can correct for up to 30 ms of lost voice. Cisco VoIP technology uses 20-ms samples of voice payload per VoIP packet. Therefore, for the codec correction algorithms to be effective, only a single packet can be lost during any given time. Excessive packet delay can cause either voice quality degradation because of the end-to-end voice latency or packet loss if the delay is variable. If the delay is variable, such as queue delay in bursty data environments, there is a risk of jitter buffer overruns at the receiving end. To provide consistent voice latency and minimal packet loss, QoS is required. The following major rules apply to QoS in LAN and WAN environments:

- Use 802.1Q/p connections for the IP phones and use the auxiliary VLAN for voice.

- Classify voice Real-time Transport Protocol (RTP) streams as "Expedited Forwarding" (EF) or IP precedence 5 (priority options in the frame and the IP header) and place them into a dedicated queue (preferably a priority queue) on all network elements.

- Classify voice control traffic with a Differentiated Services Code Point (DSCP) value of 26 or Assured Forwarding AF31 or IP precedence 3, and place it into a dedicated queue on all network elements to receive priority over bulk data traffic.

- Enable QoS in the campus if LAN buffers are reaching 100 percent utilization.

- Always provision the WAN properly, allowing 25 percent of the bandwidth for overhead, including routing protocols, network management, and Layer 2 link information.

- Use Low Latency Queuing (LLQ) on all WAN interfaces.

- Use Link Fragmentation & Interleaving (LFI) techniques for all link speeds below 768 kbps.

  Service providers (SP) can employ advanced QoS features such as Dynamic QoS (DQoS) or Resource Reservation Protocol (RSVP) to manage the dynamic nature of voice-enabled endpoints.

Network Infrastructure Services

Some of the most common network infrastructure services include Domain Name Services (DNS) for host name resolution, Dynamic Host Configuration Protocol (DHCP) for address assignment, server load balancing, and content caching. For the purpose of network readiness assessment for VoIP, we will be discussing the two most relevant services, DNS and DHCP:

- **DNS:** DNS is a critical network naming service within almost all IP networks. Network clients and servers both request connections to other devices by specifying a name. To resolve the name to an IP address, a request is sent to a configured DNS server, and from this point on, the client can use the returned IP address. DNS servers should be located centrally within core areas of the network with backup servers available. DNS should also be set up as a hierarchy with a master and secondary so that the secondary servers are updated in an appropriate time frame. Devices should also be named, and routers should have DNS entries for all ports to avoid IP address conflicts. Network devices should also be set up with backup DNS servers in case the first chosen server is down. DNS servers should also be considered critical services within the organization, requiring the highest level of security, power backup, and potential redundancy. DNS might not be required in enterprise IP communications environments because IP addresses are configured in Call Manager and IP phones for access but is useful for managing the overall network environment. In SP networks, DNS functionality is a must because it plays a crucial role in provisioning of voice endpoints such as Media Termination Adapter (MTAs). It is also required for establishing voice calls because the voice endpoints rely on DNS to resolve the Call Agent's Fully Qualified Domain Name (FQDN) to an IP address.

- **DHCP:** DHCP is typically used for client IP addressing. This allows mobility and improves IP address manageability. The only downside is that critical IP Allocation State for many nodes is kept within one file or database on the DHCP server. An organization should have adequate support for DHCP services and treat the DHCP data as highly critical data. Generally, DHCP databases should be mirrored or backed up on a continual basis. In addition, an organization should have a plan in case DHCP services fail. This can be mitigated by implementing redundant DHCP servers and distributing the load among different servers, or by configuring backup DHCP servers in case the primary server fails. DHCP server uses "options" to configure the hosts remotely when they are connected to a network. The DHCP options are defined in RFC 2132. These hosts can be VoIP endpoints as well. DHCP servers should also support option 150 for IP phone provisioning. This permits the DHCP server to pass control to the Call Manager to download phone configurations following IP address allocation. In addition, the DHCP server should also support option 43 (vendor-specific information), option 60 (vendor class identifier), option 122 (CableLabs Client Configuration), and so on for provisioning voice endpoints in SP environments.

Network Links

Network links evaluation during network readiness assessment for a VoIP domain should encompass the link redundancy and the installation of the physical media. This section discusses the most essential areas to be included as part of this assessment:

- **WAN link redundancy/diversity:** WAN link redundancy and diversity can be a consideration for distributed IP telephony deployments where WAN links are required for call setup and RTP voice traffic. The organization should determine the backup strategy when the primary WAN link is down. Distributed call processing and gateways or WAN redundancy might accomplish this. WAN redundancy and diversity can include local loop providers and long-distance providers.

- **Copper cabling installation:** Copper installation standards and testing help to ensure that the intra-building copper plant meets expected quality and performance expectations. The installation should follow standards and quality guidelines for signal attenuation, near-end crosstalk, bend radius, cable routing, distance, termination standards and components, labeling standards, patch cord routing, and building conduit requirements. The current documented standard for Category 5 testing requirements is the TIA/EIA TSB-67 standard. All verification and testing should be done following this guideline, which specifies required values for attenuation and NEXT (near-end crosstalk).

■ **Fiber cabling installation:** Intrabuilding, campus-area network (CAN), or metropolitan-area network (MAN) fiber cabling installation standards and testing help to ensure that the interbuilding fiber plant meets expected quality and performance expectations. The installation should follow standards and quality guidelines, which include parameters such as loss per connection (measured in decibel or dB), bend radius, cable routing, termination components or trays, labeling standards, patch cord routing, and organization and building conduit requirements. All fibers should be tested following termination to ensure high quality and minimal signal loss. Campus cabling should generally also offer diversity to prevent disasters caused by cable cuts.

Hardware and Software Considerations

Hardware and software evaluation is also an essential area during the network readiness assessment. This includes evaluating hardware and software resiliency that will be involved in transporting VoIP traffic:

■ **Device selection:** Network infrastructure devices identified for an IP telephony infrastructure should have the recommended features for IP telephony and faster network convergence for high availability. IP telephony features include inline power, 802.1Q/p support, hardware priority queuing, and QoS. Availability features includes spanning-tree convergence features such as uplink fast and backbone fast. Devices should also generally have improved backplane capacity, latency, and increased bandwidth. This section also looks at the Mean Time Between Failure (MTBF) of the chosen devices to determine theoretical availability. If the number of total devices is known, Cisco can also provide an expected annual failure rate for the devices.

■ **Hardware redundancy:** Redundant modules and chassis are a major contributor to network resiliency and enable normal or frequent maintenance on network equipment without service-affecting outages in addition to minimizing power, hardware, or software failure impact. Redundant chassis can also provide load-sharing capabilities used with routing protocols. The default gateway chassis can be redundant when used with HSRP. Many organizations have redundant backbone chassis and redundant distribution models. Redundant modules include power modules, supervisor modules, and interface modules. Redundant modules ensure that individual module failure does not affect network availability. It is recognized that in many cases, redundant chassis are cost-prohibitive at the access layer because of port density, especially with the introduction of IP telephony. However, the link redundancy between access layer switches and distribution layer devices becomes even more vital considering the lack of hardware redundancy at the access layer.

- **Software resiliency:** The software chosen must support the required features for IP telephony and provide the overall operational reliability. Software reliability is a factor of software configuration and software version control. Software reliability essentially means that there are no bugs with high impact such as crashes, routing errors, call control feature errors, and memory leaks. For the most part, software reliability is the responsibility of software development and testing groups within software vendors such as Cisco. However, the organization deploying VoIP must still validate whether the software is appropriate for its environment by testing or piloting the intended versions and consulting with a professional services organization or a deployment partner assisting it with the VoIP rollout. Where possible, Cisco recommends general deployment versions. The network architect must ensure that these target versions of code have been widely deployed in many customer environments and it is believed that critical and major bugs have been resolved.

- **Software version control:** Software version control is the process of testing, validating, and maintaining authorized software releases within the network. Most organizations require a handful of versions because of different platform and feature requirements. A process should be in place to choose release candidates, review potential impacting bugs, test or pilot-release candidate software, deploy authorized software, and review version accounting information to ensure that software version control is being maintained as expected. The version accounting information will contain information about the software release details, including installation date, testing information, features list, and typically when it will be retired based on the software vendor support calendar for releases. Large organizations without software version control processes can potentially end up with well over 70 software versions within the network, resulting in a higher number of software bugs, unexpected behaviors, and hardware/software incompatibility problems. Organizations requiring high availability should also weigh feature requirements with known software stability in general deployment software. Another issue is software age. Older general deployment software is considered more reliable than recently released newer versions with an untested production history.

Power and Environment

Power requirements and environmental conditions including physical security are very significant for VoIP deployment and must be considered as part of network readiness assessment:

- **Power protection:** Power protection is often a concern in IP telephony environments and might be needed to provide parity to legacy telephone systems. Power protection for IP telephony includes the use of inline power to provide backup power to phones for uninterruptible power supply (UPS)–protected LAN switching gear and power protection of all critical networking components. In addition, key networking equipment should have redundant power supplies with connectivity to separate power distribution units to prevent power loss because of a tripped circuit or Power Distribution Unit (PDU) failure. This can range from a 10-minute UPS to prevent failure because of more common short-term power outages to UPS arrays with backup generators to prevent failure because of long-term power outages. In addition, the organization should consider monitoring and servicing of the UPS equipment.

- **Environmental conditioning:** Environment is a major factor of hardware resiliency as location and temperature of network equipment affect the Mean Time Between Failure (MTBF) of network equipment. The standard operating surface temperature of most Cisco equipment is approximately 40 degrees Celsius (104 degrees Fahrenheit). When equipment fluctuates more than 10 degrees Celsius, the MTBF or hardware reliability of the component can be reduced significantly. To maintain documented MTBF estimates, the organization should ensure that proper heating, ventilation, and air conditioning (HVAC) are maintained for critical equipment.

- **Physical security:** Physical device security should ensure that unauthorized personnel do not have access to network equipment. Access to equipment allows unauthorized personnel to make unauthorized changes, obtain passwords on equipment, and even perform malicious activities. Equipment should be kept in locked rooms, preferably with card access so that all entrance is logged.

Auditing for VoIP Network Readiness

A comprehensive predeployment audit of an IP network is always recommended. It is also recommended even after a VoIP deployment for network optimization. Auditing methodology consists of analyzing deployed configurations for resiliency, modularity, QoS, high availability, IP addressing, security, and software version control, as described earlier, against established network deployment and operational standards. The auditor should also analyze operational practices for Fault, Configuration, Accounting, Performance, and Security (FCAPS) as adopted by the organization maintaining the VoIP network.

The preliminary step of this analysis includes interviewing all the stakeholders such as network administrators, architects, and Network Operations Center (NOC) staff. The interview process typically reveals high-level information including network topology and Information Technology Infrastructure Library (ITIL), International Organization for Standardization (ISO), or certain corporate standards for FCAPS in the case of enterprise or commercial deployments. SPs might have additional standards as mandated by regulatory bodies. However, this might not present an accurate and complete state of the network because compliance to standards is an ongoing process. Large networks constantly undergo changes. For example, a corporate or service provider might have the most comprehensive strategy for QoS and security. However, it might be in the initial phases of implementation at the time of auditing.

Therefore, a subsequent step should include verification of the actual status of the network as it is deployed. Network auditing tools are leveraged for this step. These tools are available from Cisco Advanced Services, such as Unified Communication Audit Tool (UCAT), Cisco Network Collector—NetAudit, and Cisco Unified Readiness Assessment Manager (CURAM). Similar tools are also available from other vendors, for example, Vivinet Assessor and AppManager from NetIQ.

For large-scale networks, it becomes necessary to use a controlled sample for expediency. It is recommended to distribute the network into logical models during the interview in the first step. For example, a large international bank might have various lines of businesses, thousands of branch locations, multiple campus locations, and several international subsidiaries. The auditor should make an attempt to categorize them logically as follows:

- Campus includes a MAN, Optical Carrier-3 (OC3 with speed of 155.52 Mbps) up-links, a three-tiered network, 6000+ employees, a contact center presence, and special considerations for emergency calling.

- Small campus location with Digital Signal 3 (DS3) with 44.73-Mbps links, a two-tiered network, up to 1000 employees, and a single building.

- Data center includes redundant OC12 links with a speed of 622 Mbps each, server farms, load-balancing servers, UPSs, and physical security.

- Branch Type A is a small-size branch with only consumer banking in grocery chains and Frame Relay circuits with speeds up to 512 kbps.

- Branch Type B represents a mid-size branch with locker facility, loan officers, commercial banking services, and up to T1 links with a data rate of 1.544 Mbps.

- Branch Type C is a large branch with high-touch services such as brokerage services, network redundancy, up to 50 employees, and multiple high-speed links.

The auditor might choose to audit in detail one campus location, two small campuses where one could be an international location, all the data centers, and three branches of each type to understand the state of the network. In this process, the auditor must keep the dialogue open with the network administrator to ensure coverage and rectify access issues commonly faced by the assessment tools.

Analyzing Configurations, Versions, and Topology

The audit tool looks for the configuration of all the devices to look for the required features as described earlier, including QoS settings and redundancy in the form of HSRP or backup links and capability of access switches to provide power over Ethernet to the IP phones in the future. The configuration data also helps validate the topology of the network as it is actually implemented. In the absence of such features, the audit tool must be able to evaluate whether the current software version allows the network architect/administrator to enable these features prior to deployment. The auditor should highlight the gaps and flag instances where either the hardware or software cannot fulfill the fundamental requirements for VoIP deployment. Cisco auditing tools are capable of analyzing the configurations based on the device role according to the established network design best practices.

Synthetic Traffic Tests

Media traffic simulation on IP networks provides engineers with a clear picture of how voice packets (RTP streams) will behave on the target network that is being prepared to carry voice. Test probes used to simulate and analyze streams use the following two methods:

- Deploying traffic generators on the edge of the network where the endpoints are intended to be installed. These remote traffic generators are controlled through a common console that instructs them to choose the appropriate codec, number of media sessions and streams, and target destination. The console then collects metrics including delay, jitter, packet loss, and Mean Opinion Scores and Perceptual Evaluation of Speech Quality (MOS/PESQ) estimates for these calls. This console should attempt to run the same number of calls as expected during the busiest hour of the day. All codec variations should be explored for this test.

- Traffic emulation can also be performed using a simpler method, which employs a central traffic generator that sets network devices, including switches or access routers, as "reflectors." These designated reflectors should be as close to the edge of the network as possible to cover the entire path taken by the RTP packet-containing media. The central traffic generator simulates media streams toward all these reflectors and records the same parameters including delay, jitter, packet loss, or other voice metrics for analysis. Cisco IOS has a feature called IP SLA that can configure a central device to generate traffic toward multiple reflectors. This method can generate a large amount of traffic to accurately emulate the expected network load. The IP SLA–based method compensates for its internal serialization delays to provide accurate statistics. Appendix A contains more information about the IP SLA–based traffic emulation test.

The transmission statistics, including voice quality metrics, are collected by means of RTP Control Protocol (RTCP), which provides information about the RTP streams related to packet statistics, reception quality, network delays, and synchronization. The information collection method can involve Simple Network Management Protocol (SNMP) MIBs, eXtensible Markup Language (XML), or even simple commands issued on the command-line interface (CLI) of the traffic generator by the central console.

Note   The expected results for delay loss and jitter are listed in Table 6-5.

Managing Network Capacity Requirements

To perform meaningful capacity planning, it is necessary to understand the traffic engineering theory and know the expected traffic flows in the VoIP network. Historical Call Detail Records (CDR) can provide this information accurately. This data can be helpful in migrating TDM voice networks to VoIP or expanding the capacity of existing VoIP networks such that the proper service-level agreement (SLA) is preserved. In scenarios where the communications networks are deployed for the first time, assumptions have to be made for anticipated call volume, and there should be plenty of room left for growth planning using the same traffic engineering principles.

Voice Traffic Engineering Theory

In traffic engineering theory, you start by determining the traffic load. Traffic load can be determined by measuring the ratio of call arrivals in a specified period of time to the average amount of time taken to service each call during that period. These measurement units are based on Average Hold Time (AHT). AHT is the total duration of all the calls completed in a specified period, which is typically the busiest time period, divided by the number of calls in that period, as shown in the following example:

(3976 total call seconds) / (23 calls) = 172.87 sec per call = AHT of 172.87 seconds

The two main measurement units used today to measure voice traffic load are Erlangs and centum call seconds (CCS). One Erlang is 3600 seconds of calls on the same circuit, or enough traffic load to keep one circuit busy for 1 hour. Traffic in Erlangs is the product of the number of calls times the AHT divided by 3600, as shown in the following example:

(23 calls * 172.87 AHT) / 3600 = 1.104 Erlangs

One CCS is 100 seconds of calls on the same circuit. Voice switches generally measure the amount of traffic in CCS. Traffic in CCS is the number of calls times the AHT divided by 100, as shown in the following example:

(23 calls * 172.87 AHT) / 100 = 39.76 CCS

Which unit you use depends highly on the equipment you use and in what unit of measurement it records. Many switches use CCS because it is easier to work with increments of 100 rather than 3600. Both units are recognized standards in the field. The following is how the two relate: 1 Erlang = 36 CCS. In this book, we use CCS. Capacity planning in voice networks is based on the busiest hour of the day. To determine the traffic load, you need to determine how many calls each voice endpoint makes during the busy hour. This number is known as the number of busy-hour call attempts (BHCA). After the BHCA and AHT are known, the resulting traffic load in CCS can be calculated using the following formula:

CCS = BHCA * AHT / 100

Example of Estimating Capacity Requirements

This section describes an example mod based on a retail chain's IP communications network supporting its nationwide retail stores. Currently, this retail chain has 225 stores and plans to add 275 additional sites to this network. The numbers are based on the Call Detail Records provided by the network administrator of this retailer during the week of Thanksgiving. In the United States, the Friday of this week in November is expected to be the busiest day of the season (also known as Black Friday). This period was chosen because it represents the busiest week of the year for this retailer by virtue of having the busiest day of the season in it, even though there is a holiday in that week. The calculation is based on the following CDR data:

- Number of sites: 225

- Calls rejected because of out of bandwidth: None

- Calls rejected because of no circuit/channel available: None

- Total call seconds: 1,804,730

- Total calls: 13,077

- Sample size: 6 days (weekly report extracted from the IP PBX covering about 148 hours)

This calculation was done for 225 stores, where the total call attempts during the busiest hour (Friday after Thanksgiving between 12 noon and 1 p.m.) were 778. The Average Hold Time (AHT) is calculated as follows:

Total call seconds / Total calls
1,804,730/13,077 = 138 sec

Following is the number of CCS per day (6 is the number of sampled data points):

CCS = (13,077 * 138) / 100/6 = 3007.7

The CCS value can be converted to Erlangs based on the fact that 1 Erlang = 36 CCS. So, the number of Erlangs in this case would be 3007.7/36 = 83.52. Cumulative call attempts are calculated as follows:

CCS / AHT * 100
1076 / 138 * 100 = 2179

As mentioned previously, the total call attempts during the busiest hour are 778.

The per-device (including phones and point-of-sales ports) Busy Hour Call Attempt (BHCA) is calculated as follows:

Cumulative call attempts / Total call attempts during the busiest hour
2179/778 = 2.8

For 500 sites, the total call attempts during the busiest hour can potentially reach up to 1600 for voice calls alone using simple extrapolation.

Each G.729 call takes about 28.6 kbps. We round it to 32 kbps to compensate for any overhead to determine the bandwidth required for calls using the G.729 codec when traversing the IP network.

A G.711 call consumes about 80 kbps. The G711 codec was used for calls from fax machines and PoS devices, constituting 18 percent of all calls.

The total bandwidth required for 500 sites with a call peak rate of 1600 is calculated as follows:

(1600 * 32 kbps * 0.82) + (1600 * 80 kbps * 0.18) = 65.024 Mbps

This retailer was expected to grow based on new product offerings and a pending acquisition to enter into new markets. The growth estimate provided by the management was 50 percent over the next two years. Based on this data, we projected the future expansion of the network that is able to support 50 percent more sites was about 500 sites. The bandwidth requirement estimate was extrapolated to be about 98 Mbps, calculated as follows:

(65.024 Mbps * 1.5) = 97.53 Mbps

The network architect or the administrator must plan for this amount of bandwidth and include any packet overhead given the protocol(s) choice and use of compression methods, as discussed in Chapter 1, "Voice over IP (VoIP) and Network Management." All the links that this traffic can traverse should be provisioned accordingly with sufficient overhead to support redundancy. The QoS scheme must take into account this capacity so that the priority queues on network devices used by RTP packets containing media are adjusted and the traffic-shaping mechanism is in effect for the correct bandwidth amount. This information can be found in the Cisco QoS solution reference network design guide at http://www.cisco.com/en/US/docs/solutions/Enterprise/WAN_and_MAN/QoS_SRND/ QoS-SRND-Book.html.

If all the calls made from the 500 retail stores have to egress to the public switched telephone network (PSTN) through a gateway at the edge of the VoIP network, Erlang formulas would be used to calculate the trunk requirements. In this example, we are calculating trunk requirements for U.S. markets, where a T1 ISDN PRI circuit has 23 bearer channels or DS0 circuits.

The business requirement includes 11 IP phones and 3 ports for a fax machine and point of sales units in every branch. The total number of devices in 225 stores is calculated as follows:

14 * 225 = 3150

Earlier, we discovered that the per-device BHCA is 2.8 calls and the AHT is 138 seconds. Because the CCS can be calculated as follows:

CCS = BHCA * AHT/100

this gives the per-device CCS of 3.062 using the following calculation:

CCS = 208 * 138 sec / 100 = 3.062

The equivalent Erlang value for every device is

Erlang = CCS / 36 or 3.062 / 36 = 0.107

The projected Erlang for 500 stores with 14 endpoints each (7000 endpoints) is

$$0.107 * 7000 = 749$$

Because the per-device BHCA value is 2.8, the BHCA value for 7000 endpoints will be 19,600.

Assuming that a desired blocking rate is no more than 1 percent, the Erlang B formula tells you that you need 778 DS0 lines. This is equivalent to 34 PRIs (778/23). Refer to the Erlang calculator at http://ict.ewi.tudelft.nl/ index.php?option=com_sections&id=164&Itemid=286.

Table 6-1 is a trunk-sizing table based on different Erlangs and service levels measured as a blocking rate of the calls as they egress and ingress from the trunks.

Table 6-1 *Erlang Calculations for Determining PSTN Trunk Capacity Requirements*

Erlangs per Phone	Total Erlangs	BHCA (7000 Users)	Blocking Rate (Service Level)	Number of Trunks Needed
0.107	749	19,600	0.05	726 DS0 or 32 PRIs
0.107	749	19,600	0.01	778 DS0 or 34 PRIs

Monitoring Network Resources

All call-processing systems offer methods to report resource utilization. In Cisco Unified Communications Manager, there are cause codes defined to monitor network resource utilization. Following is a set of minimum recommended parameters to be monitored regularly:

- Trunk resources (Cause code 34: Channel not available)
- Network availability (Cause code 38: Network out of order)
- Conferencing resources (Cause code 124: Conference full)
- Bandwidth starvation (Cause code 125: Out of bandwidth)
- Other resources such as transcoders or Music on Hold (MoH) processors (Cause code 47)

Table 6-2 lists the number of calls disconnected for abnormal reasons that need further investigation. The reasons can include oversubscribed PSTN trunks, conference bridges, lack of bandwidth, or unexpected bursts in traffic volume normally handled by this system. The analysis is done for a one-hour block of time.

An Audit for Gauging the Current VoIP Network Utilization

A periodic audit validates the assumptions in the initial design and traffic engineering and provides an opportunity to proactively address the network growth requirements and early detection of performance-related issues. This audit mainly focuses on device and link utilization.

Table 6-2 *Critical Disconnect Causes by Time of Day as Reported in CDR*

Time of Day	Cause Code 34: Channel Not Available	Cause Code 38: Network Out of Order	Cause Code 124: Conference Full	Cause Code 125: Out of Bandwidth	Cause Code 47: Resource Unavailable, Unspecified
1–2	1	1	2	0	1
2–3	3	4	2	2	2
3–4	4	4	4	4	2
4–5	4	2	5	5	5
5–6	1	1	1	1	1

Device Utilization

A baseline identifying current device resource utilization is recommended to determine the potential impact of IP telephony traffic. A baseline is normally done by monitoring peak (5-minute) utilization using SNMP over an extended period of several days to a week. This is normally done for the CPU, memory, backplane utilization, and LAN buffers. Most SNMP tools enable the organization to collect and graph the utilization over this period. The result should identify peak utilization. If peak values exceed 50 percent, the organization should consult with its Cisco representative to determine the potential impact.

Table 6-3 shows how router measurements were rated. Ratings should be based on the router result ranges for this assessment.

Table 6-3 *Router Utilization Measurements*

Measurement	Good	Acceptable	Poor
Average CPU utilization (%)	Less than or equal to 30.0	Less than or equal to 50.0	Any higher value
Average memory utilization (%)	Less than or equal to 30.0	Less than or equal to 50.0	Any higher value
Input queue drops (%)	Less than or equal to 2.0	Less than or equal to 8.0	Any higher value
Output queue drops (%)	Less than or equal to 2.0	Less than or equal to 8.0	Any higher value
Buffer errors	Less than or equal to 0	Less than or equal to 1	Any higher value
CRC errors (%)	Less than or equal to 2.0	Less than or equal to 8.0	Any higher value

Table 6-4 shows how switch measurements were rated. Ratings should be based on the switch result ranges for this assessment.

Table 6-4   *Switch Utilization Measurements*

Measurement	Good	Acceptable	Poor
Average backplane utilization (%)	Less than or equal to 50.0	Less than or equal to 75.0	Any higher value
Average CPU utilization (%)	Less than or equal to 30.0	Less than or equal to 50.0	Any higher value

Link Utilization

A baseline that identifies current trunk link utilization is recommended to determine the potential impact of IP telephony traffic. A baseline is normally done by monitoring peak (5-minute) utilization using SNMP over an extended period of several days to a week. Most SNMP tools enable the organization to collect and graph the utilization over this period. The result should identify peak utilization and busy-hour data traffic requirements. The organization can then estimate overall requirements based on estimated peak IP telephony traffic.

Table 6-5 shows how link measurements were rated. Ratings should be based on the link result ranges for this assessment.

Table 6-5   *Link Utilization Measurements*

Measurement	Good	Acceptable	Poor
Average bandwidth utilization (%)	Less than or equal to 30.0	Less than or equal to 50.0	Any higher value
Latency	0–20 ms	21–120 ms	Above 150 ms
Jitter	0–20 ms	21–120 ms	Above 150 ms
Packet loss (per-hour basis)	Less than 15	15–30c	Any higher value

Measurements for Network Transmission Loss Plan

Most VoIP networks still have to interface with the PSTN for greater reachability to all the communication devices worldwide. This exposes the TDM-IP hybrid network to even more sources for echo and other quality issues related to voice signal because of power loss, impedance mismatches, and excessive delay. Therefore, a network transmission loss plan (NTLP) must be established during the pilot phase of the VoIP implementation. NTLP maps a complete picture of signal loss in the entire path, identifying areas of improvements including signal strength adjustment at various points to help the echo cancellers (ECAN) on media/voice gateways. In an IP network, this test needs to be repeated to cover all possible permutations of routed paths. There are two primary reasons to establish a loss plan:

■   Desire to have the received speech loudness at a comfortable listening level

■   Minimize the effect of echo because of signal reflections that are caused by impedance mismatches at the 2-to-4-wire conversions in the transmission path

NTLP uses the following concepts to make loss calculations in the voice network:

■   **Send Loudness Rating (SLR):** This is defined as the loudness between the Mouth Reference Point (MRP) and the electrical interface.

■   **Receive Loudness Rating (RLR):** This is defined as the loudness between the electrical interface and the Ear Reference Point (ERP).

■   **Talker Echo Loudness Rating (TELR):** This is the loudness loss between the talker's mouth and the ear through the echo path. TELR is calculated as follows:

TELR(A) = SLR(A) + loss in the send path + ERL(A) or TCLw(B) + loss in the receive path + RLR(A), where ERL is the echo return loss of the hybrid or echo canceller and TCLw is the weighted terminal coupling loss of the digital phone set that is typically mentioned in the product manuals.

■   **Overall Loudness Rating (OLR):** The OLR of a connection is the sum of the sending terminal SLR (Sending Loudness Rating), any system or network loss, and the receiving terminal RLR (Receive Loudness Rating). The long-term goal for TELR is 8–12 dB, but because of the mix of technologies, the short-term goal is 8–21 dB. The difference between the OLR in both directions should be no more than 8 dB.

OLR = SLRtalker + [sum]attenuations + RLRlistener

Note   The NTLP (Network Transmission Loss Plan) rule of thumb is one-way loss ranges between 10 and 12 dB. This is about 66 percent of the loss at the receiving end (RX).

The degree of annoyance of talker echo depends on the amount of delay and on the level difference between the voice and echo signals.

> **Note**   The standard SLR/RLR for an analog phone is 8 dB and −6 dB, respectively. For more details, the network architect must consult Section 6.4, Item 7 of TIA 912.

Figure 6-1 shows sample calculations for on-net calls.

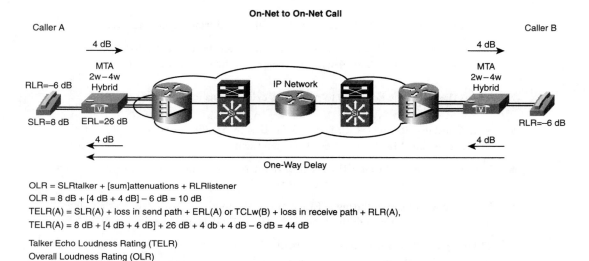

OLR = SLRtalker + [sum]attenuations + RLRlistener

OLR = 8 dB + [4 dB + 4 dB] − 6 dB = 10 dB

TELR(A) = SLR(A) + loss in send path + ERL(A) or TCLw(B) + loss in receive path + RLR(A),

TELR(A) = 8 dB + [4 dB + 4 dB] + 26 dB + 4 db + 4 dB − 6 dB = 44 dB

Talker Echo Loudness Rating (TELR)
Overall Loudness Rating (OLR)

Figure 6-1   *NTLP Calculations for On-Net Calls*

Figure 6-2 shows sample calculations for off-net calls.

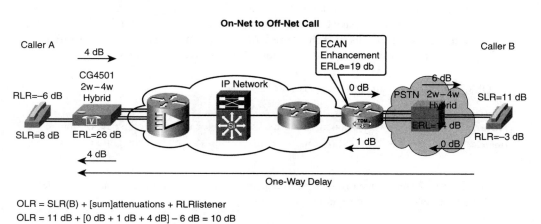

OLR = SLR(B) + [sum]attenuations + RLRlistener
OLR = 11 dB + [0 dB + 1 dB + 4 dB] − 6 dB = 10 dB

TELR(A) = SLR(A) + loss in send path + ERL(A) or TCLw(B) + loss in receive path + RLR(A) + ERLe
TELR(A) = 8 dB + [4 dB + 0 dB + 4 dB] + 14 dB + 0 db + 4 dB − 6 dB + 19 dB = 49 dB

Figure 6-2   *NTLP Calculations for Off-Net Calls*

Any voice quality test set should be able to measure the previously mentioned parameters necessary to develop an accurate NTLP prior to a wide-scale IP telephony deployment. A Sage Instruments test set was used to conduct the tests in the previous two examples.

The value of TELR would be plotted against the one-way delay measured for the test calls on the chart with the Echo Tolerance Curve specified by the ITU-T G.131 standard for echo, as illustrated in Figure 6-3.

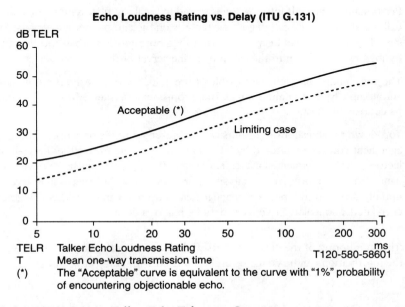

Figure 6-3   *ITU-T G.131—Talker Echo Tolerance Curves*

An attempt should be made to bring the coordinates, containing the TELR and the one-way delay values on the *x*- and *y*-axes, respectively, into the acceptable range as shown in the graph by modifying the gain setting on the TDM voice gateway. Digital signal processors (DSP) on Cisco gateways are equipped with echo cancellers providing echo canceling enhancement (ECANe) to conceal the echo in the speech path.

Effectively Monitoring the Network

Effective monitoring of the network is based on the philosophy of proactive or preventive maintenance. Preventive maintenance is based on proactive monitoring and consists of performance monitoring of voice quality metrics.

Every network is prone to faults and problems. Faults can be detected by operations personnel, by users of the service equipment, by performance monitoring and testing within elements, by trend analysis, and so on. The troubleshooting of these problems can be done using the following two approaches:

- **Bottom-up troubleshooting:** Fault indication and performance information generated at the endpoint flow upward through a hierarchy of levels beginning at the bearer element level through the signaling element level to the network management level and finally to reach the system's operation level.

  Performance and fault information is stored in local log files, remote local databases, call detail records, or call diagnostic records on the call-processing servers. This stored information can be analyzed later by a bottom-up troubleshooting approach by logging directly on to the call-processing server or the suspect device.

  The voice endpoint might be capable of preliminarily troubleshooting the fault autonomously, or it might always expect top-down assistance from one or more of its controlling entities.

- **Top-down troubleshooting:** Faults can be presented to the network's central managemnent console monitored by NOC personnel arising from either lower-level collectors, such as Element Management System (EMS), or from user or device maintenance trouble reports, trend analysis reports, and so on. When a system-level fault is initially detected by a network management station, it can be further categorized, correlated, and isolated to a more definite link or a device.

Effective monitoring of the network starts with discovering the network in its entirety so that the information can be collected and correlated with complete context.

Discovery—Complete Picture

The completeness of coverage and accuracy of device identification are keys to effective network management. Network discovery must be performed only when the network is not undergoing major changes such as device additions or upgrades.

The length of time required to discover a network is dependent on various factors that are not necessarily tied to the device population size. For example, you cannot assume that 200 devices can be discovered in 2 hours simply because the initial 100 devices detected in a session might have been discovered in 1 hour. The underlying processes of a discovery session can result in requiring 1 hour for the initial 100 devices and only 10 minutes for the remaining 100. The discovery process as implemented by most discovery engines as part of a network management system employ a multithreaded process for discovery, significantly reducing the discovery of a large network.

Typical factors that impact the amount of time to run a discovery process on a network include the following:

- **Link speed:** When deploying a discovery session across WAN links, slower link speeds and latency result in slower discovery.

- **Bandwidth controls:** The discovery process throttles itself by regulating the amount of traffic it creates on the network in order not to disturb the production traffic.

- **Device focus:** Looking only for specific network elements results in a faster discovery time as compared to wider probes. Firewalls and access control lists (ACL) often block SNMP traffic. This can be mitigated by ensuring that the security control points allow SNMP and Internet Control Message Protocol (ICMP) traffic by allowing the discovery engine's IP address and these protocols specifically by way of configuring a firewall conduit or modifying the access control list. Alternatively, including a seed device from the other side of the security control point can bypass this problem.

Seed Devices for Network Discovery

When selecting seed devices, it is helpful to remember that the start of the discovery process relies on complete Cisco Discovery Protocol (CDP) and/or routing table information from one or several key devices. These devices need not be core devices but should be devices that contain complete routing tables and/or CDP neighbor tables. Aggregate devices, rather than core devices, might be the wiser choice for use as seed devices. This potentially avoids overwhelming the core devices that are primarily responsible for fast routing and switching traffic between a large number of devices compared with distribution layer devices.

Concern about the effect of SNMP queries on the seed device can be settled by observing the effect on the SNMP process or a test device, running the same version of software, and containing SNMP tables loaded using a test tool to generate IP route or ARP entries.

A good option for seed devices might be a redundant distribution layer device. This is because the hot standby devices have full knowledge of the network routing tables but are not actually handling the majority of the network traffic. This reduces potential SNMP network traffic on the active devices and links.

Cisco Discovery Protocol (CDP) Discovery

A Cisco device that supports CDP can both transmit discovery requests and respond to CDP queries from other Cisco devices. As a result, each Cisco device is aware of its immediately connected neighbors. The Cisco Network Collector (CNC) discovery engine collects the CDP information from devices by using SNMP queries to form a list of all neighbors of the queried device. This list contains all devices that have advertised their presence on the network and provides clues about other devices to be discovered.

Limitations of the CDP-based discovery method include the following. Not all Cisco devices support CDP, for example, content networking devices. Some transmission media, such as ATM, do not support CDP. The network administrator can disable CDP on some parts of the network or even the entire network.

CDP does not have to be enabled on all the devices for it to work. As long as the core, distribution, and some of the access layer devices have CDP enabled, the majority of the network can be discovered using CDP.

Routing Table Discovery

The routing table method uses the routing table from seed devices and retrieves the subnet address and subnet mask from the routing table MIB. It then compares each subnet against the list of subnets already discovered. If the connection point for the subnet is not found, it then uses SNMP to retrieve the next-hop address for that particular subnet, and then it compares the next-hop address with the IP addresses already discovered. If the next-hop address is not discovered, it is added to a list of devices to be discovered. Routing protocol neighbor lookup is also effective but does not find Layer 2 devices and only currently supports the OSPF protocol and Border Gateway Protocol (BGP). This discovery method is available as part of CNC Discovery Engineer and Unified Communication Audit Tools (UCAT), and can be implemented using a set of scripts.

ARP Discovery

The ARP method looks at devices discovered and retrieves the ARP table using the "at" address table MIB. This method retrieves the list of all IP addresses the device has in its cache and compares the MAC address to a list of known Cisco MAC address prefixes. If the MAC address prefix matches, the IP address is added to the list of devices to be discovered.

ARP discovery is not an efficient discovery tool. Devices for which ARP entries have timed out will not be discovered. However, it becomes useful in finding devices that do not route, do not support CDP, or have CDP disabled.

Routing Protocol—OSPF Discovery

Open Shortest Path First (OSPF) is an internal gateway protocol. If OSPF is active in a network, OSPF's inherent discovery feature is the preferred method to determine neighbor information. OSPF discovery uses the OSPF MIB information, which maintains a list of all Layer 3 neighbors. This list contains clues for further device discovery.

Routing protocol neighbor lookup is also effective but does not find Layer 2 devices and only currently supports the OSPF protocol and BGP (EIGRP does not have a neighbor MIB). Of course, discovery is ineffective if OSPF is not running on the network. Therefore, it is recommended to use a generic routing table instead of a protocol-specific method.

Ping Sweep Discovery

Ping sweep discovery generally provides two types of ping sweep mechanisms using either a specified IP address range or a specified starting IP address. Both methods issue sequential pings to the IP addresses. If an address responds, discovery then attempts to communicate with the device at the IP address by SNMP using a host-resource MIB. If SNMP communication is successful, the device is then considered to be manageable.

The Cisco network management tools, including CNC, CiscoWorks, and UCAT, provide the following two ping sweep methods:

- **Ping sweep with hop:** This method starts from an IP address and continues pinging new addresses up to the given hop count.

- **Ping sweep range:** This method uses a range of IP addresses, from the starting IP address to the ending IP address for the given IP address and netmask. Ping sweep might not be an efficient method, but given enough time and if ICMP messaging is not blocked, it will find everything on the network. Depending on how network devices have been configured and the addressing scheme in the network, the network might start responding with ICMP unreachable and redirect messages.

Seed Files

Seed files contain device information including IP addresses, SNMP community strings, and passwords to log in to the command-line interface (CLI) or access other APIs such as XML. Seed files guarantee complete device access, leading to full discovery and manageability of the network. Even though loading a seed file can result in quick network discovery, the creation of a seed file (generally by a network administrator) can itself be a time-consuming process and is prone to human errors. It is recommended that multiple methods for discovery be employed for greater coverage. The results of the discovery process should be verified with the network administrator to ensure the accuracy of the network topology as discovered by the Network Management System (NMS) using these methods.

Voice Quality Metrics

Media has always been predisposed to quality degradation as it traverses through any network, including both circuit-switched and packet-switched networks. IP networks offer greater flexibility to manage media streams and supplementary applications compared to dedicated circuit-switched networks. However, packet-switched or IP networks can further exacerbate some problems or introduce newer issues that need to be managed, as discussed in Chapter 1. There are multiple factors working independently and in concert to degrade the perceived quality of the voice signal as contained in RTP packets traversing from point A to point B on an IP network. These can be categorized into the following three major areas:

- **Environmental issues:** Environmental issues include acoustic problems caused by handset, headset, analog-to-digital convert, and impedance mismatch. Other issues can result from poor cabling or network clock synchronizations, resulting in crackled voice, clicking sounds, or the presence of crosstalk.

- **Signal processing:** Signal processing includes speech compression, voice activity detection (VAD), silence suppression, and signal gain variations. Issues related to VAD include front-end clipping and incorrect comfort noise levels, and the presence of static, hissing, and often underwater sounds.

■ **VoIP network issues:** An IP network introduces significant propagation and serialization delays compared to circuit-switched networks. This creates network jitter and packet loss, which often require error concealment procession. Delay also makes an inherently present echo more perceivable to human ears. Delay, jitter, and packet loss are manageable through the proper implementation of QoS end to end in the IP network. However, if this is not managed properly, it can result in robotic or synthetic voice sounds because of periods of silence (caused by packet drops) or a choppy voice.

Metrics are needed for sustaining quality of voice in an IP network. There are three comprehensive quality metrics: the Mean Opinion Score (MOS), the Perceptual Speech Quality Measurement (PSQM), and the Perceptual Evaluation of Speech Quality (PESQ).

MOS or K-factor

MOS is a subjective measure of voice quality. An MOS is generated when listeners evaluate prerecorded sentences that are subject to varying conditions, such as compression algorithms. Listeners then assign the scores to the received voice signal, based on a scale from 1 to 5, where 1 is the worst and 5 is the best. The test scores are then averaged to a composite score. The tests are also relative, because a score of 3.8 from one test cannot be directly compared to a score of 3.8 from another test. Therefore, a baseline needs to be established for all tests using G.711 with no compression so that the scores can be normalized and compared directly.

Cisco engineering has devised a computerized method to estimate an MOS, called K-factor. K-factor (*klirrfaktor* is "clarity measure" in German) is a clarity, or MOS-LQ (listening quality), estimator. It is a predicted MOS based entirely on impairments because of frame loss and codecs. K-factor does not include impairment because of delay or channel factors (echo or levels). K-factor MOSs are produced on a running basis, with each new MOS estimate based on the previous 8–10 seconds of frame loss data. That is, each K-factor MOS is valid over the past 8 seconds. The computation of new scores can be performed at any rate (every second, for example, with the score based on the past 8 seconds), but the computation window of the MOS is a constant. This method is used in the UC Manager that is able to provide a meaningful value for voice quality in a rather objective manner in its call detail records.

PSQM

Perceptual Speech Quality Measure (PSQM) is an automated method of measuring speech quality "in service," or as the speech happens. The PSQM measurement is made by comparing the original transmitted speech to the resulting speech at the far end of the transmission channel. PSQM systems are deployed as in-service components. The PSQM measurements are made during real conversation on the network. This automated testing algorithm has more than a 90-percent accuracy compared to the subjective listening tests, such as MOS. Scoring is based on a scale from 0 to 6.5, where 0 is the best and

6.5 is the worst. Because it was originally designed for circuit-switched voice, PSQM does not take into account the jitter or delay problems that are experienced in packet-switched voice systems. PSQM software usually resides with IP call-management systems, which leverage SNMP for collecting PSQM statistics.

PESQ

Perceptual Evaluation of Speech Quality (PESQ) is the current standard for voice quality measurement and is documented in ITU Standard P.862. PESQ can take into account codec errors, filtering errors, jitter problems, and delay problems that are typical in a VoIP network. PESQ combines the best of the PSQM method along with a method called Perceptual Analysis Measurement System (PAMS). PESQ scores range from 1 (worst) to 4.5 (best), with 3.8 considered "toll quality" (that is, acceptable quality in a traditional telephony network). PESQ is meant to measure only one aspect of voice quality. The effects of two-way communication, such as loudness loss, delay, echo, and sidetone, are not reflected in PESQ scores.

Many equipment vendors offer PESQ measurement systems. Such systems are either standalone or they plug into existing network management systems. PESQ was designed to provide an alternative to the MOS measurement system. So, if a score of 3.2 is measured by PESQ, a score of 3.2 should be achieved using MOS methods. PESQ measures the effect of end-to-end network conditions, including codec processing, jitter, and packet loss. Therefore, PESQ is the preferred method of testing voice quality in an IP network. When this metric is available on a call-processing system through XML, SNMP, or CDRs, it should be used for monitoring voice quality.

Approaches to Measure Jitter, Latency, and Packet Loss in the IP Network

The previously mentioned voice quality metrics provide an overall picture of the perceived voice quality. It is still important to look at the individual factors that impact the voice quality, including jitter, latency, packet loss, and various aspects of voice signal such as signal strength and bandwidth. This provides explicit information about what is contributing to the voice quality degradation in addition to getting an overall quality score in the form of K-factor or PESQ, for example. This section explains approaches to measure these parameters in the VoIP network interfacing the TDM network.

Using Call Detail Records for Voice Quality Metrics

A most common and the easiest approach to measure voice quality metrics, including jitter, latency, packet drop, and other statistics, is to analyze Call Detail Records (CDR), because they are already managed on central call-processing devices such as the Cisco BTS10200, Cisco Unified Communications Manager (CUCM), and Cisco PGW call agent. CDRs can also be collected on voice gateways that provide connectivity to the PSTN, as discussed in Chapter 3, "VoIP Deployment Models in Service Provider Networks," and

Chapter 4, "Internet Telephony," by leveraging RADIUS. Remote Authentication Dial-In User Service (RADIUS) is a networking protocol that provides centralized Authentication, Authorization, and Accounting (AAA) management for gateways.

These call-processing devices rely on the RTP Control Protocol (RTCP), as defined by RFC 3550, with VoIP metrics and collection and processing details described in RFC 3611. RTCP defines a method to report voice quality metrics, including such parameters as packet loss rate, packet discard rate, burst density and duration, gap density and duration, jitter round-trip delay, end system delay, signal level, noise level, residual echo return loss (RERL), and estimation of MOS listening quality. RTCP is mandated by G.799.1 VoIP Trunking Gateway specifications and Packet Cable 1.5 U.S. Cable Industry specifications.

Call control protocols, including Session Initiation Protocol (SIP), Media Gateway Control Protocol (MGCP), and H.323, can also provide statistics on voice quality metrics at the end of the call. In the case of MGCP, the call quality metrics will be included in DeleteConnection or a DLCX message. H.323 extensions as defined in H.460.9 Annex B define call control statistics. These call quality metrics will be pushed to the call control devices at the end of the call and recorded in CDRs for analysis and reporting purposes.

Using IP-SLA and RTTMON for Voice Quality Metrics

Cisco IOS Software features IP Service-Level Agreement (IP-SLA) and Round Trip Time Monitor (RTTMON) provide methods for measuring delay, jitter, and packet loss on the IP network. The IP-SLA and RTTMON features perform measurements by deploying small Cisco IOS routers as agents to simulate customer end stations. The routers are referred to as delay and jitter probes. Alternatively, existing IOS gateways at the edge of the network can be enabled with IP-SLA and RTTMON features. The delay and jitter probes can be configured with the remote monitoring (RMON) alarm and event triggers after baseline values are included in Table 6-5. This allows the delay and jitter probes to monitor the network for predetermined delay and jitter service levels and alert Network Management System (NMS) stations when a threshold is exceeded.

After they are configured and activated, the delay and jitter probes begin collecting data and place it in SNMP MIB tables. The rttMonStats table provides a 1-hour average of all the jitter operations. It is recommended that the NMS collector poll the rttMonStats table every hour. A higher polling frequency for the rttMonLatestJitterOper table will provide more granular statistics. If the delay and jitter probe is calculating jitter every 5 minutes, it is recommended to not poll the MIB at an interval less than 5 minutes.

IP-SLA can also be used to simulate voice calls and measure the voice metrics. This also makes it a tool to perform call load testing for readiness assessment, as discussed in the section "Synthetic Traffic Tests," earlier in this chapter. Appendix A discusses in detail how to leverage the IP-SLA feature for comprehensive network readiness assessment. More details on configuring IP-SLA and RTTMON for monitoring jitter, delay, and packet drop can be found at http://www.cisco.com/en/US/tech/tk869/tk769/ technologies_white_paper09186a00801b1a1e.shtml#saarttmon.

Using Cisco NetFlow for Measuring Voice Quality Metrics

NetFlow is a feature in Cisco IOS that enables network administrators and architects for network utilization monitoring, application profiling, security analysis, and accounting by caching the information about the traffic flows on the network. The flow is comprised of these seven key fields:

- Source IP address

- Destination IP address

- Source port number

- Destination port number

- Layer 3 protocol type

- Type of service (ToS) byte

- Input logical interface

The NetFlow feature enables the Cisco IOS device to create a cache entry that contains the information about all the active flows by processing the first packet of the flow. This flow record is created in the NetFlow cache for packets with similar characteristics defined by these seven key fields. Flow record provides very accurate counting or tracking of packets and bytes per flow. NetFlow operation is highly efficient and scaleable; it can provide a condensed yet detailed view of all network traffic while collecting only 1.5 percent of the switched traffic in the router.

In earlier chapters, you learned that voice call-signaling protocols use specific port numbers. The voice traffic contained in RTP packets also uses a well-known port range that is between 16384 and 32767. In addition to this, the IP address information for voice endpoints, call control devices, and any other application is known to the network administrator. A properly designed QoS scheme will be marking the ToS byte in the voice signaling and media packets. All of this information can accurately determine the voice-specific flows on a converged IP network to enable the operations staff to generate the reports for voice monitoring, profiling, security analysis, accounting, and capacity planning.

NetFlow version 9 has some voice-specific parameters that provide explicit information on RTP jitter, delay, and packet loss statistics for the relevant flows. Table 6-6 includes this list.

Cisco NetFlow collector periodically collects data from the NetFlow cache from multiple NetFlow export-enabled devices and performs data volume reduction through selective filtering and aggregation in addition to very comprehensive and customizable reporting. However, this is an ongoing process that needs to be constantly fine-tuned for greater accuracy and coverage as the VoIP network in particular and the overall network in general grows. Other third-party collectors are also available that can additionally provide traffic analysis, billing, security, and monitoring for NetFlow-enabled devices in the network. The collection can also be performed through SNMP by using NetFlow MIB. SNMP-based collection is especially useful when troubleshooting VoIP issues or when export is not possible from a specific network.

Table 6-6   *VoIP Metrics Available in NetFlow Version 9*

SIP	RTP
SIP_CALL_ID	RTP_FIRST_SSRC
SIP_CALLING_PARTY	RTP_FIRST_TS
SIP_CALLED_PARTY	RTP_LAST_SSRC
SIP_RTP_CODECS	RTP_LAST_TS
SIP_INVITE_TIME	RTP_IN_JITTER
SIP_TRYING_TIME	RTP_OUT_JITTER
SIP_RINGING_TIME	RTP_IN_PKT_LOST
SIP_OK_TIME	RTP_OUT_PKT_LOST
SIP_ACK_TIME	RTP_OUT_PAYLOAD_TYPE
SIP_RTP_SRC_PORT	RTP_IN_MAX_DELTA
SIP_RTP_DST_PORT	RTP_OUT_MAX_DELTA

The following link provides a list of these third-party collNetFlow collector vendors, NetFlow versions, Cisco IOS, and devices that support the NetFlow feature and configuration details: http://www.cisco.com/en/US/prod/collateral/iosswrel/ps6537/ps6555/ps6601/prod_white_paper0900aecd80406232.html.

Round-Trip Delay Measurement

The MOS readings indicate the speech transmission quality or the listening clarity. The delay has no direct impact on the MOS readings, although it indirectly affects the real phone conversation in following ways:

■ Long delay affects the natural conversation interactivity and causes hesitation and overtalk. A caller starts noticing delay when the round-trip delay exceeds 150 ms. ITU-T G.114[9] specifies the maximum desired round-trip delay as 300 ms. A delay greater than 500 ms makes the phone conversation impractical.

■ Long delay exacerbates echo problems. An echo with a level of –30 dB is not "audible" if the delay is less than 30 ms. But if the delay is greater than 300 ms, even a –50-dB echo is audible. The echo delay and level requirements are specified in ITU-T G.131[10].

Voice Jitter/Frame Slip Measurements

A frame slip or voice jitter is defined as a sudden delay variation at the audio signal side. The audio signal requires continuous and synchronous playout. The packet-switched network is inherently jittery where each packet arrives asynchronously and might be out of order. To compensate for the jittery nature of the packet-switched (IP) network, jitter buffers are used on voice gateways or MTAs. A large jitter buffer can minimize packet loss, but induces a longer delay. To balance the conflicting need for a shorter delay and less packet loss, the jitter buffer can be dynamically resized depending on the network traffic situation. Whenever the jitter buffer resizes, the audio signal experiences a sudden delay variation (jitter or frame slip) in an amount (in ms) that matches the voice frame size (6, 10, or 30 ms). You should measure two types of frame slips to get an accurate idea of the nature and amount of jitter:

■ **Positive (+) frame slip:** The total amount of compressive jitters (shortening of delays) that correspond to the downsizing of the jitter buffer or the deletion of packets.

■ **Negative (–) frame slip:** The total amount of expansive jitters (lengthening of delays) that correspond to the upsizing of the jitter buffer or the insertion of packets.

A good system should maintain a total jitter of less than 3 percent of the test duration. For a 10-second test, the total amount of positive and negative slips should be within –300, 300 milliseconds. If the measured value shows a higher amount of jitter, the network should be reconfigured for better traffic engineering and prioritization as part of call admission control and QoS settings on call control and network devices. Chapter 8, "Trend Analysis and Optimization," discusses this subject in detail by use of examples.

Measurement of Effective Bandwidth

A test should measure the attenuation distortion by analyzing the frequency response of the communications network under test by analyzing the 300–3400-Hz band (which is the frequency band for voice perceivable to the human ear). For Pulse Code Modulation (PCM) using G.711 and Adaptive Delta Pulse Code Modulation (ADPCM) using G.726 waveform coders, the effective bandwidth largely reflects the attenuation distortion caused by analog or digital filtering. If a network under test uses PCM or ADPCM waveform coders, its measured effective bandwidth should be higher than 0.9 in the range of 0 to 1. Anything below 0.85 signifies either excessive loop attenuation distortion (for analog circuits) or excessive band-limiting digital filtering on the IP leg for the test calls. This test can be run during network readiness assessment prior to deployment or during on-demand audits after VoIP deployment. They need not to be enabled on regular basis.

Voice Band Gain Measurement

A test should measure the overall voice band (300 to 3400 Hz) signal level change (attenuation or gain). Flat gain change is not reflected in the MOS reading. However, an excessive level change (too loud or too faint) does affect human perception. A VoIP network with a balanced network loss plan should maintain the change in voice level (gain) in the range of [–10, 3] dB.

Silence Noise level Measurement

The silence noise level in a VoIP network measures the comfort noise level generated by a Comfort Noise Generator (CNG). The level should not be too high (sounds too noisy) or too low (sounds like a dead line). The noise level is expressed in dBrnC, which stands for decibels above reference noise in reference to C-message weighting. C-message weighting is a noise spectral weighting used in a noise power-measuring test set to measure noise power on a line. An ideal system should maintain a silence noise level of [10, 30] dBrnC. This is based on IEEE Standard 743, "IEEE Standard Equipment Requirements and Measurement Techniques for Analog Transmission Parameters for Telecommunications." Anything above 30 dBrnC sounds too "noisy" and below 10 dBrnC might sound too "quiet."

Voice Clipping

The intention of voice-clipping measurement is to quantify the voice quality degradation caused by VADs (voice activity detectors). VADs help reduce bandwidth requirements through the silence suppression scheme. An overly aggressive VAD, however, can cause the leading or trailing edges of an active signal burst to be clipped. The voice clipping will severely effect modem and fax tone transmission over a VoIP network because both of these devices uses tones of specific frequencies to send information. The human ear might be able to compensate for a distorted voice, but these devices can lose data or the connectivity altogether if affected by voice clipping. Voice clipping is difficult to measure accurately and can be confirmed by looking at the symptoms on a reactive basis. It is advisable to simply turn off VAD features on voice gateways, especially if they are going to be processing modem or fax calls.

Echo Measurements

In a VoIP network, echo is an inherent issue because of the analog 2-wire loop presence, which causes an impedance mismatch at the hybrid junction (linking a 2-wire analog loop with a 4-wire trunk). The echo becomes perceivable because of the network delay, as explained earlier in this chapter. The higher level of the echo signal and the significant network delay will make the echo perceivable to the human ear. Echo cancellers are employed to cancel the echo by sampling the original signal in the receive path to cancel it from the reflected or the transmitted signal, hence suppressing the echo from the originally transmitted signal. The sampling time is referred to as *tail coverage*.

Figure 6-3 shows the minimum requirements for TELR as a function of the mean one-way transmission time T (half the value of the total round-trip delay from the talker's mouth to the talker's ear), as discussed in section "Measurements for Network Transmission Loss Plan," earlier in this chapter. In general, the values for TELR and one-way delay must fall above the "acceptable" curve to avoid echo-related problems. Only in exceptional circumstances should values for the "limiting case" be allowed; otherwise, all such cases should be compensated for by enabling echo cancellers and properly adjusting tail coverage, as shown in the TELR calculations in Figure 6-2. Test equipment from test set vendors such as Sage Instruments, Agilent, or Spirent should be able to calculate the echo level against the delay to characterize the echo present in the network. This provides guidance in setting the echo cancellers (ECAN) with the appropriate tail coverage. This is done by adjusting the TELR value against the given delay of the network by adding the appropriate tail coverage. The tail coverage value is directly related to ECAN enhancement that determines Echo Return Loss, which is an additive component of TELR, by bringing the final TELR and delay values into the acceptable range, as shown in the echo/delay tolerance curve of Figure 6-3.

Voice-Signaling Protocol Impairments in IP Networks

Signaling connections are implemented with protocols that allow the detection of packet losses and the retransmission of lost packets. As such, they are better equipped than voice media connections to survive packet losses. For example, Skinny Call Control Protocol (SCCP) used by Cisco IP Phones, uses TCP as a transport protocol. Similarly, MGCP and SIP implement their own retransmission scheme because their underlying transport protocol, UDP, does not provide retransmission services for lost packets. SIP can also use TCP instead of UDP. In either case, similar to MGCP, it has its own retransmission mechanism.

Even though lost packets are retransmitted, it is the network conditions that determine the success or failure of the retransmission attempts. Even successful retransmissions can have negative effects if the period needed to complete the transaction (from the initial attempt, through the retransmission attempts, to final success) delays system response by a user-perceivable amount of time. We can classify the IP communications system behavior according to the relative severity of the interruption to the signaling link connectivity, as follows:

- **Light packet drops, with short duration and low frequency of drops:** In this case, the system appears to be generally unresponsive to user input. The user might experience effects such as delayed dial tone, delayed ringer silence upon answer, and double dialing of digits because of the user's belief that the first attempt was not effective (thus requiring hang-up and redial).

- **More frequent, longer-duration packet drops:** In this case, the system alternates between seemingly normal and deteriorated operation. Packet drops cause endpoints to activate link failure measures, including reinitialization. This link interruptions caused by continuous packet drops for long durations can reach the point of causing phone or gateway reset, resulting in media teardown as well. Users might experience survivable remote site telephony (SRST) activation, whereby all active calls are dropped when the link is interrupted and again when the link is reestablished. Phones can also appear unresponsive for several minutes.

- **Complete link interruption:** Although most likely caused by an actual network failure, link blackouts can be the result of a congested network where end-to-end QoS is not configured. For example, a high degree of packet loss can occur if a signaling link traverses a network path experiencing large, overprovisioned, sustained traffic flows such as network-based storage/disk access, file download, file sharing, or software backup operations. In such cases, the IP communications system interrupts calls, and the initiation of a backup mechanism—for example, SRST in enterprise networks—provides continued telephony service for the duration of the link failure. However, the switchover to the backup system might be associated with a delay where the endpoints might have to reregister to the alternate system or the advanced telephony features might become unavailable.

These effects apply to all deployment models, as shown in Figure 6-4. However, single-site (campus) deployments tend to be less likely to experience the conditions caused by sustained link interruptions because the larger quantity of bandwidth typically deployed in LAN environments (minimum link speed of 100 Mbps) allows some residual band-width to be available for the IP communications system. In any WAN-based deployment model and any service provider–managed residential services model, as shown in Figure 6-4, traffic congestion is more likely to produce sustained and/or more frequent link interruptions because the available bandwidth is much less than in a LAN (typically less than 2 Mbps for WAN links), so the link is more easily saturated. The effects of link interruptions impact the users, whether or not the voice media traverses the packet network.

How to Effectively Poll the Network

To recap, a VoIP network is a large and complicated solution that encompasses many integrated technologies. This presents a problem for VoIP infrastructure managers because each technology brings its own network management challenges. A device-polling strategy needs to be implemented for the network so that all VoIP functional segments have coverage.

Polling involves utilizing the existing device management mechanisms like the EMS/NMS to perform device polls. These polls, when done in regular periods, in essence act as health audits for VoIP segments. In reality, custom probes need to be implemented through scripts for achieving full polling coverage. A best practice is to create or have a dedicated in-house Linux- or UNIX-based system to develop these probes. Open-source tools like Nagios accommodate these types of probes easily.

The following sections describe the processes that should be in place to allow a VoIP SP to effectively manage its network. The key fact is that the development and deployment of specialized tracking systems require up-front cost and man-hours. This can be implemented through the assistance of vendor services groups.

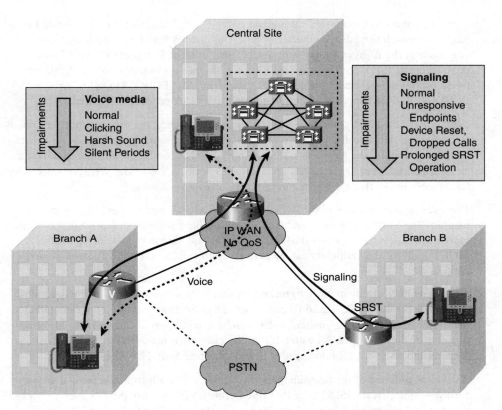

Figure 6-4 *Effects of Packet Loss on Voice Media and Signaling Protocols*

Polling Strategy

VoIP segments are driven by certain protocols, and all these protocols ride over IP. Thus, an organized and a layered approach needs to be developed to effectively poll the VoIP network for key information through the corresponding protocol metrics. A layered approach implies that the polling is done in such a way that all protocols riding IP are covered. Figure 6-5 reflects this concept. All the VoIP-related segments need to be polled for key metrics. The base connectivity to the network components is tested at the IP layer, and then the next layer depicts the various segments and functional components.

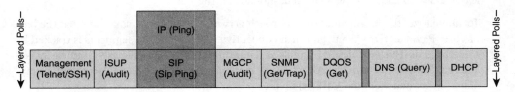

Figure 6-5 *Polling of Network Elements a Layered Approach*

The base connectivity is tested through an IP ping poll. The entire network should be mapped through this basic connectivity test. This creates a connectivity health knowledge base at the IP layer. If ping is disabled, other custom TCP probes should be developed to create this map. There are open-source alternatives available to the classic ping-based probes like *echoping*, which do not use the ICMP_ECHO_REQUEST or ECHO_REPLY packets but can communicate using other protocols like http.

The SNMP connectivity needs to be verified. At a minimum, the trap functionality needs to be verified periodically. The traps are critical because they map to alarms and events being generated by the Network Elements (NE). The NE should be configured to trigger an informational trap to the SNMP manager on a periodic basis. The SNMP connectivity validation ensures that the alarm and key event stream is flowing.

Considering the example of a cable deployment, MGCP polling needs to be implemented. The VoIP deployment architecture for a cable environment has MGCP as a protocol used in several segments (PSTN, CPE/MTA, and Announcements); a periodic MGCP audit for the endpoints can be performed to validate the MGCP-based device connectivity.

SIP polling needs to be in place to make sure that SIP-supporting devices are functional. The devices could be SIP-based voicemail servers or Session Border Controllers (SBC) that terminate SIP trunks. Custom probes can be used to periodically test the SIP functionality of these devices. A custom SIP probe example is included in Appendix A, "Scripts and Tools for Monitoring and Troubleshooting VoIP Networks."

A periodic poll test of the Signaling System 7 (SS7) link needs to be performed, along with verifying that the ISDN User Part (ISUP) functionality is up. ISUP is used to backhaul the SS7 information to the Call Management Switch (CMS). Chapter 7, "Performance Analysis and Fault Isolation," covers ISUP key performance indicators (KPI) that need to be tracked. The reported SNMP alarms also need to be tracked and mapped back to SS7 and ISUP segments to create a comprehensive SS7 health view.

Management connectivity needs to be verified, making sure that the devices can always be reached through Telnet or Secure Shell (SSH). In some cases, out-of-band connectivity through modem dialups needs to be verified as well.

A periodic polling of DNS and DHCP functionality needs to be in place. All the DNS servers, including the primary and secondary DNS servers, need to be polled for DNS queries. Similarly, the DHCP functionality needs to be verified on a periodic basis; this can be done by tracking the DHCP statistics (such as DISCOVER messages sent and leases granted) and creating a visual dashboard for the stats.

To summarize, the idea behind this approach is twofold: One is to make sure that the base or core IP connectivity is up, and then connectivity to each of the segments is tracked in some fashion. This is achieved by polls at the protocol layer. This allows you to easily isolate the down segment in case of issues by mapping the failed polls to alarms and events beings generated from the VoIP segments.

Key Alarms and Events Monitoring

An exercise of identifying the key alarms and events should be done across the VoIP network. It is imperative that this exercise be done on a periodic basis, especially when major software upgrades are performed on the network. The new software would deprecate some old key alarms and introduce potentially new critical and more verbose alarms to better track the network. This information is available in the release notes of the new software.

Table 6-7 depicts the general buckets of alarms as collected from the Cisco BTS 10200 product. This can be used as a high-level guideline for making sure that the operation centers are including such alarm categories in their network-monitoring tools.

The general buckets do have alarms that get reported on a frequent basis and can cause the Network Operation Centers (NOC) to drop them in an ignore bucket. Typically, end devices going into and out of service trigger these audit alarms. A scenario can be that of DNS functionality failing for a particular market, thus causing a flood of audit endpoint failure notifications. This should be tracked immediately and cannot be ignored.

A best practice would be to go through the vendor services groups and classify key and chatty alarms on new major software releases. At the same time, understand the behavior of the chatty alarms so that the critical ones do not get ignored.

Table 6-7 *Alarm/Event Groups*

Alarm/Event Groups	Type
OSS	Alarms
DATABASE	Alarms and Events
AUDIT	Alarms and Events
MAINTENANCE	Alarms
SYSTEM	Alarms and Events
BILLING	Events
CALLP	Alarms
SIGNALING	Alarms

SNMP Configuration and Setting

SNMP configurations and connectivity are at the core of network operations. It is among the first set of configurations being pushed on the NEs. This section gives an overview of some basic configurations and then describes key SNMP trap-related configuration settings.

Basic Configuration

The basic SNMP configurations involve setting up the following:

- SNMP community read string and write string with passwords need to be updated. Typically, the default is set to "public" and should be overwritten.

- SNMP trap destination configurations need to be made. This basically tells the SNMP agent running on the NE the location (IP address) of the NMS to send the traps to. Most of the time, the default port also needs to be overridden for security reasons. There can be multiple NMS devices listening for the traps, so all these devices need to be accounted for.

SNMP Trap Settings

It is important to configure the SNMP trap settings correctly. The monitoring of the VoIP network is affected if the key traps are not generated.

The following categories of traps need to be generated at a minimum: CRITICAL, MAJOR, MINOR, and WARNING. These categories constitute the alarms. The other category is of events; it includes INFO and DEBUG types. Often, the events are also critical as they could include audit information.

The NMS needs to be optimized to handle all alarms effectively such that the chatty alarms and even events go in other buckets but can still be tracked. The chatty alarms can have a custom threshold trigger alarm mechanism on the NMS, thus catching systematic issues such as the DNS failure example mentioned earlier.

Traps Use Case BTS 10200 Cisco Softswitch

Softswitches enable even greater flexibility in tracking the type of traps. In a centralized model, where they perform the switching and handling of all VoIP segments, it becomes critical to generate and subscribe to relevant categories and types of traps. All types might need to be included. The BTS 10200 enables the following detail types:

- BILLING
- CALLPROCESSING
- CONFIGURATION
- DATABASE
- MAINTENANCE
- OSS
- SECURITY
- SIGNALING
- STATISTICS

- SYSTEM

- AUDIT

Thus, subscribing to all the traps related to all these types, and then in turn effectively monitoring them, is the key to VoIP network management success. We expand on the usefulness of this extensive subscription in the upcoming section "Alarm and Event Audit Correlation."

Standard Polling Intervals and Traps

The minimum polling interval depends on the type of the SNMP object(s) being polled, the number of devices you are polling, and how much network bandwidth you want to devote to network management. Most critical SNMP objects (for example, ifOperStatus, ifInErrors, and so on) should be polled every 5 minutes. Other SNMP objects might require more frequent polling (for example, nl-ping-response).

Most performance SNMP objects should be polled at 30-minute intervals. This is a fairly conservative polling interval that provides 48 data points per 24-hour reporting period. Forty-eight data points provide enough granularity to establish general performance baselines.

Traps from the managed devices are sent to Network Management Systems (NMS) unsolicited on a reactive basis as the problem occurs. Traps notify the problems such as link down when there is an outage. However, there is no trap to identify link congestion. For that, we have to rely on polling using the SNMP object identifier (OID). There is a specific set of OIDs or MIB for network statistic gatherings such as QoS traffic-shaping packet discards, forward explicit congestion notifications (FECN), backward explicit congestion notifications (BECN), and so on. This poses a challenge because excessive polling to increase the time resolution of problem notification through polling can increase management traffic on a production network.

Because the polling responses arrive at a certain time delay, the problems encountered for a short duration during the polling intervals can go unnoticed. Moreover, the traps need to be correlated to the polled results or the other traps that might be related. Also, some tools and certain SNMP OID tables calculate and store values that are representative of an average number rather than instantaneous values. This might not give an accurate idea of the severity of the problem as the counters show an average value over a spread of a time period.

The scenarios described in the following sections illustrate this challenge.

Scenario 1: Phones Unregistering from Unified CM and Reregistering to SRST Router Because of WAN Link Outage

In this scenario, as shown in Figure 6-6, the NMS is polling the WAN for a Frame Relay congestion (monitoring for FECN or BECN or QoS TS discards or packet drops) with an interval of 30 minutes. Assume that the polling intervals are 30 minutes apart and that a poll occurs at 9:00 a.m. The successive polls reoccur at 9:30 a.m., 10:00 a.m., and 10:30 a.m. Around 9:35 a.m., the Frame Relay network close to the aggregation site encounters an outage that causes the Frame Relay link connecting the branch to go down. This causes the IP phones to unregister from the CallManager cluster and register with the SRST gateway local to the branch. This unregistering is notified by a trap from the Unified CM originally hosting those IP phones. At the same time, the aggregation router sends an ifDown (link down) trap to the NMS. Because of the close time proximity of these traps, an NOC staffer is able to correlate the IP phone registration with the SRST gateway to the Frame Relay link outage.

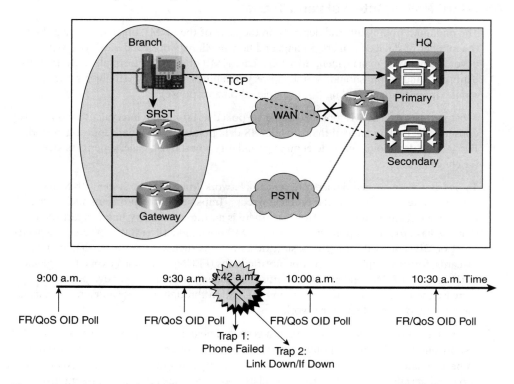

Figure 6-6 *Phones Unregistering from the Unified CM and Reregistering to the SRST Router Because of a WAN Link Outage*

Scenario 2: Phones Unregistering from the Unified CM and Reregistering to the SRST Router Because of WAN Congestion

In this scenario, the NMS is polling the WAN for Frame Relay congestion (monitoring for FECN or BECN or QoS TS discards or packet drops) with an interval of 30 minutes. Assume that the polling intervals are 30 minutes apart, and a poll occurs at 9:00 a.m. The successive polls reoccur at 9:30 a.m., 10:00 a.m., and 10:30 a.m. Around 9:35 a.m., the Frame Relay network close to the branch starts to experience network congestion, and at about 9:40 a.m., the IP phones cannot get the keepalives (heartbeat signals) serviced by the UC Manager cluster. By 9:42 a.m., they unregister themselves from the CallManager cluster and register with the SRST gateway local to the branch. This unregistering process will be notified by a trap from the CallManager originally hosting those IP phones. However, the underlying problem in the Frame Relay network is reported by the next polling cycle occurring at 10:00 a.m. There is a possibility that the Frame Relay network congestion is relieved around 9:45 a.m. In that case, the network administrator can correlate the phone unregistering and registering problem with the actual cause. The troubleshooting efforts might be directed toward the source of the trap, which is the CallManager.

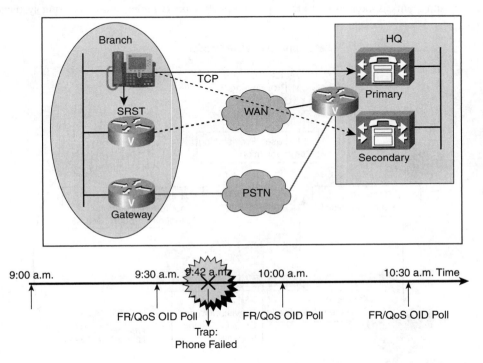

Figure 6-7 *Phones Unregistering from the UC Manager and Reregistering to the SRST Router Because of WAN Congestion*

These challenges can be addressed by adopting a layered approach, as discussed earlier. Chapter 7, "Performance Analysis and Fault Isolation," further elaborates using statistical data collection, which is instantaneous rather than average, to get a more accurate profile of the network. Chapter 7 also dicusses the use of syslog analysis to bridge gaps in polled data.

Using eXtensible Markup Language (XML) for Polling and Extraction of Key Information

eXtensible Markup Language (XML) is used extensively in the industry these days to address a variety of needs. Some of the key XML advantages in the scope of our discussion are as follows: XML simplifies data sharing, XML simplifies data transport, and

XML is used as building blocks for new protocols. We cover these aspects briefly and then describe how they are applied in the VoIP network management and polling.

XML Overview

NEs, Softswitches, and EMSs provide communication and reporting interfaces through XML. Figure 6-8 depicts most of these interfaces. This XML capability thus allows generic and seamless integration to third-party reporting provisioning and monitoring systems.

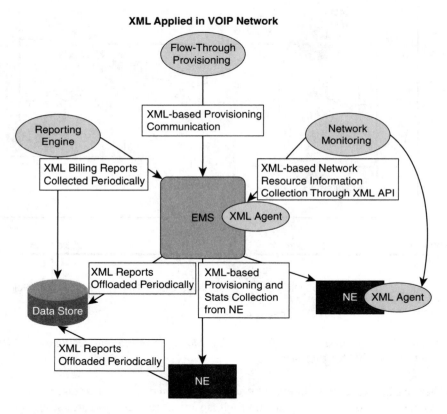

Figure 6-8   *XML Application VoIP Networks*

A flow-through provisioning system from a certain vendor can be integrated to another vendor's EMS implementation through a generic XML interface. There are many reporting engines that take in XML data reports and transform them to manageable information. These reporting engines facilitate trouble ticket tracking, manage billing records, and provide data-mining information for capacity planning through performance measurement reports. In some cases, there are explicit XML agents present on the NEs and EMSs that listen for XML-based queries over a TCP socket. The XML agents handle the request from the client. Typically, these requests are otherwise made through the CLI. The returned result is a well-formed XML report.

In short, the XML usage in VoIP network operations is crucial. It allows the VoIP provider to scale and integrate with third-party vendors.

XML APIs

Generally, the XML APIs can be broken down into categories of exporting data, communication payload, and as agent interfaces. These aspects enable the reduction of interoperability challenges and facilitate the adaptability to a changing industry.

Data Exported as XML Reports

VoIP network EMSs, in particular the EMS for BTS 10200, provides interfaces to periodically generate Call Detail Record reports, Billing Record reports, and Performance measurement reports in XML format. Each of these reports can be imported into a third-party Billing and Operation Sub-Systems (OSS) system.

SOAP Communication Utilizing XML

In VoIP networks, provisioning applications are used for flow-through configuration. These systems communicate through the NMS and EMS and utilize the XML over Simple Object Access Protocol (SOAP) frequently. The well-formed aspect of XML enables strict syntax checking. The XML capability of incorporating new tags facilitates third-party provisioning systems to adapt to changes with VoIP vendors' provisioning API interfaces. This makes the interoperability within a slew of VoIP-related applications manageable.

Thus, XML-based communication (which basically implies that the communication protocol is utilizing XML to format its payload) enables the VoIP SP to better manage its flow-through provisioning systems.

Querying XML Agent for Information Retrieval

Almost all the mainstream VoIP hardware and software vendors are moving toward providing an XML-based agent interface on their products; they also expose XML/SOAP web services APIs. Network operators utilizing web services or through the well-defined XML agent interfaces can now easily integrate their strategic applications with those of their equipment vendors over IP networks without interoperability issues typically experienced with different SNMP version and Management Information Base (MIB), COBRA implementations, or Remote Monitoring (RMON) methods.

The XML agent is present at the NE or in an EMS. This agent enables the following:

- A mechanism using XML to transfer, configure, and monitor objects in the NE or the EMS.

- XML capability that allows the user to easily shape or extend the CLI. The XML query can query for NE objects in various data structure formats to support single and multiple CLI queries, thus as an example creates a superset of CLI commands.

- Query and reply of data in XML format to meet different specific business needs.

- Transfer of **show** command output from the CLI interface in XML format for statistics and status monitoring. This **show** command output transfer capability would allow the user to query and extract data from the NE/EMS.

- The NE/EMS XML Document Type Definition (DTD) scheme for formatting CLI queries or parsing the XML results from the NE/EMS to enable third-party software development through XML communications.

- Remote user authentication through AAA. The interface is exposed in the XML DTD.

- Communication through HTTP or HTTPS.

- A set of return error codes to easily diagnose the NE/EMS faults and also XML query syntax issues. Errors can be as simple as the wrong XML form or a syntax issue with the XML NE/EMS query.

- Other NE/EMS specific-feature support through the DTD.

To summarize, a VoIP service provider needs to understand and implement XML-based applications and devices that support XML interfaces. This results in flexibility to adapting to change and easily accommodating growth.

Using the Syslog/Trace Logs for Deep Analysis

Syslogs or trace logs contain information that was typically used during deep debugging and root cause analysis sessions. There is a lot of valuable information embedded in the syslogs, which if tapped, on a real-time basis can take the VoIP network monitoring to another level. It allows creating in-depth views of the network elements.

The basic idea is to stream the syslogs to a dedicated server for deep analysis, where a continuous process should be introduce to parse the syslogs for key trends. Key exercises that need to be performed are as follows:

Step 1.   Identify systemic metrics that can be tracked through the logs.

Step 2.   Identify frequent service-affecting issues that can be tracked through logs.

Step 3.   Capture the frequency of the systemic metrics.

Step 4.   Group the metrics into functional buckets, thus allowing easier functional debugging and tracking.

Table 6-8 highlights some key metrics that are captured from the Cisco Softswitch BTS 10200. These metrics are basically derived from the log text. Each piece of text acts as a signature for the metric. The syslog is periodically parsed for these signatures. The signatures then map to event types, based on their criticality. They can fall into an ERROR or a WARNING category. These events can then trigger traps or other notifications to monitoring systems.

Table 6-8   *Sample Key Metrics from Cisco Softswitch Trace Logs*

Log Text	Severity	Explanation
MGW admin state not allow subscriber maint request	Error	Cannot perform maintenance request (administrative and diagnostic module related).
ANM_process state_waitcrxresp Connection failed event type	Error	Announcement connection failed (Announcement Manager).
Failed SRV lookup and A record lookup while attempting add port and Domain name	Error	DNS issue for softsw_tsap_addr. BTS could not resolve domain name during an audit of the SIP table.
KA timer expired for aggrIdx	Error	Cops protocol (BTS to CMTS) related error. Keepalive timer expired for AGGR.

These metrics should be used for tracking overall system health through periodic summary reports along with triggered event notifications.

The smart analysis of the syslogs enables the service provider to develop an extra layer of monitoring that might not be covered through the typical alarm notification functionality and might even drive toward improving the reported alarms.

Alarm and Event Audit and Correlation

Root cause analysis (RCA) is important to incident management. RCA can be accelerated by following a steady alarm and event analysis practice. The practice can be developed through the following best practices:

- Have a vigorous exercise of understanding the alarm and alarm groupings for all the network elements. As a result, a key alarm summary list should be produced for all the products.

- A periodic collection of alarm history should be introduced and then monitored actively through a Dashboard.

- The Dashboard should be used for history trending of alarms by market and functional groups. For example, two functional groups of SIP- and PSTN-related alarms should have their own visual mentoring groups.

- In a VoIP network, the VOICE switch (or corresponding EMS) sees all the protocols and most NE-reported anomalies. It can be a good starting point for creating an alarm summary report.

Table 6-9 shows a summary of alarms for a particular customer site. The table has been generated from the alarm history report for a Cisco BTS 10200. The table is periodically generated for many customer sites. It allows identification of the hotspots and infrequent anomalies.

Table 6-9   *Summary Alarm Table from Cisco Softswitch*

Alarm Count	Alarm Group Type	Alarm Explanation
1	CALLP	Country code dialing plan error
1	DATABASE	EMS database alert.log alerts
1	OSS	SNMP authentication error
1	SIGNALING	SS7 message decoding failure
1	SIGNALING	Unanswered REL
2	BILLING	FTP/SFTP transfer failed
2	DATABASE	Daily database backup completed successfully
2	SIGNALING	AGGR connection down
2	SIGNALING	Feature server is not up or is not responding to Call Agent
2	SIGNALING	Continuity recheck successful
3	SIGNALING	AGGR gate set failed
3	SIGNALING	Continuity recheck is performed on specified CIC
6	SIGNALING	RLC received in response to RSC message on the specified CIC
10	AUDIT	Start or stop of SS7-CIC audit
11	CALLP	No route available for carrier dialed
14	DATABASE	Errors in the EMS database DefError queue
27	MAINTENANCE	Admin state change failure
56	MAINTENANCE	Admin state change successful with warning
76	AUDIT	Call exceeds a long-duration threshold
112	SIGNALING	Continuity recheck failed
215	SIGNALING	Media gateway/termination down
379	MAINTENANCE	Admin state change
469	SIGNALING	Timeout on remote instance
803	CALLP	Invalid call
981	SIGNALING	Unexpected message for the call state is received: Clear CA
1166	CALLP	Call failure

Table 6-9 *Summary Alarm Table from Cisco Softswitch   (continued)*

Alarm Count	Alarm Group Type	Alarm Explanation
1194	SIGNALING	COT message received on the specified CIC
1267	BILLING	Message content error
1391	SIGNALING	General MGCP signaling error between MGW and CA
5257	SIGNALING	Trunk locally blocked
5495	SIGNALING	Trunk remotely blocked
1391	SIGNALING	General MGCP signaling error between MGW and CA
5257	SIGNALING	Trunk locally blocked

The table shows the general group buckets and the detailed breakdown of the alarm frequency within the groups.

The most prevalent alarm buckets surface up. The periodic report can be easily used to first identify the problem, and then the results are fed into the systemic root cause analysis (RCA) engine. In Table 6-9, you can see the signaling alarm count is high. Alarms like "Trunk remotely blocked" were ignored because a large number of trunks were being turned up, which triggered a high number of maintenance "Trunk locally blocked" alarms, therefore masking other alarm categories.

Thus, the results are obvious for performing these alarm audits on a periodic basis and analyzing them at the same time through a Dashboard. The alarm count and the alarm categories help isolate the trouble segments. The resolution can be prioritized based on the issues that are most critical.

Effectively Monitoring the PSTN Bearer Traffic

Tracking a VoIP service provider's PSTN bearer trunk utilization is growing rapidly with the fast growth of the customer base. Trunks represent T1s that carry basically 24 DS0s/CICs (Circuit Identifier Codes).

These CICs represent the state of the voice channel and basically the voice capacity. If these CICs are not available to carry voice traffic, the SP's voice network is critically impaired as OFFNET or PSTN calls will not go through. Again, this depends on the percentage of CICs affected.

Another important need to monitor these CICs is because the SP routes certain categories of OFFNET calls like emergency, long-distance calling, toll-free calling, directory assistance, and so on to specific set of trunks. If these particular trunks are affected, that entire category of service is down.

Summary email reports are one form of output for the abnormal CIC states. The same data can be used to generate web/HTML reports that are updated on a pseudo-real-time basis. These reports are to be used to create a Trunk Monitoring Dashboard. This monitoring Dashboard would greatly improve voice network operations, and the Dashboard would give a view into the health of the voice trunks on a ready basis. Figure 6-9 shows this report.

```
1018   92   ADMIN_INS   TERM_DOWN                 LBLK   IDLE   TERM_FAULT   DES_LOC_FO
1018   93   ADMIN_INS   TERM_DOWN                 LBLK   IDLE   TERM_FAULT   DES_LOC_FO
1018   94   ADMIN_INS   TERM_DOWN                 LBLK   IDLE   TERM_FAULT   DES_LOC_FO
1018   95   ADMIN_INS   TERM_DOWN                 LBLK   IDLE   TERM_FAULT   DES_LOC_FO
1018   96   ADMIN_INS   TERM_DOWN                 LBLK   IDLE   TERM_FAULT   DES_LOC_FO
1021   96   ADMIN_INS   TERM_ACTIVE_CTRANS_BUSY   ACTV   TRNS_OBSY   NON_FAULTY
1024   95   ADMIN_INS   TERM_ACTIVE_CTRANS_BUSY   ACTV   TRNS_OBSY   NON_FAULTY
1031   23   ADMIN_INS   TERM_ACTIVE_CTRANS_BUSY   ACTV   TRNS_OBSY   NON_FAULTY
1050    1   ADMIN_INS   TERM_DOWN                 LBLK   IDLE   TERM_FAULT   DES_LOC_FO
1050    2   ADMIN_INS   TERM_DOWN                 LBLK   IDLE   TERM_FAULT   DES_LOC_FO
1050    3   ADMIN_INS   TERM_DOWN                 LBLK   IDLE   TERM_FAULT   DES_LOC_FO
1050    4   ADMIN_INS   TERM_DOWN                 LBLK   IDLE   TERM_FAULT   DES_LOC_FO
1050    5   ADMIN_INS   TERM_DOWN                 LBLK   IDLE   TERM_FAULT   DES_LOC_FO
1050    6   ADMIN_INS   TERM_DOWN                 LBLK   IDLE   TERM_FAULT   DES_LOC_FO
1050    7   ADMIN_INS   TERM_DOWN                 LBLK   IDLE   TERM_FAULT   DES_LOC_FO
1050    8   ADMIN_INS   TERM_DOWN                 LBLK   IDLE   TERM_FAULT   DES_LOC_FO
1050    9   ADMIN_INS   TERM_DOWN                 LBLK   IDLE   TERM_FAULT   DES_LOC_FO
1050   10   ADMIN_INS   TERM_DOWN                 LBLK   IDLE   TERM_FAULT   DES_LOC_FO
1050   11   ADMIN_INS   TERM_DOWN                 LBLK   IDLE   TERM_FAULT   DES_LOC_FO
1050   12   ADMIN_INS   TERM_DOWN                 LBLK   IDLE   TERM_FAULT   DES_LOC_FO
1050   13   ADMIN_INS   TERM_DOWN                 LBLK   IDLE   TERM_FAULT   DES_LOC_FO
1050   14   ADMIN_INS   TERM_DOWN                 LBLK   IDLE   TERM_FAULT   DES_LOC_FO
```

Figure 6-9   *Summary Anomaly CIC Report*

The highlighted section in the sample report reflects the CIC's status. It shows the CICs as administratively in service (ADMIN_INS) but locally blocked (LBLK); similarly, other CICs are in the ADMIN_INS state but stuck in the transient state (TERM_ACTIVE_CTRANS_BUSY). Both of these CIC states need to be investigated.

An alerting mechanism can be set up to send an email or SMS when certain configured thresholds of trunks go local block (LBLK) or remote block (RBLK) in between two monitoring periods.

Also, this captured data can be used to generate XML reports that can plug into web browsers or other aggregation tools to show the monitoring information.

The ideas presented so far have the same key theme: Create specialized reports and monitor them through Dashboards. This specialization costs time and money, but the positive impacts are immediate to the overall VoIP service as the anomalies are identified quickly.

QoS in VoIP Networks

As discussed earlier, VoIP is most commonly deployed over converged IP networks carrying data, voice, and video traffic. When network resources are congested, they can severely affect the quality of VoIP traffic and cause a poor user experience for the subscribers. This can result in increased customer calls (trouble tickets) for the voice SP and loss of revenue because of customer turnover.

Therefore, it is important for the Voice SP or an enterprise to implement QoS for VoIP traffic in its networks. This can help guarantee good voice quality when network resources are congested.

A number of factors can affect the quality of VoIP traffic as perceived by the end user. Some of the common factors include delay, jitter, and packet loss. These factors have been discussed in detail in Chapters 1 and 2. This section does not cover the various methods of configuring and troubleshooting QoS to prevent delay, jitter, and packet loss in VoIP networks. It describes (at a high level) the methodology of how to use these key indicators to implement and manage a QoS policy in the network. This can help the voice SP or an enterprise isolate problems in the network more effectively and prevent them from happening in the future.

Defining a QoS Methodology

The QoS policy implemented for VoIP traffic should encompass the end-to-end voice network. It is recommended to take a layered QoS approach that makes it easier to implement and manage the QoS policy for VoIP.

The QoS policy for VoIP traffic should cover Layer 2, Layer 3, and the application layer. This helps guarantee that the VoIP traffic is given preferential treatment when it is transported from one endpoint to another. QoS at the application layer is especially useful when end users are using PC-based VoIP applications to place and receive voice calls. In this case, the VoIP traffic might receive the desired QoS as it traverses the network, but the end user's PC-based application might not prioritize VoIP over other applications demanding CPU resources. This can result in poor voice quality because of delay, jitter, or packet loss, as described previously.

One thing to keep in mind is that QoS might only help when resources are congested. If there is no contention for bandwidth or other network resources, applying QoS might not provide any additional benefits.

Differentiated Services (Diff Serv) for Applying QoS

A good QoS policy involves marking or classifying the VoIP traffic at the edge of the network so that intermediate devices in the network can differentiate voice traffic from other traffic and process them according to the defined policy. This marking or classification can be done using Differentiated Services Code Point (DSCP) values or by using the IP precedence bits in the type of service (ToS) byte in the IP header.

Diff Serv (RFC 2474 and RFC 2475) defines the required behavior in the forwarding path to provide quality of service for different classes of traffic. An important aspect in the definition of forwarding path behavior for QoS is the method of doing packet classification. Packet classification is required for quality of service to determine which treatment a particular packet gets for shared resource allocation. The Diff Serv model also defines boundaries of trust in a network and the associated functions that occur at the edges of a region of trust. A DSCP specifies a Per-Hop Behavior (PHB) for forwarding treatment. A PHB specifies a scheduling treatment that packets marked with the DSCP will receive. A PHB can also include a specification for traffic conditioning. Traffic conditioning functions include traffic shaping and policing. Traffic shaping conditions traffic to meet a particular average rate and burst requirement. Policing enforces an average rate and burst requirement. Actions to take when traffic exceeds a policing specification can include remarking or drop.

The precedence level for IP packets carrying for voice bearer traffic in a PHB model is commonly 46, which is also known as the Expedited Forwarding (EF) PHB. The PHB precedence value commonly used for call signaling is 26, also known as the Assured Forwarding 31 (AF31) PHB. Figure 6-10 illustrates a Differentiated Services–based QoS model.

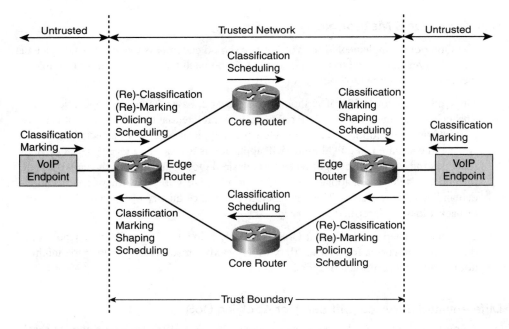

Figure 6-10 *Differentiated Services–Based QoS Model*

In SP environments, endpoints are typically untrusted devices. This means that endpoints might not mark or classify the VoIP traffic correctly. Therefore, this traffic would need to be remarked and reclassified at the edge of the network. After the VoIP traffic is reclassified at the edge, it can be scheduled into appropriate queues to receive the desired level of service per QoS policy.

In enterprise networks, endpoints such as IP phones are considered trusted devices while PC-based soft clients are conditionally trusted. Trusted devices are supposed to classify the VoIP traffic correctly, while traffic from conditionally trusted devices is trusted only if it meets a defined criteria. This criteria is typically defined at access layer switches that are directly connected to these conditionally trusted devices. If a device is compromised and starts sending misclassified VoIP traffic, it can be policed at the edge of the trust boundary and put into a scavenger queue (a very-low-priority queue that gets less than best-effort treatment and is allocated a very small bandwidth percentage, typically 1 percent of interface bandwidth). This queue is meant for mitigating worm traffic, denial of service (DoS) attacks, and so on. This queue can be monitored periodically to discover any undesired network activity, and the data can be used for trending to predict failures as well as linked to the trouble ticketing system to correlate to any network issues.

Using Bandwidth/Resource Reservation and Call Admission Control (CAC) for Providing QoS

Another approach for providing QoS to VoIP traffic is to reserve the required network resources before setting up the voice call and using CAC for rejecting calls that might not be able to receive the desired QoS because of congestion or high utilization of network resources. Although this approach can definitely guarantee QoS to VoIP traffic, it does have its disadvantages.

One of the problems with using this approach is that network resources need to be reserved end to end to guarantee QoS to VoIP traffic from one endpoint to another. This can be challenging because resources might not be available on certain network segments because of congestion, which results in the call setup failing. This also means that after the resources are reserved, they cannot be used for any other traffic; hence network resources might not be efficiently utilized.

Even with the downsides mentioned previously, this approach is still used in some deployment models in SP environments. The approach is slightly modified though to make better use of network resources. Instead of reserving network resources ahead of time, they are only reserved when a voice call needs to be set up and are released after the voice call is torn down. This enables more efficient use of network resources because they can be used for other traffic if they are not being utilized for VoIP traffic. This approach is preferred especially in cases where VoIP is deployed in converged networks.

Managing QoS

QoS management helps to set and evaluate QoS policies and goals. A common methodology entails the following steps:

Step 1.   Establishing a network baseline. This helps in determining the traffic characteristics of the network.

Step 2.   Deploying QoS techniques when the traffic characteristics have been obtained and an application(s) has been targeted for QoS.

Step 3.   Evaluating the results by testing the response of the targeted applications to see whether the QoS goals have been reached.

To effectively manage QoS policies in a VoIP network, it is important to use a layered approach. Information needs to be gathered from different points in the network and at various layers (Layer 1, physical layer; Layer 2, data link layer, Layer 3, network layer; and Layer 7, application layer). This information needs to be correlated to different events occurring in the network, such as degraded voice service in certain network segments or complete voice outage in a specific location.

It is important to establish a baseline for the voice endpoints as well. For example, a baseline can be established for PacketCable MTAs based on their state (In-Service, Out-of-Service, and so on), registration status (registered, unregistered), and so on. If a mass deregistration occurs, this can be correlated to a provisioning server failure, or if a large number of MTAs go into the Out-of-Service state, this event can be correlated to a CMS failure.

For monitoring QoS, look for PHB as defined in the Diff Serv model for QoS. Look for QoS policy violations, queue drops, interface statistics (traffic rate, queue depth, drops, errors), and resource overutilization (memory, CPU) on routers, switches, voice gateways, and endpoints. This information can be correlated to alarms and syslog messages stored on management servers.

The previously mentioned information can be gathered using the command-line interface (CLI) or by polling through SNMP or XML, as mentioned in earlier sections of this chapter. To poll information from various network devices, different MIBs can be used. Examples of the QoS MIB "CISCO-CLASS-BASED-QOS-MIB" and relevant Object IDs (OID) are as follows:

- **cbQosPoliceExceededBitRate (1.3.6.1.4.1.9.9.166.1.17.1.1.14):** The bit rate of the nonconforming traffic.

- **cbQosQueueingDiscardByteOverflow (1.3.6.1.4.1.9.9.166.1.18.1.1.3):** The upper 32-bit count of octets, associated with this class, that were dropped by queuing.

- **cbQosQueueingDiscardPkt (1.3.6.1.4.1.9.9.166.1.18.1.1.7):** The number of packets, associated with this class, that were dropped by queuing.

- **cbQosTSStatsDropPktOverflow (1.3.6.1.4.1.9.9.166.1.19.1.1.10):** This object represents the upper 32-bit counter of packets that have been dropped during shaping.

- **cbQosTSStatsDropPkt (1.3.6.1.4.1.9.9.166.1.19.1.1.11):** This object represents the lower 32-bit counter of packets that have been dropped during shaping.

If the problem is occurring because of network congestion, this can be diagnosed by monitoring QoS at different network elements at different layers. To explain this concept, we take an example of a PacketCable network, as discussed in Chapter 3.

PacketCable Use Case

In a PacketCable environment, QoS is provided using dynamic quality of service (DQoS) architecture (PKT-SP-DQOS-I12-I05-050812), which ensures that VoIP traffic gets better treatment than other types of traffic and focuses on the access part of the network between the MTA and the Cable Modem Termination System (CMTS). Resources are assigned to the MTA at the time of call setup after performing admission control, and QoS is assigned based on the information received from the CMS (through gate messaging). If the call setup fails, it can be because of any of the following reasons:

■ Lack of resources on the MTA.

■ Layer 2 messaging is dropped between the MTA and the CMTS. This can be caused because of Layer 1 events such as noise on the cable plant or Layer 2 events such as DOCSIS queues filling up.

■ Lack of resources on the CMTS. This can be either at the DOCSIS layer (Layer 2), at the IP layer (Layer 3), or at upper-layer protocols like COPS (used for carrying DQoS messages between the CMTS and the CMS).

■ Call-signaling failure because of network congestion, causing delayed or dropped packets by intermediate devices between the MTA and the CMS.

Similarly, if the quality of the voice call is degraded after being set up, the problem might be related to the following issues:

■ Packet drops between the MTA and the CMTS because of physical layer (Layer 1) issues (degraded signal-to-noise ratio [SNR], uncorrectable errors, and so on).

■ Proper QoS not assigned to the voice call. The voice call might be set up over best-effort service flows instead of a dedicated service flow with guaranteed QoS for voice.

■ The voice service flows might be impacted because of resource overutilization (high CPU utilization, DOCSIS scheduler issues, and so on) on the CMTS. This can cause voice packets to be delayed or dropped on the service flows.

■ Packets getting dropped by an intermediate device between the two VoIP endpoints (Layer 3).

The layered approach for monitoring the previously mentioned issues is illustrated in Figure 6-11.

In the approach mentioned previously, we start at Layer 1 by monitoring the physical parameters of the cable plant like SNR, power levels, and correctable and uncorrectable errors caused by noise. These parameters can be monitored by using the DOCS-IF-MIB. If there are issues at the physical layer that can affect VoIP traffic (degraded SNR, power levels, errors, and so on), we correlate this data to network events or alarms to see whether they are causing any VoIP-related issues. The following OIDs in the "DOCS-IF-MIB" MIB table can be used to collect the Layer 1 statistics from the CMTS:

■ **docsIfSignalQualityTable (1.3.6.1.2.1.10.127.1.1.4):** At the CM, describes the PHY signal quality of downstream channels. At the CMTS, describes the PHY signal quality of upstream channels.

The following OID can be queried for more specific data on a per-cable modem basis. This data can be useful when trying to troubleshoot problems on specific cable modems:

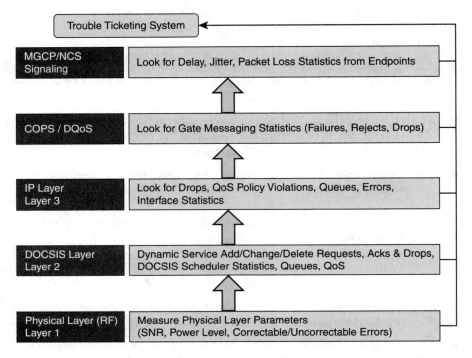

Figure 6-11 *Layered Approach for Monitoring QoS*

■ **docsIfCmtsCmStatusTable (1.3.6.1.2.1.10.127.1.3.3):** A set of objects in the CMTS like SNR, power level, correctable and uncorrectable errors, and so on, maintained for each cable modem connected to this CMTS.

Next, we look at the DOCSIS layer (Layer 2) to see whether the DOCSIS layer messaging between the MTA and CMTS is working as expected. We need to look at the DSX messages (Dynamic Service Add [DSA], Dynamic Service Change [DSC], and Dynamic Service Delete [DSD]) to ensure that requests being sent by the MTA are not being rejected or dropped by the CMTS. Failure in DSX messaging also needs to be correlated to any VoIP events in the network to make sure that service is not getting impacted. The DSX messaging on the CMTS can be monitored by using the DOCS-QOS-MIB. The following OID in the "DOCS-QOS-MIB" MIB table can be used to acquire this data:

- **docsQosDynamicServiceStatsTable (1.3.6.1.2.1.10.127.7.1.6):** This table describes statistics associated with the Dynamic Service Flows in a managed device.

  Next, we look at the DOCSIS QoS parameters on the CMTS to make sure that the VoIP traffic is getting the appropriate QoS when it is transported over the cable network. This information can be monitored using the DOCS-QOS-MIB. The following OID can be used to acquire this data:

- **docsQosParamSetTable (1.3.6.1.2.1.10.127.7.1.2):** This table describes the set of DOCSIS 1.1 QoS parameters defined in a managed device.

The next thing we look at is the IP layer (Layer 3) information to make sure that packet drops under the cable (RF) interfaces or the WAN links are not affecting VoIP traffic. We would also look at the queues under these interfaces to make sure that packets are not backing up in the queues, which can cause delay and jitter for VoIP traffic. The interface statistics can be monitored using the IF-MIB. The following tables can be used to collect the relevant data: IF-MIB, ifTable, and the (1.3.6.1.2.1.2.2) table, which contains a list of interface entries. The number of entries is given by the value of the ifNumber.

Another thing to check on the intermediate devices between the MTA and the other VoIP endpoint is the QoS policy defined to make sure that it is operating as designed. This information can be monitored using the CISCO-CLASS-BASED-QOS-MIB, as described previously.

Additional checkpoints can include any other devices such as firewalls, Layer 2 switches, and so on to make sure that they are not interfering with the quality of the VoIP traffic. On Layer 2 devices, make sure that they are configured to trust the incoming class of service (CoS) values for the VoIP signaling and bearer traffic. This allows the Layer 2 devices to queue VoIP traffic based on its CoS values and prioritize it over other types of traffic. On the Layer 3 devices connected to the Layer 2 devices, the CoS values can be mapped to the right DSCP values to provide appropriate QoS and apply relevant queuing techniques.

Typically, VoIP bearer traffic is classified with a CoS value of 5, while the VoIP signaling traffic is classified with a CoS value of 3. Table 6-10 shows the recommended Layer 2 and Layer 3 classification values for various types of traffic.

Table 6-10 *Traffic Classification Recommendations*

Traffic Type	Layer 2 CoS	Layer 3 IP Precedence	Layer 3 DSCP (PHB)
VoIP bearer	5	5	46 (EF)
VoIP signaling	3	3	26 (AF31)
Data	0–2	0–2	10–22 (0–AF23)
Video	4	4	34 (AF41)

Finally, we also need to look at the information from signaling protocol (MGCP/NCS) counters from the endpoints to track delay, jitter, and packet loss. This can also help explain issues contributing to degradation of the voice quality. These performance counters can also be used for trend analysis and capacity planning, as explained Chapter 8.

So far, we have explained the high-level methodology to monitor QoS and correlate this information to network events to help isolate problems more effectively. The details of this approach are explained in Chapter 7.

As mentioned earlier in this chapter, it is important to group different network elements in the VoIP network (group endpoints, aggregation and core devices, provisioning and management servers, CMS and voicemail servers, and so on) to isolate problems caused because of certain device types. This can also help in establishing a baseline for each device type and for collecting periodic information from these devices that can be used for analysis and trending. This is discussed in more detail in Chapter 8.

After issues are categorized by different device types or groupings, they can be correlated to the trouble ticketing system so that problems can be tied back to specific vendors and so on. This is discussed in more detail in the next section.

Trouble Ticketing (TT) Systems

A trouble ticket system is one of the initial problem-tracking systems deployed in a VoIP network. It is crucial that the system be utilized to categorize issues so that they help in tying them back to specific vendors and partner SPs. A logging and alerting mechanism should be in place so that systemic issues can be identified and rectified. Also, the TT systems should help drive better customer satisfaction and uptime.

Identifying and Streamlining the Categories of Trouble Tickets

The TT system should be developed so that it captures the relationship between the user-reported problems or proactive trouble tickets and the service uptime.

The geographic reporting should be captured in the trouble ticket along with detailed information about the type of problem being reported. If it's a voicemail-related problem being reported for market A or a particular campus, the location and the type of problem should be captured as a tag. This tagging would facilitate postanalysis for correlation to reported alarms and the vendor product.

Correlating the TT to the Service Uptime

The customer experience can be deduced through effective tracking of trouble tickets. The following key practices centered on a TT system help reduce downtime and improve customer satisfaction:

■   A continuous network availability tracking in the trouble ticketing system, by logging user downtime.

- A proactive alarm creation and association to track volume trouble tickets.

- Tracking reported faults to trouble tickets on periodic basis, thus allowing diagnosis of resolution time for reported customer issues. As an example, trending this relationship would map the increasing faults to customer downtime. Similarly, if there is fast resolution to issues, faults could be high but the downtime is low, a good indication of fast resolution of issues.

- The TT systems are polled to generate reports that tie them back to vendor products and partner SPs. This can only be achieved through a detailed tagged trouble ticketing systems, which facilitates a query mechanism based on these tags. Thus, the tag-based queries can generate focus-specific TT reports. These reports can be correlated to the faults reporting during the same periods.

- Based on effective correlation of TT to network uptime, a process should be introduced for improving service availability. Basically, this should be a dynamic process derived through automated querying of TT systems and fault-reporting systems.

Summary

Network management starts even before VoIP deployment. Prior to VoIP deployment, the transport network is assessed for its resiliency, high availability, performance, and capacity. This process involves analysis of the IP network done by the tools, traffic engineering for capacity planning, verification using voice traffic simulation, and network transmission loss planning to ensure that voice quality is preserved throughout the IP network and its interfaces with public land and mobile network (PLMN) through TDM voice gateways.

Voice quality metrics, including MOS/K-factor, PSQM, and PESQ, are monitored for proactive management on the call-processing entities involved in coding and decoding of voice signals. All the contributing factors, including latency, jitter, packet drop, and signal levels, should be analyzed on the network devices through which media traffic traverses. This is only possible if the NMSs are accurately and completely able to discover all the devices in the network using seed devices, CDP, routing tables, Address Resolution Protocol (ARP) cache analysis, or ping sweeps.

Network management systems can employ various methods, including SNMP polling, subscribing to alarms/traps, and syslog analysis, to track KPIs. All of this information is correlated and tracked on customized Dashboards to provide meaningful metrics in the proper context.

In essence, VoIP guidelines include transmitting voice the fastest way possible by keeping the delay less than 150 ms. Excessive delay worsens the echo and causes awkward conversations. VoIP packets must be transmitted as a steady, smooth stream to minimize jitter without dropping packets. This requires end–to-end QoS implementation covering all the network layers.

The proactive management approach might still have some fallout covered by trouble ticketing systems. These TT systems should be tied with NMS so that the problems can be resolved quickly by virtue of correlation with underlying cause(s).

References

1. *Voice over IP (CVoice)*, Second Edition. Copyright 2006, Cisco Systems, Inc., Reproduced by permission of Pearson Education, Inc., 800 East 96th Street, Indianapolis, IN 46240.

2. "Introduction to Cisco IOS NetFlow," October 2007, Cisco, http://www.cisco.com/en/US/prod/collateral/iosswrel/ps6537/ps6555/ps6601/prod_white_paper0900aecd80406232.html.

3. "Understanding Call Detail Records," January 2009, Cisco, http://www.cisco.com/en/US/docs/voice_ip_comm/cucmbe/service/6_0_1/car/carcdrdef.html#wp1061813.

4. Dai, Renshou, "A White paper on SMOS," November 15, 2001, Sage Instruments, http://sageinst.com/downloads/935at/smoswp.pdf.

5. "TIA 912—IP Telephony Equipment: Voice Gateway Transmission Requirements," April 12, 2007. Telecommunications Industry Association, http://standardsdocuments.tiaonline.org/tia-912-b.htm.

6. "ITU-T G.107—The E-model, a computational model for use in transmission," Feburary 12, 2003, International Telecommunications Union, http://www.itu.int/itudoc/itu-t/aap/sg12aap/history/g107/g107.html.

7. "ITU-T G.131—Control of Talker Echo," October 15, 2003, International Telecommunications Union, http://www.itu.int/itudoc/itu-t/aap/sg12aap/history/g131/g131.html.

8. "PKT-SP-DQOS-I12-050812—PacketCable Dynamic Quality-of-Service Specification," August 12, 2005, CableLabs, http://www.cablelabs.com/packetcable/downloads/specs/PKT-SP-DQOS-I12-I05-050812.pdf.

9. T. Li, B. Cole, P. Morton, D. Li, "RFC 2281—Cisco Hot Standby Router Protocol (HSRP)," March 1998, Network Working Group IETF, http://www.ietf.org/rfc/rfc2281.txt.

10. S. Alexander, R. Droms, "RFC 2132—DHCP Options and BOOTP Vendor Extensions," March 1997, Network Working Group IETF, http://www.ietf.org/rfc/rfc2132.txt.

11. H. Schulzrinne, S. Casner, R. Frederick, V. Jacobson, "RFC 3550—RTP: A Transport Protocol for Real-Time Applications," July 2003, Network Working Group IETF, http://www.ietf.org/rfc/rfc3550.txt.

12. T. Friedman, Ed., R. Caceres, Ed., R. Caceres, Ed., "RFC 3611—RTP Control Protocol Extended Reports (RTCP XR)," November 2003, Network Working Group IETF, http://www.apps.ietf.org/rfc/rfc3611.html.

13. K. Nichols, S. Blake, F. Baker, D. Black, "RFC 2474—Definition of the Differentiated Services Field (DS Field) in the IPv4 and IPv6 Headers," December 1998, Network Working Group IETF, http://www.ietf.org/rfc/rfc2474.txt.

14. S. Blake, D. Black, M. Carlson, E. Davies, Z. Wang, W. Weiss, "RFC 2475—An Architecture for Differentiated Services," December 1998, Network Working Group IETF, http://www.ietf.org/rfc/rfc2475.txt.

Chapter 7

Performance Analysis and Fault Isolation

This chapter discusses an approach for proactive monitoring of the Voice over IP (VoIP) network for performance analysis and fault isolation of problems caused by anomalies in the network. It starts with explaining the VoIP network-monitoring aspects, including collection, categorization, and correlation of performance counters for both enterprise and service provider (SP) networks. It also discusses different ways of gauging the performance of a large-scale VoIP network by looking at various counter-based key performance indicators (KPI).

Performance monitoring includes monitoring certain aspects of the transit network by tracking its effect on the signaling protocols and the packets containing the media (Real-time Transfer Protocol [RTP]). Collecting the performance counter data for the call-processing devices, such as Cisco Unified Communications Manager (CUCM) or Microsoft OCS in enterprise and BTS 10200 or PGW in SP networks, and analyzing data from VoIP endpoints including both customer premises equipment (CPE) devices and central trunking gateways provide the mechanism to the network administrator for gauging the performance of the network.

A comprehensive monitoring strategy involves looking at syslog and trace log information from the various nodes in the network. It is important to manage this information by evaluating security, storage strategy, the impact on the processing power of the devices, and link utilization when extracting this information for correlation of the data.

This chapter also discusses tools and scripts that aid in performance management using some of the best practices using case studies from enterprise and SP perspectives.

For optimal network operations, this chapter also covers software maintenance and auditing of VoIP network practices.

Proactive Monitoring Through Performance Counters

This section talks about how information from performance counters can be used to proactively monitor the network and identify and isolate problems before they impact the VoIP service.

First, the VoIP provider needs to go through an important exercise of identifying the devices that need to be monitored and then classifying the information that needs to be collected from these devices. These KPIs fall into various classifications and are defined in the following section.

Second, the VoIP provider needs to determine the frequency of monitoring these devices and the related performance counters. It would make sense to periodically monitor the network performance to create a baseline, instead of only taking these measurements when a problem occurs.

Measurements should be taken throughout the day to include data points during peak-utilization hours and nonbusy (low-utilization) hours. This can help identify whether the problem impacting VoIP service is dependent on network resource utilization or whether it occurs because of other factors. An effective polling strategy is discussed in Chapter 6, "Managing VoIP Networks."

The third and one of the most important tasks the VoIP provider must do is correlate the different data points to identify problematic areas of the VoIP network. This correlation helps to quickly identify the source of the problem and isolate issues when the VoIP service is impacted. An approach for data correlation is mentioned later in this chapter.

Classification of Performance Counters

Managing and optimizing a large VoIP network requires a methodology to be implemented for classifying the key performance metrics. The classifications and then collection and tracking of these KPIs enable a measurable system to be put in place. This system helps guide the network operator to be ahead of the curve on capacity bottlenecks, systemic issues, and vendor product anomalies.

Figure 7-1 illustrates that the available KPI set can be segmented into various groups. Furthermore, a KPI-layering approach is applied within these segments to help resolve an issue. The classification is elaborated in the following sections.

Network Device KPIs

The network device layer metrics are the main building block KPIs. Network device KPIs are comprised of the following parameters:

- Delay, jitter, packet loss, and so on
- Link and CPU utilization on relevant network components such as routers and switches that act as transit devices for the VoIP signaling and data traffic,

Softswitch, or Session Border Controller (SBC) that handles call control messages and trunking gateways

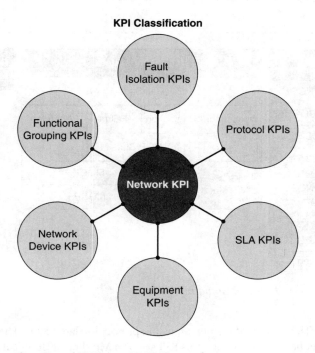

Figure 7-1 *Classification of KPIs in a VoIP Network*

- Errors and packet drops at the link interfaces
- Physical layer characteristics of the link (signal-to-noise ratio [SNR], transmit and receive power level, and so on)

Functional- or Services-Based Grouping of KPIs

The goal of the functional grouping or the service-related grouping of KPIs is to enable the network operator to develop a proactive service monitoring strategy. As an example, the VoIP network public switched telephone network (PSTN) call functionality provides the capability for Off-net call service and needs to be tracked with KPIs. The service monitoring strategy enables the network operator to effectively and expeditiously restore the service. For a VoIP network, some of the major services can be comprised of On-net and Off-net call legs. The VoIP operator has to then keep on top of the call functionality for both the IP leg and the PSTN leg. We cover some use cases to describe how the signaling KPIs can be functionally grouped to track the PSTN and the IP legs of the call. The VoIP use cases are based on a cable-based deployment and are discussed in the following sections.

On-net Call Use Case

This use case covers the call that originates on the IP network and also terminates on the IP network. Figure 7-2 is illustrates the On-net use case.

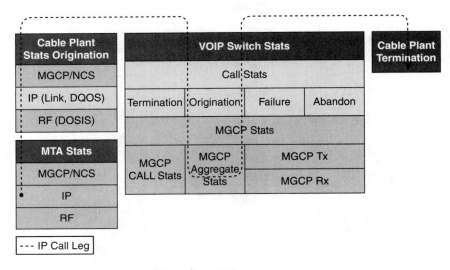

Figure 7-2   *A Layered Map of Signaling KPIs*

Figure 7-2 reflects a layered map for signaling protocol–related KPIs. The On-net call capability can be tracked through this KPI set. The Multimedia Terminal Adapter (MTA) originates the call, the radio frequency (RF)–to–IP gateway Cable Modem Termination System (CMTS) sitting in the cable plant that handles the L2 Data Over Cable Service Interface Specification (DOCSIS)–to–IP conversion, and the packets are forwarded off to the VoIP switch over IP. The switch handles the Media Gateway Control Protocol (MGCP)/Network-based Call Signaling (NCS) protocol commands sent by the MTA to establish the call leg with the termination side, thus setting up a call. The operator can trace the entire call by tracking MGCP performance counters from the MTA (Layer 3), the RF (SNR and link errors), the IP (link counters), and the signaling (dynamic quality of service (DQoS)—Gate set, attempt, commit) performance counters at the CMTS. Lastly, detailed call-related counters for call origination, termination, failures, and abandon calls—along with the MGCP-related counters (for both ingress and egress sides)—determine whether the call service is successful. The call capability is determined by tracking both failure and success counters for each of the protocols; an increase in failure counters can be symptoms of call failures.

The approach involves tracking the KPIs on the protocol layer stack across these VoIP components, as described in Figure 7-2. Network operators must set up a visual tracking and threshold-based alerting system for these KPIs. More guidance is provided in Chapter 8, "Trend Analysis and Optimization," which discusses the concept of a dashboard. The system is instrumental in improving service uptime, providing data to keep ahead of the capacity issues, and reducing any customer experience impacts because of network serviceability.

> **Note**   The case study illustrated in Figure 7-2 is a typical VoIP On-net call flow and can be extrapolated for noncable technology–based types of VoIP deployments as discussed in Chapter 3, "VoIP Deployment Models in Service Provider Networks," Chapter 4, "Internet Telephony," and Chapter 5, "VoIP Deployment Models in Enterprise Networks."

Off-net Call Use Case

In an Off-net call scenario, the call is made to a PSTN number; thus the call has an On-net and an Off-net leg. Two scenarios are covered in this section: one where the SP manages the PSTN leg and the other where the SP maintains just the IP network piece by utilizing Session Initiation Protocol (SIP) connectivity through the SBC.

Off-net Call with PSTN Managed by the SP

Figure 7-3 represents the managed PSTN scenario.

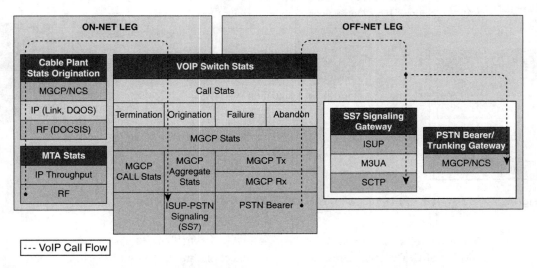

Figure 7-3 *Layered KPI Depiction for Off-Net PSTN Call*

Figure 7-3 shows the VoIP switch or a Call Management Switch (CMS) counter set for the PSTN leg. The VoIP switch initiates a Signaling System 7 (SS7) signaling session with the IP-to-SS7 gateway called the IP Transfer Point (ITP). The ISDN User Part (ISUP) protocol is used by the VoIP switch to communicate over IP to the ITP. It basically encapsulates the SS7 messages. The ISUP counters include SS7 Initial Address Message (IAM), Answer

Message (ANM), and Release (REL) message counters. The far-end PSTN switch to which the telephone number belongs responds, and during this communication, a bearer trunk is also allocated to the call. The key metrics in the PSTN leg are tracking the ISUP counters and the bearer traffic counters (ingress and egress) through the trunking gateway.

Off-net Call Leveraging SIP Proxy (SBC)

In the second scenario, the VoIP SP routes the PSTN call to an SBC through an SIP trunk. This SBC resides in the VoIP SP's network at the edge of its IP network. This is where it has SIP connectivity to a PSTN SP's peer SBC through a SIP trunk. The use case is shown in Figure 7-4.

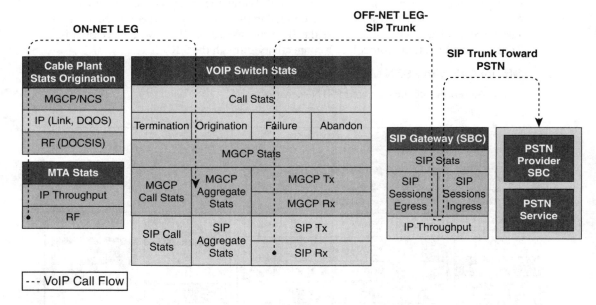

Figure 7-4   *Layered KPI Depiction for an Off-net SIP Call*

In this figure, we see the VoIP call being routed to an SBC. There is an SIP trunk from the CMS to SBC. The SBC ingress and egress connections are the key statistics to be monitored. Tracking these KPIs provides accountability for the PSTN service to the VoIP SP, because they can track the health of the inbound and outbound VoIP traffic flow, thus allowing them to enforce service-level agreements (SLA) on the PSTN SP.

Similarly, services such as voicemail, announcements, and general SIP functionality, such as presence, can be tracked by extracting a key set of KPIs.

Thus, the functional grouping of KPIs as captured in the use cases of On-net and Off-net calls can be taken as methodology guidelines for analyzing each unique VoIP network. Through an elaborate set of dashboards that track these service KPIs, the operator can track the services flowing through the network. The elaborate set of dashboards allow capturing the views from an aggregate level to a drilled-down level for specific components.

Fault Isolation–Based Grouping of KPIs

Fault isolation is the Achilles' heel of all network operators. To quickly get to the root cause of a problem is a big challenge. It takes many hours to achieve a proper diagnosis.

Following the concept described in Figure 7-1, we extract the key KPIs to track issues. The functional grouping dashboard strategy mentioned earlier provides the means of isolating the faults by tracking failure and success counters. Figure 7-5 illustrates this scenario.

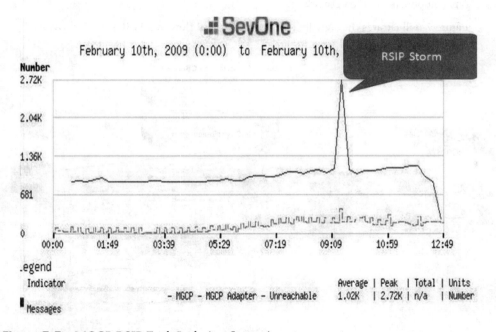

Figure 7-5 *MGCP RSIP Fault Isolation Scenario*

Here the MGCP Restart in Progress (RSIP) event notifications are tracked through a dashboard. MGCP RSIP events are generated by a device that communicates through MGCP when it reboots. The issue illustrated in Figure 7-5 is a CMS where the calls were failing. This MGCP KPI dashboard confirmed a potential issue in the MGCP statistics that was caused by a single faulty MTA that was stuck in a constant RSIP state, creating an RSIP storm. The RSIP storm caused by the single MTA was overwhelming the CMS, thus causing it to not handle calls.

The section "Trace Log Monitoring on Softswitch and Network Devices," later in this chapter, covers how the syslogs helped in locking down the final piece of the root cause.

The other key indicators for isolating the problems are the reported trouble tickets and the alarms. Most of the time, these notifications are trigger points for Network Operations Center (NOC) operators to become aware of a system issue. The time taken to isolate the fault is a key for the network operator. The KPIs tracked to isolate the VoIP network faults need to derive a reduction in this time. The key idea is to go through an exercise to identify the KPIs, correlate and create visual representation and threshold-based alerting mechanisms for the KPIs, and then through trained staff, track them.

Protocol-Based Grouping of KPIs

The protocol-based grouping of KPIs and in turn the visualization and threshold-based alerting of this grouping enables the network operator to track the performance of various components in the network.

Figure 7-6 illustrates the signaling protocol for a call flow in a VoIP service provider solution.

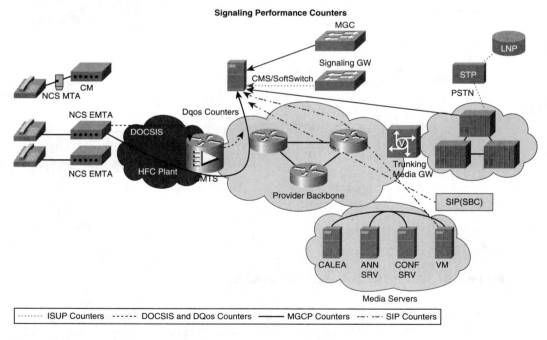

Figure 7-6 *Overall Signal Flow for a Centralized VoIP Switch Deployment*

The topology captures the various VoIP network segments and what signaling protocols they use for communication. The CMS has visibility into all the signaling communication and typically has extensive collection and reporting of these signaling counters. The signaling counters can be tracked to generate a comprehensive view. Any performance bottlenecks and anomalies can be easily isolated and worked on by classifying, collecting, correlating, and visually representing these KPIs. Use cases were discussed earlier in Figures 7-2, 7-3, 7-4, and 7-5, where we tracked the flow of signaling protocols through MGCP counters from the MTA (an endpoint) to the CMS to the MGW (a Trunking Media Gateway). MGCP is also used to perform signaling communication with bearer traffic trunking gateways. Collecting the MGCP KPIs from the MTAs, CMSs, and MGW controllers at fixed intervals gives good health and performance indicators for the call functionality.

To further clarify this point, when the MGCP error is reported, it is handled by the CMS for taking appropriate action on the call flow. Because one of the collection points is the CMS for the MGCP error counters, the CMS MGCP counter set provides a baseline of the MGCP signal network. If the errors are above normal or above the expected baseline range, it should trigger an alert for the network operator. The trigger might be a visual one, as observed by NOC personnel, or it might be through an alerting system. The NOC personnel would then further isolate the MGCP flow that is causing the errors by deeply looking at the collector path. A signaling counter collection path basically represents the dashboard of all the MGCP-based KPIs that can affect the MGCP flow. As shown in Figures 7-2, 7-3, and 7-4, the MGCP signaling path has several possible legs.

Note   Network operators also use probe-based external collectors for the protocol statistics. These collectors connect to the switches and sniff IP and other protocol traffic. The probes listen on the IP wire through spanned ports for different protocol traffic and report it back to the collector. The collector is protocol-intelligent because it has protocol stacks built in; thus a comprehensive counter set is reported. This approach enables the operator to create a nonvendor-biased protocol statistics view, where they are not dependent on the vendor's product for error and success reporting.

SLA Tracking Through KPIs

The SLA is critical for the VoIP operator's survival as it helps keep the operator's vendor quality in check. The resiliency of the network can be improved by having a methodical SLA strategy. The strategy should encompass

■   Third-party service providers

■   Equipment vendors

■   Internal SLA metrics for various lines of services

The VoIP networks are dependent on SPs by other providers for their PSTN support and their long-haul support. Our focus is on PSTN support. VoIP providers offload the in-house PSTN support to typical legacy telcos. They have an SIP trunk tied to their SBC, which is connected to the PSTN provider's SBC. The key metrics that the VoIP SP has to track are the ingress and egress SIP sessions going across the SBC, and the PSTN provider keeps up with the egress side without any session drops. The session throughput is the key SLA metric.

VoIP equipment vendors also need to be held accountable through some relaxed SLAs. It's called relaxed because VoIP technology is so dynamic that the equipment often becomes bleeding edge and the product cannot meet very stringent SLA requirements. However, service uptime is the key, so the strategy here is to track the protocols (SIP, MGCP, and H.323) that the VoIP equipment communicates through, along with the corresponding hardware's defect resiliency. Through real-time dashboards and periodic static comprehensive reports, the vendor must be held accountable.

A comprehensive SLA strategy should be developed for tracking the services the VoIP operator provides. This encompasses the base call uptime, quality, and availability of the call features set. Following the use-case scenarios of earlier sections and tying them to Figure 7-6, the VoIP operator can develop a set of SLA KPIs. These SLA KPIs can be used to measure performance and uptime data for the services a VoIP provider gets from their SPs for access to PSTN and the Internet. Tracking the SLA KPIs can generate accountability reports for service impacts and performance for these external SPs. Analyzing the KPI-based reports and taking corrective action on a periodic basis would allow the VoIP SP to improve the quality of their leased services.

The layered approach described so far in this chapter is summarized in Figure 7-7.

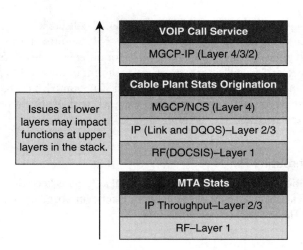

Figure 7-7   *KPI Layred Stack Concept*

Figure 7-7 shows that a lower networking layer issue in the VoIP network has broad impacts as the protocols within the stack act as building blocks for the VoIP network communication. If there is an impact at the DOCSIS layer (Layer 2), the problem has a cascading effect, as the overall call flow stops. If the impact is at a higher layer, only a select group of components that communicate using those protocols and the corresponding functionality are impacted. It is critical that the KPIs be tracked on a real-time basis for all the layers, as depicted in Figure 7-7. This facilitates the narrowing of the time to resolve the critical faults.

Equipment-Based Grouping of KPIs

Another key grouping of performance counters is equipment. Figure 7-6 illustrates protocols flowing through the various VoIP network components. The strategy that needs to be used here is to create a tag system for each KPI collection point. This tag would basically reflect the type of equipment and possibly the vendor's initials. Cataloging the type of equipment would allow the differentiation in a multivendor VoIP equipment network environment. It might be that not all the vendors that have equipment in the protocol flow are the cause of the issues and are reporting the errors, but rather only one of the vendors is the source of the problem. Tallying these source root problems by the vendor components and types enables the operator to develop good metrics for vendor accountability and in turn create a resilient network.

Collection

Collection involves receiving unsolicited messages in real time, such as Simple Network Management Protocol (SNMP) traps, syslog messages, or events containing information about a problem with a device in the infrastructure, such as a CMTS, router, or a switch and processing servers including Unified CM, BTS, or an SIP proxy. Collection typically happens through a network of distributed collectors in NMS in large networks. Network and device performance statistics are generally polled by the collectors as well.

Element management systems (EMS) for the call-processing devices, including Softswitches and application servers, proactively retrieve performance data from these devices to review their overall health, including CPU loads broken down by processes; memory utilization and its details, such as paging information; disk utilization, including free space and input/output (IO) activity; and software service statuses running on these managed elements. An EMS is also responsible for processing Call Detail Records (CDR) or the platform logs. The EMS-managed server-based applications typically have a large storage capacity. Hence, the performance data or CDRs can be extracted by the EMS less frequently compared to other performance data using Secure File Transfer Protocol (SFTP), Trivial File Transfer Protocol (TFTP), or simple File Transfer Protocol (FTP). An EMS might use the Simple Object Access Protocol (SOAP) or NetFlow MIBs for link utilization for real-time monitoring of performance data. A network management system might have multiple EMS instances for collecting performance data depending upon the type of devices in the network, such as Cisco Unified Operation Manager (CUOM), EPOM for BTS, and CiscoWorks for routers and switches.

Similarly, infrastructure devices, including routers and switches, are monitored for their CPU utilization levels, memory utilization, and link utilization. SNMP MIBs or more detailed Remote Network Monitoring (RMON) MIBs for performance data can be used by allowing Cisco IOS Software to send traps and alarms based on threshold crossings directly to a alarm/event parser on a collector.

For example, the following router configuration commands set up a scheme called "cpmCPUTotal5min" on the router to monitor its CPU every 300 seconds. This router fires event 1 if the CPU exceeds 75 percent and fires event 2 when the CPU falls back to 50 percent. In both cases, an SNMP trap message is sent to the network management station with the community private string.

```
rmon event 1 trap private description "cpu hit 75%" owner tsiddiqu
rmon event 2 trap private description "cpu recovered" owner tsiddiqu
rmon alarm 10 cpmCPUTotalTable.1.5.1 300 absolute rising 75 1 falling 50 2 owner
  tsiddiqu
```

Alarm Processing

Alarm processing is focused on summarizing network behavior, root cause analysis, and impact analysis. Some of the key aspects of alarm processing involve

- Event filtering
- Event correlation
- Propagation paths discovery

Figure 7-8 presents the multilevel alarm filtering hierarchy.

Syslog messages are first parsed by an event filter by analyzing the structure of these messages. An event filter can identify the severity from the descriptive text of the occurred event recorded in the syslog. Event notification can also come in the form of SNMP trap event filtering and can also involve managing event storms by suppressing irrelevant or redundant information. For example, a "link flapping," where a network link is unstable and changes its operational state frequently, will generate a large number of SNMP traps. This link instability will also be recorded in syslog messages. The event parser should be able to identify the link, analyze SNMP traps showing repeated "link up" and "link down" events, reduce them to a single event based on the close proximity of the timestamps on these traps, and prioritize the event. The event parser will perform the same function when analyzing the syslog messages for the same incident.

Alarm processing can occur at an EMS, which converts threshold crossings of performance counters into alarms. For example, when the CPU load crosses a 75 percent threshold as defined by the network administrator or preset in the EMS by the software/equipment vendor, an alarm will be generated by the EMS and sent to the collector, where this alarm can be parsed by an alert parser. The alert parser then updates the central management console about this event with appropriate priority.

If a central CUCM loses connectivity to a remote site with a Survivable Remote Site Telephony (SRST) gateway, there will be multiple SNMP traps generated by the CUCM that will include IP phones and digital signal processors (DSP) on the local SRST gateway in that remote location losing their registration status. All of these traps and other notifications, including event log messages, will be first collected by CUOM that plays a role of an EMS, as shown in Figure 7-8. The CUOM console can present that event more concisely as a network outage and present the event as such. If the CUOM is able to reach the SRST, it can potentially show that the IP phones have registered with the SRST gateway as part of the remote site survivability design. The CUOM can also relay this event to the Cisco Info Center (CIC by Netcool) either directly or through one of its distributed collectors, as shown in Figure 7-8.

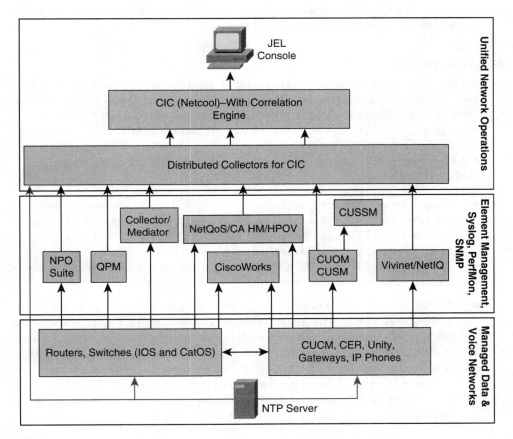

Figure 7-8   *Framework for Monitoring a Converged Network*

Correlation

In a voice-oriented network, faults can occur in any part of the network because of undesired behavior of one or many network elements or their interaction. The fault management system must be able to perform alarm correlation to determine the root of the problem.

Alarm correlation provides mechanisms to identify relationships among various alarms, favoring root cause determination. Additionally, network discovery and network topology updates are two instrumental key features for fault management. Fault management must have a view on both physical and logical topology of the network to perform advanced correlations. There are three levels of complexity related to fault management: simple, advanced, and complex. For example, a link outage can generate multiple alarms, including the link outage itself from multiple devices (if that link connects multiple devices), routing protocol–related alarms, and a loss of VoIP endpoint registration with the central call control server. A correlation engine in the fault management system should be able to establish the relationship between all these alarms so that the corrective action is directed to the right device on the network.

Simple Correlation

Simple correlation does not require knowledge of network topology. These correlations concern one device at a time. As an example, a simple correlation would entail correlating "link up" and "link down" alarms or traps related to a single network link "flip-flop" behavior, where the same link repeatedly changes its status from up to down to up again. Simple correlation will not extend correlating this event to other affected areas such as routing protocols and route resolution services associated with link disruption. It can perform the following:

- **Compression of similar alarms:** Multiple alerts should not be generated for duplicated events received within a configurable amount of time. For example, if a remote site with 400 VoIP IP endpoints loses its connectivity with the central call processing server, all 400 endpoints will loose their registration. This deduplication process should convert all of these 400 alarms into one alarm stating that 400 phones lost registration with the call agent.

- **Recognizing state changes:** Often the alarms of the same nature are produced by the same network node, and they have similar message contents, such as link status (up or down) except timestamps. Compression is the task of reducing multiple occurrences of identical alarms to a single representative of the alarm. As mentioned in an earlier example, if a "link down" and "link up" event from the same device for the same interface is received in a short amount of time, it will be compressed to a single event stating the link disruption for the total time period where all the actual events occurred.

Advanced Correlation

Advanced correlation requires network topology awareness. Advanced correlation is applied for aggregated links (trunks) or interlayer connections. These types of correlations relate to multiple devices, various zones, and domains. Commonly, they refer to a given plane (signaling, routing, and so on). Constant awareness of both physical topology and logical topology is required to perform this correlation.

Advanced correlation can be observed in a VoIP network managed using a framework, as described in Figure 7-8. Consider an enterprise VoIP network with a central CUCM with several remote sites. Each site will have a gateway and multiple switches to which IP phones will be attached. The WAN links will be terminated on an aggregation router in the data center. The IP phones, local gateway and switches, and WAN link can be grouped logically in a domain and tied with a "location" or "device pool" configuration on CUCM. A location or device pool pointing to a set of IP phones and DSPs can be logically tied to an SRST gateway in the remote site on a CUOM. If all the phones and DSPs belonging to the same device pool lose their registration statuses, the CUOM will receive alarms that it can relate to a common device pool. This will indicate an issue with the device pool or the remote site rather than the IP phones, thus localizing the issue.

The CUOM can further relay this as a single event associated with a particular device pool and SRST gateway to one of the CIC's distributed collectors. If this is related to a link outage, the aggregation router will also send a trap containing the link information either through a local collector or CiscoWorks to CIC's event correlation engine. CIC's correlation engine contains a "rules" file that has the correlation rules. These rules can be configured to match the SRST gateway's IP address to the affected link's subnet. This will enable the CIC rule engine to match IP phones and DSPs losing connectivity to a specific link. The two parsed events can be grouped for presentation on the "JEL console" monitored by the NOC staff. This will save time for the CUCM administrator, who would otherwise have to look in the CUCM debug logs to find the cause of the registration status change and coordinate with the administrator responsible for the IP infrastructure to find the root cause by manually looking for the links status that serve the same area code belonging to the affected phones during the same time frame.

The section "Case Study of Link Congestions," later in this chapter, also reiterates the importance of complex correlation by use of an example.

Complex Correlations

Complex correlations can pinpoint specific voice fault routing or call quality issues caused by

■   Network configuration problems

■   Misrouting/setup of voice calls

■   Network bandwidth, or performance problems associated with a particular segment

The CUCM generates Call Detail Records (CDR) and call diagnostic records that contain call quality information. This information can be archived by Cisco Unified Service Manager (CUSM) and analyzed by Cisco Unified Service Statistics Manager (CUSSM), as shown in Figure 7-8. To analyze calls experiencing voice quality issues, we will begin by analyzing call diagnostic records and CDRs in the CUSSM. The CUSMM will provide information about these issues:

■   Date: 19-Jan-09

■   Origination time: 9:35:42 a.m.

- Termination time: 9:35:42 a.m.

- Duration (ms): 83,000

- Origination number: 8082936200

- Destination number: 4793

- Origination codec: G711Ulaw

- Destination codec: G711Ulaw

- Origination device: 10.200.19.112

- Destination device: 10.200.14.4

Troubleshooting this problem will require complex correlation. The first step in troubleshooting will involve establishing a pattern of calls with bad call quality or K-factor values with the help of the CUSSM. If they tend to belong to the same originating or terminating gateway, NetQoS logs can be consulted. NetQoS, as shown in Figure 7-8, is responsible for collecting QoS and link utilization data from routers. If NetQoS data eliminates the QoS configuration and suggests link congestion at a time that matches the time frame of call quality issues as observed on the CUSSM, it might require further analysis of events related to link congestion on the CIC. Without this complex correlation, the origination or termination device common to the calls with a lower K-factor (bad call quality) will be considered the culprit. But there will be no explicit corrective action that can be taken against these devices or gateways in the absence of complete knowledge of the root cause.

Note that synchronizing all the managed devices with single NTP source is crucial for advanced and complex correlations. Using a common NTP source for clock synchronization on all the devices in the network will ensure accurate correlation of events and alarms among several devices involved in the VoIP call setup.

Recommendations for VoIP-Centric Network Management Framework

A network management framework might already exist in an enterprise or SP environment. When VoIP applications are introduced, the framework must accommodate the new elements introduced in the network related to VoIP while focusing on the fact that their management must be coordinated with the existing infrastructure that provides the transport to VoIP traffic. Here are some basic guidelines to achieve this objective:

- Establish a network utilization baseline, including devices (CPU and links), trunk usage, and call completion rates. A network baseline is established by collecting data for some initial amount of time determined to be sufficient to provide a representative sample of overall network behavior. This can also be done by performing periodic network audits.

- Synchronize all the managed devices in the network, the element management systems (EMS), collectors, and central rules engine using the same NTP source.

- Identify element management systems that can be interfaced to a common network management system that might have several workstations.

- Ensure that the EMSs provide complete coverage for all the devices and the applications with the features enabled.

- Ensure that any fallout from the scope of complete coverage through SNMP or performance counters is covered by syslog or other alternate methods such as SOAP/XML.

- Explore the EMS options so that the statistical data collection is instantaneous rather than average to get a more accurate profile of the network.

- To the extent possible, perform correlation at collectors level to alleviate the processing burden on the central correlation engine.

- Evaluate the possibility of correlation, for example, leveraging a rules file on the CIC without impacting the performance. Evaluate external correlation engines such as Cisco Info Center or Netcool.

Performance Analysis from a Transit Network Perspective

The VoIP signaling and bearer traffic typically traverses several network components and goes through several hops while it is transported from one VoIP endpoint to another. To maintain good voice quality and guarantee reliable service, the VoIP provider needs to ensure that the transit network that carries the VoIP traffic is optimized specifically for this service and is performing accordingly. In general, a multiservice network should take into consideration all the services that it supports in addition to VoIP.

It is important that VoIP signaling and bearer traffic receive the appropriate QoS end to end and is treated better than best-effort data traffic in the network. The VoIP operator needs to proactively monitor the transit network using the KPIs mentioned in the previous section to ensure that the network is performing optimally to guarantee reliable VoIP service.

Chapter 6 describes the voice-signaling protocol impairments in IP networks because of packet loss and excessive delay. To discuss how to monitor the behavior of voice protocols and derive conclusions regarding the health of the endpoints and the call control servers, we first establish the recommended configuration for QoS parameters in the devices in the network infrastructure. With the assumption that the QoS is configured correctly for proper treatment of voice packets, we can look at detecting and correlating performance-related issues.

Signaling Protocol Transport Optimization

Often, the VoIP network operators provide the feedback to the engineering team to provision the network to provide the appropriate QoS to VoIP bearer traffic but might not provide any special treatment to the VoIP signaling traffic. This can result in call setup failures and interruption to the VoIP service, as discussed earlier in Chapter 6.

Typically, the VoIP bearer traffic is treated better than best-effort data traffic by using one or more of the following approaches:

- Separating VoIP bearer traffic from best-effort data traffic

- Creating separate virtual local-area networks (VLAN) for data and voice traffic

- Transporting VoIP bearer traffic over dedicated links

- Reserving bandwidth for VoIP bearer traffic using the Resource Reservation Protocol (RSVP) and modular QoS (as discussed in Chapter 6)

- Marking the VoIP bearer traffic with a higher IP precedence or Differentiated Services Code Point (DSCP) value and classifying this traffic into special queues supported by Cisco IOS.

A similar strategy needs to be implemented for VoIP signaling traffic to minimize the impact to the VoIP service because of signaling failures. The following sections describe how the transit network can be optimized for transporting VoIP signaling traffic in enterprise and SP environments.

Enterprise Networks

IP packets containing a signaling protocol must be uniquely identified. Signaling traffic's classification happens by setting either the class of service (CoS) bits at the 802.1Q/p frame level or at an IP packet level by setting the IP type of service (ToS) byte. In this section, we focus on the DSCP value that is represented by the six most significant bits in the ToS byte. Here are the classification principles:

- Whenever there is a choice, QoS policies should always be performed in hardware rather than software.

- Classify and mark voice traffic as close to their sources as technically and administratively feasible.

- Use DSCP markings whenever possible.

The signaling traffic should be marked as a DSCP value of "CS3" per the recommendations in RFC 2474 to get the appropriate level of priority over other data traffic. However, some endpoints, gateways, and call-processing servers might still use the DSCP value AF31. Therefore, it is recommended to honor both markings when queuing packets in the QoS configurations.

After proper classification, the signaling traffic should get queued correctly. Real-time media, especially voice, is always queued in the priority queue. Queuing is mandatory at any node that has the potential for congestion to provide service guarantees. Here are the queuing design principles:

■ Limit the amount of strict-priority queuing to 33 percent of a link's capacity to minimize the impact on the CPU of the routing and the switching device. This is because the priority queuing is an interrupt-driven process that impacts the CPU in most of the routing and switching devices.

■ At least 25 percent of a link's bandwidth should be reserved for the default best-effort class.

■ Configure consistent queuing policies in the LAN and WAN, including VPN (over Internet), according to platform capabilities to ensure consistent per-hop behavior (PHB).

■ Enable the weighted random early detect (WRED) queuing method on all TCP flows, whenever supported. DSCP-based WRED is preferred because it is able to prioritize different traffic flows so that important data flows get better QoS services.

QoS Configuration Example on Access Switch

The following configuration is based on a best-practice recommendation for QoS configuration. The packets are assigned the voice VLAN (VVLAN) based on their DSCP markings at the edge on the access layer switch. All real-time media traffic (RTP) that is marked with DSCP 46 and the signaling traffic that is either marked with DSCP values AF31 or CS3 are matched and placed in the VLAN, which preserves these markings for transport through the transit network. Then, the policy is bound to the access interface that is connecting to the voice endpoints.

```
CAT3550(config)# class-map match-all VOICE
CAT3550(config-cmap)# match ip dscp 46                          ! DSCP EF (voice)
CAT3550(config-cmap)# class-map match-any CALL-SIGNALING   ! Need 'match-any' here
CAT3550(config-cmap)# match ip dscp 26                          ! DSCP AF31 (old)
CAT3550(config-cmap)# match ip dscp 24                          ! DSCP CS3 (new)
CAT3550(config-cmap)#
CAT3550(config-cmap)# class-map match-all VVLAN-VOICE
CAT3550(config-cmap)# match vlan 110                       ! VLAN 110 is VVLAN
CAT3550(config-cmap)# match class-map VOICE               ! Matches VVLAN DSCP EF
CAT3550(config-cmap)#
CAT3550(config-cmap)# class-map match-all VVLAN-CALL-SIGNALING
CAT3550(config-cmap)# match vlan 110                       ! VLAN 110 is VVLAN
CAT3550(config-cmap)# match class-map CALL-SIGNALING         !Matches VVLAN
  AF31/CS3
CAT3550(config-cmap)#
CAT3550(config-cmap)# class-map match-all ANY
CAT3550(config-cmap)# match access-group name ANY    ! Workaround ACL
CAT3550(config-cmap)#
```

```
CAT3550(config-cmap)# class-map match-all VVLAN-ANY
CAT3550(config-cmap)# match vlan 110                    ! VLAN 110 is VVLAN
CAT3550(config-cmap)# match class-map ANY          ! Matches other VVLAN traffic
CAT3550(config-cmap)#
CAT3550(config-cmap)# class-map match-all DVLAN-ANY
CAT3550(config-cmap)# match vlan 10                     ! VLAN 10 is DVLAN
CAT3550(config-cmap)# match class-map ANY          ! Matches other DVLAN traffic
CAT3550(config-cmap)# policy-map VOICE+SIGNALLING
CAT3550(config-pmap)# class VVLAN-VOICE
CAT3550(config-pmap-c)# set ip dscp 46                          ! DSCP EF (Voice)
CAT3550(config-pmap-c)# class VVLAN-CALL-SIGNALING
CAT3550(config-pmap-c)# set ip dscp 24                          ! DSCP CS3 (Call-
  Signaling)
CAT3550(config-pmap-c)# class VVLAN-ANY
CAT3550(config-pmap-c)# set ip dscp 0
CAT3550(config-pmap-c)# class DVLAN-ANY
CAT3550(config-pmap-c)# set ip dscp 0
CAT3550(config-pmap-c)# exit
CAT3550(config)#
CAT3550(config)# interface FastEthernet0/1
CAT3550(config-if)# switchport access vlan 10              ! DVLAN
CAT3550(config-if)# switchport voice vlan 110             ! VVLAN
CAT3550(config-if)# mls qos trust device cisco-phone   ! Conditional Trust
CAT3550(config-if)# service-policy input VOICE+SIGNALLING
CAT3550(config-if)# exit
CAT3550(config)#
CAT3550(config)# ip access-list standard ANY          ! Workaround ACL
CAT3550(config-std-nacl)# permit any
CAT3550(config-std-nacl)# end
CAT3550#
```

QoS Configuration Example on an Access Router

In the following configuration from an access router, any packet with DSCP = 46 (PHB = EF, or Expedited Forwarding) gets assigned to a class that will get a high-priority queue with 33 percent bandwidth. The signaling traffic queue is assigned 2 percent of the interface bandwidth.

```
UCP-2821-2(config)# class-map match-all voice
UCP-2821-2(config-cmap)# match ip dscp ef
UCP-2821-2(config-cmap)# Class-map match-all voice-control
UCP-2821-2(config-cmap)# match ip dscp 26
UCP-2821-2(config-cmap)# match ip dscp 24
UCP-2821-2(config-cmap)# !
UCP-2821-2(config-cmap)# policy-map WAN
UCP-2821-2(config-pmap)# class voice
```

```
UCP-2821-2(config-pmap-c)# priority percent 33
UCP-2821-2(config-pmap-c)# class voice-control
UCP-2821-2(config-pmap-c)# bandwidth percent 2
UCP-2821-2(config-pmap-c)# class class-default
UCP-2821-2(config-pmap-c)# fair-queue
UCP-2821-2(config-pmap-c)# !
UCP-2821-2(config)# int Serial0/1
UCP-2821-2(config-if)# ip address 10.1.6.2 255.255.255.0
UCP-2821-2(config-if)# bandwidth 128
UCP-2821-2(config-if)# no ip directed-broadcast
UCP-2821-2(config-if)# service-policy output WAN
```

Cisco IOS QoS Recommended SNMP Polling Guidelines

It is imperative to gauge the QoS policy settings to ensure that the media traffic is getting proper treatment and adjust as necessary. Chapter 8 elaborates on QoS adjustments as a result of monitoring the QoS performance in the network. Table 7-1 shows the minimum recommended MIB set for CISCO-CLASS-BASED-QOS-MIB MIB, and the corresponding Object IDs (OID) can be used to monitor the QoS behavior in the network. It also shows the polling intervals and the threshold. The QoS counters are cumulative; hence they do not require frequent polling in a stable network.

Case Study of Link Congestions

This case study discusses a scenario in an enterprise environment where a centrally located Unified CM supports multiple branches. The branches are connected through Frame Relay circuits to the main site, where the Unified CM and the aggregation router are located. The branches also have a Cisco Unified Communications Manager Express (CUCME) and a Survivable Remote Site Telephony (SRST) router to offer some capabilities locally, such as DSP resources for transcoding and conferencing as well as a backup for the branch IP phones in case of loss of connection to the main site.

The CISCO-SRST-MIB feature allows remote site status data for the managed devices on your network management system to be retrieved by SNMP. You can specify the retrieval of CISCO-SRST-MIB information from a managed device. However, most importantly, the SRST device must be set to send the SNMP traps to the NMS. In this example, we focus on these two OIDs, which are sent as a trap from the CUCME/SRST router to the NMS:

- **csrstStateChange:** This trap indicates the SRST system state changes (either up or down).

- **EphoneUnRegistrationThresholdExceed:** This trap is sent by the SRST router when devices start to unregister and cross a user-defined threshold. It is recommended to set this threshold using an EphoneUnRegistrationThreshold OID through the network management system that has CISCO-SRST-MIB downloaded and compiled on it. It should be set to 10 percent of the total device that SRST is configured to support. For example, if the branch has 30 IP phones and all of them are backed up by SRST, the threshold should be set to 3.

Table 7-1   *SNMP Polling Recommendations for QoS Statistics*

Object Name	Object Description	OID	Poll Interval	Threshold
cbQosPolice-ExceededBitRate	The bit rate of the nonconforming traffic.	1.3.6.1.4.1.9.9.166.1.17.1.1.14	24 hrs	> 1
cbQosQueueing-DiscardByteOverflow	The upper 32-bit count of octets associated with this class that were dropped by queu-ing.	1.3.6.1.4.1.9.9.166.1.18.1.1.3	1 hr	> 1
cbQosQueueing-DiscardPkt	The number of packets associated with this class that were dropped by queuing.	1.3.6.1.4.1.9.9.166.1.18.1.1.7	1 hr	> 1
cbQosTSStatsDrop-PktOverflow	This object repre-sents the upper 32-bit counter of packets that have been dropped dur-ing shaping.	1.3.6.1.4.1.9.9.166.1.19.1.1.10	1 hr	> 1
cbQosTSStatsDropPkt	This object repre-sents the lower 32-bit counter of packets that have been dropped dur-ing shaping.	1.3.6.1.4.1.9.9.166.1.19.1.1.11	1 hr	> 1

Now we analyze the phones' registration on the Unified CM. It is recommended to moni-tor the status of all the following registered devices using the performance counters:

```
SUB01_Cisco CallManager_RegisteredHardwarePhones
SUB01_Cisco CallManager_RegisteredAnalogAccess
SUB01_Cisco CallManager_RegisteredMGCPGateway
SUB01_Cisco CallManager_RegisteredOtherStationDevices
SUB01_Cisco CallManager_AnnunciatorResourceTota
SUB01_Cisco CallManager_HWConferenceResourceTotal)
SUB01_Cisco CallManager_MTPResourceTotal
SUB01_Cisco CallManager_PRISpansInService
```

The SUB01 prefix indicates the individual node in the Unified CM cluster. All the call-processing nodes need to be monitored for these counters. These counters can be tracked by either the Real-Time Monitoring Tool (RTMT) or CUOM. For this example, we will just concentrate on "SUB01_Cisco_CallManager_RegisteredHardwarePhones," which tracks the status of the IP phones' registration on SUB01, as shown in Figure 7-9.

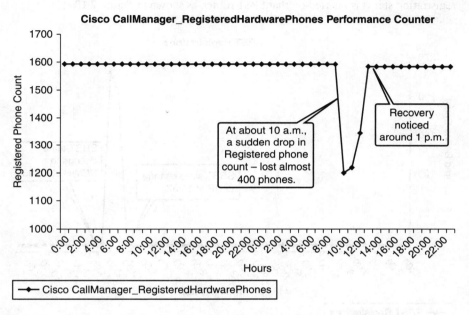

Figure 7-9 *Hardware Phones' Registration Status Tracking on Unified CM*

The graph in Figure 7-9 shows a sudden drop in the hardware phones' registration count from a steady 1592 from the last day until about 11:00 a.m. to 1201 phones. This loss of about 390 phones points to a network anomaly because 75 percent of the phones are still registered with this particular subscriber. Other subscriber nodes in the same cluster do not show any out-of-the-ordinary behavior. At about the same time, an NOC engineer notices a csrstStateChang trap on the central management console coming from one of the branch SRST routers, which had its state changed to "up."

```
*Apr 10 10:57:23.207: SNMP: V1 Trap, ent ciscoMgmt.441, addr 10.45.196.10,
 gentrap
6, spectrap 1
ciscoMgmt.441.2.2.2.1.2.1 = 2
ciscoMgmt.441.1.3.1.2.1 = 1
ciscoMgmt.441.2.2.2.2.2.1 = SRST system state change up
```

Notice the timestamp that indicates 10:57:23.207, which is about the same time when there was a drop in phone registration on the centrally located Unified CM cluster. The NOC engineer then turns to the SRST router identified by its IP address contained in the trap as 10.45.196.10. This SRST router is also the WAN access device for the branch. The first order of business is to ensure that the lost phones are recovered by the SRST. The registration status is queried on that SRST router, as shown in Figure 7-10.

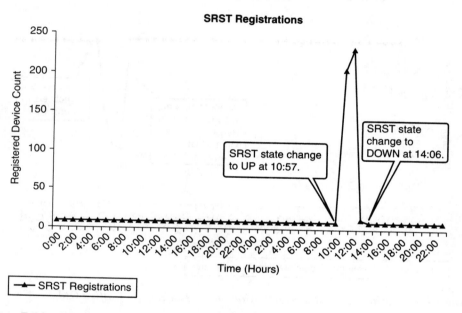

Figure 7-10   *Hardware Phones' Registration Status Tracking on SRST*

In Figure 7-10, it is apparent that the phones are registered to the SRST router coinciding with the trap about its state change to up. Also, the SRST router returns to its normal state at 14:06 as follows:

```
csrstStateChange - SRST status change trap (Down)
*Apr 10 14:06:23.199: SNMP: V1 Trap, ent ciscoMgmt.441, addr 10.45.196.10,
 gentrap
6, spectrap 1
ciscoMgmt.441.2.2.2.1.2.1 = 2
ciscoMgmt.441.1.3.1.2.1 = 2
ciscoMgmt.441.2.2.2.2.2.1 = SRST system state change down
```

This is around the same time that the Unified CM node recovers most of the phones, and they are registered to SUB01 again. The NOC engineer noticed about eight missing phones because the device count was 1585 compared to 1593 before the incident. The NOC engineer did not find any link outage traps either from the branch or the aggregation router. The NOC engineer then turned to QoS to ensure that all the signaling packets were serviced because they were prioritized per the QoS scheme. The phones maintain their registration status using heartbeat signals that are 2 minutes apart. Detailed IP phone registration processes, the keepalive method, the deregistration process, and failover to SRST are described in Chapter 5. The NOC engineer discovers at the time of the incident queue drops around the same time as shown on the graph in Figure 7-11. The graph in Figure 7-11 is based on the polling of the cbQosQueueingDiscardByteOverflow OID that is part of CISCO-CLASS-BASED-QOS-MIB. Also, frames were received on the Frame Relay circuit coming into the branch SRST router with the Backward Explicit Congestion Notification (BECN) bit set indicating congestion in the backward direction.

Figure 7-11 shows queue drops as observed on an SRST router in the branch.

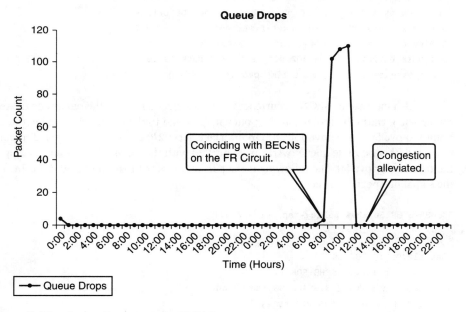

Figure 7-11   *Queue Drops on an SRST Router*

Note   The Backward Explicit Congestion Notification (BECN) bit is set by a Frame Relay network in frames traveling in the opposite direction of frames encountering a congested path. The data termination equipment (DTE) receiving frames with the BECN bit set can request that higher-level protocols take flow control action as appropriate. Frames where the BECN bit is set are tracked by polling the cfrExtCircuitBECNOuts (equivalent to 1.3.6.1.4.1.9.9.49.1.2.2.1.12) OID, which is part of the CISCO-FRAME-RELAY-MIB MIB. The recommended polling interval for this OID is 30 minutes.

The BECN counters increased on the SRST router around the same time that the SNMP trap csrstStateChange was received from 10.45.196.10. The NOC engineer verified the information received through SNMP polling of the **cfrExtCircuitBECNOuts OID** using the **show frame-relay pvc 50** command on that specific branch router.

```
RCDN3825-BR25# show frame-relay pvc 50


PVC Statistics for interface Serial2/0 (Frame Relay DTE)
DLCI = 50, DLCI USAGE = LOCAL, PVC STATUS = ACTIVE, INTERFACE = Serial2/0.1
 input pkts 172116     output pkts 318342     in bytes 11704519
 out bytes 20067856     dropped pkts 0        in pkts dropped 0
 out pkts dropped 0        out bytes dropped 0
 in FECN pkts 0       in BECN pkts 15260     out FECN pkts 0
 out BECN pkts 0      in DE pkts 1929      out DE pkts 0
 out bcast pkts 35      out bcast bytes 4016
 5 minute input rate 243000 bits/sec, 432 packets/sec
 5 minute output rate 393000 bits/sec, 765 packets/sec
 pvc create time 2d04h, last time pvc status changed 00:01:53
```

The sudden increase in BECN occurrences was then traced back to the queues dedicated for signaling traffic by looking at the policy map bound to the frame-relay subinterface bound to pvc 50. The **show policy-map interface serial2/0.1** command was used to correlate it with a possible queue overflow associated with the serial interface terminating the frame-relay pvc 50. The following is the partial output of that command, focusing on the signaling-specific queue:

```
RCDN3825-BR25# show policy-map int se2/0.1


 Serial2/0.1: DLCI 50 -
   Service-policy : 5PHB.FOR.T1
    Class-map: VOICE.SIGNALLING (match-any)
     75134 packets, 3465272 bytes
     30 second offered rate 107000 bps, drop rate 12 bps
     Match: ip precedence 3
      75134 packets, 3465272 bytes
      30 second rate 107000 bps
     Match: ip precedence 4
      0 packets, 0 bytes
      30 second rate 0 bps
     Queueing
      Output Queue: Conversation 73
      Bandwidth 460 (kbps)
      (pkts matched/bytes matched) 0/0
     (depth/total drops/no-buffer drops) 0/110/80
       exponential weight: 9
       mean queue depth: 0
```

class	Transmitted pkts/bytes	Random drop pkts/bytes	Tail drop pkts/bytes	Minimum thresh	Maximum thresh	Mark prob
0	42823/2912224	4944/336192	27833/1892644	20	40	1/10
1	0/0	0/0	0/0	22	40	1/10
2	0/0	0/0	0/0	24	40	1/10
3	75735/3492918	0/0	110/6693	26	40	1/10
4	0/0	0/0	0/0	28	40	1/10
5	0/0	0/0	0/0	30	40	1/10
6	0/0	0/0	0/0	32	40	1/10
7	0/0	0/0	0/0	34	40	1/10
rsvp	0/0	0/0	0/0	36	40	1/10

The incident report was generated along with a root cause analysis. The diagram in Figure 7-12 shows a consolidated graph correlating the problem.

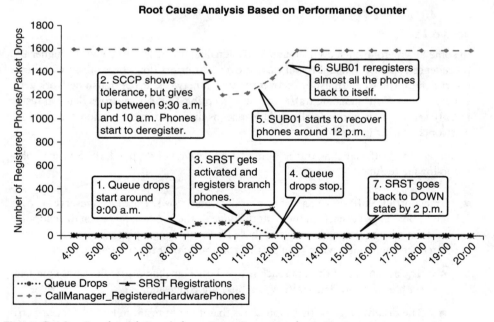

Figure 7-12   *Combined Graph from Root Cause Analysis Report*

The NOC engineer can quickly establish the root cause of the problem associated with the SRST activation trap at 10:57 a.m. The registered hardware phone counter indicates that it is not a systemic issue on the Unified CM because a majority of the phones remained registered to the SUB01 subscriber node in the cluster during the same time interval. Moreover, the overall health of the Unified CM—including CPU, processes, and memory—was good, and no threshold-crossing alarms were noticed on the central management console. However, the SRST router in the affected branch showed the BECN counter increasing based on cfrExtCircuitBECNOuts earlier around 9:00 a.m. This went unchecked because of the Skinny Call Control Protocol (SCCP) protocol, which is responsible for maintaining the IP phone registration status on Unified CM and has tolerance built into it. It can retransmit keepalives. Also, the fact that it is TCP based shows some resiliency for almost one hour until 11:00 a.m., when the phones ultimately lost registration status and registered themselves to the SRST router located within the branch. The trap csrstStateChange from the SRST router has information that enabled the NOC engineer to trace back the problem by looking at its link. This problem was not apparent because link congestion does not generate a trap automatically, compared to a link down situation that would have generated a trap. The actionable item from this incident was to monitor closely the congestion occurrence on the Frame Relay circuit and possibly set up a threshold-crossing alarm/trap for cfrExtCircuitBECNOuts and/or cbQosQueueingDiscardByteOverflow OIDs in the Frame Relay or QoS MIBs, respectively.

Chapter 8 further discusses how to optimize the network to avoid situations like this.

SP Networks

In some SP environments, even though VoIP signaling traffic is marked with a higher IP precedence or DSCP value, it is still carried over best-effort flows or is mixed with best-effort data traffic when transported over the transit network. If the transit network gets congested, the VoIP signaling traffic can be susceptible to delay or drops that can result in VoIP service interruption in the same manner as discussed in the section "Enterprise Networks," earlier in this chapter.

To protect the VoIP signaling traffic from getting delayed or dropped, the SP can deploy the following techniques:

- Transporting VoIP signaling traffic over dedicated flows. The SP can reserve <64-kbps flows for each endpoint at the edge of the network (between the endpoint and the edge router) and aggregate these flows into bigger pipes (10 percent of the interface bandwidth) when transporting this traffic over the transient network.

  - The advantage of this approach is that VoIP signaling traffic does not compete with best-effort data traffic when links are congested.

  - The disadvantage of this approach is that it might result in loss of network efficiency because other traffic might not be able to use these dedicated flows even if bandwidth is available. This might require additional links to be deployed in the network, driving up the cost for the SP.

- Marking the VoIP signaling traffic with a higher IP precedence or DSCP value as compared to best-effort data and classifying this traffic on special queues based on the QoS recommendations in Chapter 6.

  - The advantage of using this approach is that it does not require pre-allocated fixed bandwidth or dedicated network resources specifically for VoIP signaling traffic. This results in better utilization of network resources and improves efficiency.

■ The disadvantage of this approach is that markings from VoIP endpoints either need to be trusted or the packets have to be remarked at the edge of the network. Based on the marking, packets would be classified into special queues. This classification and treatment would need to be applied on all the transient network devices, instead of edge devices only (edge routers, trunking gateway, and so on) to provide end-to-end QoS to VoIP signaling traffic. This increases the complexity of the configuration on transient devices and requires a consistent QoS policy to be defined throughout the network. A comprehensive end-to-end QoS policy can increase the load on CPUs in the core network elements, as well as increase complexity in the core network configuration, which is highly undesirable.

To apply these recommendations, we need to consider the two different deployment models in SP environments discussed in Chapters 3 and 4. In the first deployment model, voice SP owns and manages the transit network. The second deployment model illustrates the case where the voice SP offers VoIP service to its customer over a public IP infrastructure; thus it does not own or manage the transit IP network.

Figure 7-13 illustrates the managed VoIP deployment model.

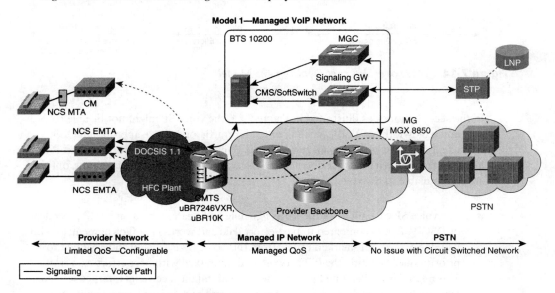

Figure 7-13 *Managed VoIP Deployment Model*

In the first model, as illustrated in Figure 7-13, the voice SP can provide special treatment to VoIP signaling traffic using the techniques mentioned previously. This can help the voice SP provide consistent and reliable VoIP service to its customers.

Figure 7-14 illustrates the Voice over Internet deployment model.

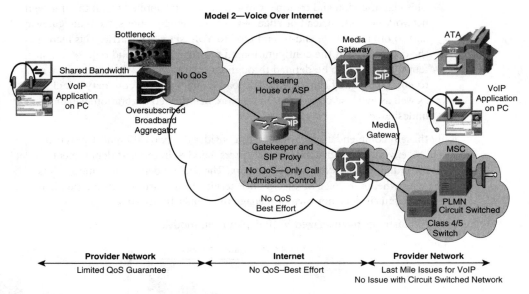

Figure 7-14   *Voice over Internet Deployment Model*

In the second model, as illustrated in Figure 7-14, the voice SP might not be able to dedicate any bandwidth for VoIP signaling or bearer traffic or provide appropriate QoS because it does not own or manage the public IP network that transports this VoIP traffic between endpoints (two subscribers for this service). In this case, the following options are still available to the voice SP:

- The voice SP can still mark the VoIP signaling and bearer with a higher IP precedence or DSCP value as compared to best-effort data and work out a QoS agreement with its ISP. So, the VoIP-marked traffic from this voice SP is trusted and assigned to priority queues within the ISP network and not mixed with best-effort data traffic. This might be challenging because the voice SP might receive its Internet service from multiple providers, so it might need to work out such agreements with several entities.

■ The VoIP SP can purchase dedicated bandwidth (links or a fat pipe) from its ISP to carry the VoIP traffic. This can help ensure that the VoIP traffic for this voice SP is not mixed with traffic from other providers using Virtual Private Networks (VPN). The VoIP SP still needs to rate-limit the traffic as it egresses from its network to match the dedicated bandwidth guaranteed by its ISP.

■ The Voice SP can work out different SLAs with its ISP, either based on QoS guarantee or guaranteed bandwidth.

The previously mentioned approaches can help the voice SP provide high-quality and reliable VoIP service to its customers.

Performance Data in an Enterprise VoIP Environment

It is recommended to monitor the critical call-processing servers such as call agents for performance. These central servers generally have a comprehensive picture of the VoIP network because they register and control the VoIP endpoints and hence represent the central brain of the VoIP network. Metrics data from these servers can provide an estimate of the overall network performance. This performance data can offer clues to other network issues, as discussed earlier in section "Case Study of Link Congestions," where monitoring the status of phone registration on the CUCM lead to the discovery of WAN link congestion. Here we are presenting an example of Cisco Unified Communications Manager (CUCM) that is used in enterprise VoIP environments for central call-processing features. This discussion is intended to provide guidelines on what type of information is critical to track performance. This section has some specific details for CUCM performance data collection, including specific OIDs, alarm conditions, and the polling frequency. Similar information must be collected for performance monitoring on similar central call-processing servers in SP or enterprise VoIP networks.

A CUCM is a central call-processing server that has all the VoIP-related devices registered to it, including all the endpoints (IP phones and analog ports), trunking gateways, intercluster trunks, gatekeepers, SIP proxies, DSP resources supporting transcoding and conference calls, Music on Hold ports, voicemail integration ports, and CTI route points for connectivity to applications like voicemail systems and Unified Call Center platforms that are all registered to it. Therefore, it is much easier to manage the unified communications network in an enterprise environment by focusing on a UC manager that provides a full view of its managed devices and offers Call Detail Records and call diagnostic records.

A CUCM supports application (CISCO-CCM-MIB) and platform MIBs such as HOST-Resource-MIB. Monitoring through SNMP might be desirable if a CUCM is being integrated with CiscoWorks or Cisco Info Center for SNMP-based monitoring. This section also discusses Perfmon counters that can be viewed through the Real-Time Monitoring Tool (RTMT) that is available as part of the CUCM. SNMP community strings on a Unified CM can now be configured through the Serviceability page.

> **Caution**   The following recommendations are for critical items to be monitored. Polling intervals need to be adjusted based on bandwidth availability, low-speed WAN links, or the capacity of the server.

CPU Status

CPU status is a critical item to monitor to determine the overall health and the capacity of the system to process calls. It can point to potential network anomalies and abrupt changes such as sudden surges in endpoint registration requests:

- Periodicity of the polling: Every 2 minutes

- Alarm trigger condition: On high CPU

- Trending recommendation: Maximum and average

- SNMP OID: CpqHoCpuUtilFiveMin (1.3.6.1.4.1.232.11.2.3.1.1.3)

- Perfmon counter: Processor\% CPU Time (Total)

- Threshold for alarm: This is determined by establishing a baseline depending on the load on the server. The typical value is 70 percent. Chapter 8 will elaborate on establishing this baseline. If a high CPU spike is seen during a polling period and the CPU keeps at that level during the next polling interval, only one event should be generated so that the system is not overwhelmed, especially when the CPU utilization is already high.

Physical Memory

Physical memory utilization pertains to the health of the system as well as capacity. It can also point to potential software issues that might slowly leak memory over time:

- Periodicity of the polling: Every 12 hours

- Alarm trigger condition: Based on baseline data

- SNMP OIDs:

  - CpqHoPhysicalMemorySize: (1.3.6.1.4.1.232.11.2.13.1)

  - CpqHoPhysicalMemoryFree: (1.3.6.1.4.1.232.11.2.13.2)

- Perfmon counter: Memory\% Mem Used

- Memory/Cache kbytes

- Memory/Pages

- Trend analysis: Minimum and average values are used for deciding an alarm threshold and establishing system growth needs. Maximum free memory and cached bytes by processes along with paging frequency values are used as a reference to detect memory leaks. As a rule of thumb, a maximum memory utilization of 90 percent should trigger critical alarms. Anything that is about 30 percent above the average utilization but less than 90 percent of the total memory usage should trigger major alarms.

Hard Disk Status

Although there is no SNMP MIB or performance counter available for automated monitoring, hard disk status must be checked every 12 hours manually. This includes ensuring the status of the different disks belonging to a RAID array. (The array might be in proper working condition although one of the physical drives is not.)

High Utilization of Disk Space

The hard disk on the call-processing server maintains debugs, syslog messages, and Call Detail Records in addition to static configuration files or databases. Because this information grows based on call-processing rate, the hard drive space must be managed by the administrator:

- Periodicity of the polling: Every 12 hours

- SNMP OIDs: HrStorageSize: (1.3.6.1.2.1.25.2.3.1.5)

- HrStorageUsed: (1.3.6.1.2.1.25.2.3.1.6)

- Trap: LogPartitionHighWaterMarkExceeded

- Trap: LogPartitionLowMarkExceeded

- Perfmon Counter: Partition\% Used

- Alarm trigger condition: Above 75 percent usage or less than 1 GB free (depending on partition usage)

Utilization of disk space of the CUCM servers should be closely monitored, especially if detailed log collection is activated. Most often the application on the CUCM will automatically start to delete older files. But this still needs to be managed more proactively because the trigger-based automatic file deletion can cause high input and output (I/O) counts that can potentially degrade the performance of the system.

Virtual Memory

Virtual memory consists of physical memory (RAM) and swap memory (swap partition). When virtual memory runs out, the operating system (OS) kills process(es) in an unpredictable manner to free virtual memory. Therefore, it is critical to proactively take care of the situation before the OS starts killing processes. For example, if you suspect a memory leak where a particular process's memory usage keeps increasing or a process is using a lot more than it should, you might want to restart that process and log the incidence for further investigation:

- Periodicity of the polling: Every 2 minutes

- Alarm trigger condition: Above 75 percent usage

- SNMP OIDs: CpqHoPagingMemorySize (1.3.6.1.4.1.232.11.2.13.3)

- CpqHoPagingMemoryFree (1.3.6.1.4.1.232.11.2.13.4)

- Perfmon counter: Memory\%VM Used and Memory\%PageUsed

Number of Active Phones

It is critical to monitor the phone registration status from the CUCM using the following guidelines. This information will help in proactively notifying potential network outages that can impact the CUCM's reachability to these phones even when a disaster recovery scheme such as the Survivable Remote Site Telephony (SRST) feature is activated. This information is also helpful in measuring the SLA for the dial tone availability despite network outages:

- Periodicity of the polling: Every 15 minutes

- Alarm trigger condition: Decrease in phones' registration by 10 percent in 15 minutes. An alarm should be set to notify the system administrator when the number of registered phones decreases drastically.

- SNMP OIDs: CCM-MIB

- ccmActivePhones: (1.3.6.1.4.1.9.9.156.1.5.1)

- Perfmon counter: Cisco CallManager\RegisteredHardwarePhones

- Gives the total number of active (registered and sending keepalive messages) phones for that Unified CM

Gateway Registration (MGCP)

VoIP gateways in VoIP networks are often central points for a remote site, an entire campus, or even an entire install base to interface with the PSTN to not only reach outside the VoIP network but also for important services such as police, fire, or ambulance. This resource is often oversubscribed. This further reiterates the importance of the gateway's availability. Therefore, the following guidelines are based on more frequent polling and alarm settings for any change in gateway registration status:

- Periodicity of the polling: Every 5 minutes

- Alarm trigger condition: *Any* change

- Perfmon counter: Cisco CallManager\RegisteredMGCPGateway

Gatekeeper Registration (H.323 RAS)

Gatekeepers are responsible for call admission control and often address resolutions between two or more clusters of CUCMs. If a CUCM loses its gatekeeper registration, it will not be able to complete a call destined for another directory number belonging to a different CUCM cluster over the IP network, despite bandwidth availability. If this call gets completed using an alternate route, which is typically the PSTN, toll charges might apply. Therefore, registration status monitoring is critical using the following guidelines:

- Periodicity of the polling: Every 5 minutes

- Alarm trigger condition: *Any* change

- Perfmon counter: Cisco CallManager\RAS retries

- Threshold-based trigger: If this counter is increasing rapidly, it is indicative of failed registration attempts. Check whether the Cisco Gatekeeper\ARQs Attempted counter has shown a spike. Alternatively, the Route List exhaustion can be monitored for a Routelist that has Gatekeeper-controlled trunk(s) by monitoring for a ccmRouteListExhausted trap from the CUCM.

There is no SNMP MIB or direct performance counter available for this parameter, but Cisco recommends a periodic check if the CUCM is registered with the gatekeeper. This can be done with an automated script for the **show gatekeeper end-points** command on the IOS gatekeeper.

Calls in Progress

Tracking calls in progress provides an estimation of the VoIP network's capacity utilization. Proactively monitoring calls in progress will assist the network administrator with capacity planning to absorb growth before any SLAs are violated:

- Periodicity of the polling: Every 5 minutes (real-time monitoring in the NOC)

- Perfmon counters: Call Manager\CallsInProgress and Call Manager\Cisco SIP\CallsInProgress

- Alarm trigger condition: If over 95 percent of system capacity

- Perfmon trending: Real-time graphing of this parameter, compared with expected values based on historical data, is useful in detecting subtle system performance degradation (generally by detecting that the real-time number of calls in progress is below expected values compared to same time-of-day/day-of-week baseline values). It is recommended that any drop that represents 30 percent less calls than the established baseline for that time of the day is investigated.

Calls Active

Active calls represent the number of streaming connections that are currently active (in use). In other words, "Call Active" represents the number of calls that actually have a voice path connected. Tracking active calls will provide another level of granularity to calls-in-progress monitoring by accurately determining the bandwidth utilization associated with active calls. This performance data can help in adjusting call admission control (CAC) parameters on gatekeepers, the CUCM, or SIP proxies:

- Periodicity of the polling: Every 5 minutes (real-time monitoring in the NOC)

- Perfmon counters: Call Manager\CallsActive, Call Manager\Cisco SIP\CallsActive, and Call Manager\Cisco SIP\VideoCallsActive

- Alarm trigger condition: If over 90 percent of system capacity

- Perfmon trending: This value can be analyzed in a fashion similar to calls in progress.

Calls Attempted

A sudden increase in call attempts in reference to an established baseline for this parameter will proactively notify the NOC to investigate potential VoIP-related security threats such as theft of services or distributed denial of service (DDoS) attacks. The potential VoIP-related security threats have been mentioned in the section "Safeguarding Network Resources from Potential Security Threats" in Chapter 8.

Similarly, an abrupt drop in call attempts can be indicative of network-related problems such as VoIP endpoints' inability to place calls because of network congestion:

- Periodicity of the polling: Every 15 minutes

- Alarm trigger condition: Outside the preestablished (through periodic audits) baseline. This can indicate a potential system anomaly requiring further investigation.

- Perfmon counters: Call Manager\CallsAttempted and Call Manager\Cisco SIP\CallsAttempted

- Perfmon trending: This value must be collected over time and used to compute the Busiest Hour Call Attempt (BHCA) value. This is explained in Chapter 6 in the "Managing Network Capacity Requirements" section and in Chapter 8.

Calls Completed

Tracking and correlating calls attempted and calls completed are helpful in determining specifically the CUCM's and generally the VoIP endpoints' (including IP phones, gateways, and DSPs) ability to successfully process calls. This is discussed in detail in a case study in the section "Safeguarding Network Resources from Potential Security Threats" in Chapter 8:

- Periodicity of the polling: Every 15 minutes.

- Alarm trigger condition: Outside the preestablished baseline.

- Perfmon counters: Call Manager\CallsCompleted, Call Manager\Cisco SIP\CallsCompleted, and Cisco CallManager\Cisco SIP\VideoCallsCompleted.

- Perfmon trending: This value must be collected over time and used to compute the Busiest Hour Call Completed (BHCC) value that is used in traffic engineering in Chapter 6.

PRI Channels Active

Tracking active ISDN PRI channels serves a similar purpose as tracking gateway registration status but at a more granular level by looking at the individual PRI trunks. Because PRI channels are often oversubscribed, it is critical to ensure that they are always available for incoming and outgoing calls from and to the PSTN. PRI channel status tracking through "PRI Channels Active" monitoring can provide feedback to the network administrator to adjust the dial plan or increase the trunk capacity if the trend based on this Perfmon counter suggests an increase in call volume on the PRI channels that consistently reach the maximum trunk capacity:

- Periodicity of the polling: Every 5 minutes (real-time monitoring in the NOC).

- Alarm trigger condition: Underutilization of the preestablished baseline.

- Perfmon counters: Cisco CallManager\PRIChannelsActive (for MGCP-controlled PRIs only) or SNMP OID: CISCO-CALL-TRACKER-MIB on IOS.

- Perfmon counters trending: Collection of this data over time is used to understand call patterns and busy-hour peak calls. Baseline data can be used to detect real-time underutilization of circuits, which is an indication of possible system performance degradation (including otherwise hard-to-detect PSTN call routing or circuit-down conditions). Data trending allows circuit growth and provisioning planning.

Conferencing/Transcoding DSP's Depletion

DSP resources are used to perform transcoding and provide conferencing capabilities to the end users. DSP resources can be distributed on IOS gateways or reside locally on the CUCM subscriber servers. The CUCM registers and controls these DSP resources. This allows the NOC to centrally manage and track DSP resources' utilization by monitoring their status from the CUCM. Like PSTN trunks, DSP resources are also oversubscribed. Their unavailability means that the end users will no longer be able to use features like conferencing and transcoding. These DSP resources are often used to broker SIP call features and provide call setup normalization functions. Therefore, it is critical to monitor DSP resource depletion:

- Periodicity of the polling: Every 5 minutes (real-time monitoring in the NOC).

- Alarm trigger condition: Whenever this performance counter increments, it should be set to generate an alarm.

- Perfmon counters: Cisco CallManager\UnicastHardwareConfResourceOutOfResources, Cisco CallManager\UnicastSoftwareConfResourceOutOfResources, and Cisco CallManager\TranscoderOutOfResources

Available Bandwidth of a Location (CAC)

The CUCM centrally controls the amount of calls it can send and receive between "locations." Locations represent the various remote locations that are typically connected through WAN links. Based on the bandwidth reserved for VoIP calls, the "location" and "region" settings on the CUCM will provide call admission control (CAC) to preserve the call quality by avoiding overloading the WAN links. The CUCM can redirect the overflow calls using the Automated Alternate Routing (AAR) feature. AAR can redirect the overflow calls to use a PSTN trunk resource. Monitoring for the bandwidth availability Perfmon counter, especially when it reaches zero, can provide critical feedback to the network administrator to make decisions about increasing the network bandwidth or adjust the settings to use compressions codecs to accommodate more calls on existing WAN links:

- Periodicity of the polling: Every 5 minutes (real-time monitoring in the NOC).

- Perfmon counters: Cisco Locations\CurrentAvailableBandwidth.

- Alarm trigger condition: When it reaches zero.

- Perfmon counters trending: This information should be collected over time for capacity planning, including bandwidth increase and call capacity increase by use of compression codecs.

Sometimes a Resource Reservation Protocol (RSVP)–based CAC scheme is used to manage bandwidth. RSVP reserves bandwidth for every flow, where the flow can be a voice or video call. An RSVP-based CAC design includes all the nodes in the call path that support RSVP or act passively to RSVP resource reservation requests. RSVP-related Perfmon counters on the CUCM must be tracked in a similar fashion to monitor bandwidth usage on the IP network. This can be achieved by tracking the following using the same criteria as prescribed for the Cisco Locations\CurrentAvailableBandwidth Perfmon counter:

Cisco CallManager\Cisco Locations\RSVP TotalCallsFailed and Cisco CallManager\Cisco Locations\RSVP VideoCallsFailed

Recommendations for Categorizing Performance Measurements

The amount of information collected in a large-scale VoIP network, including KPIs and other performance data, can be overwhelming. This section provides recommendations for categorization of performance data into three categories: green, yellow, and red. Events, alarms, or performance counters categorized as green would indicate normal operations where no action is required. The performance data in this category can be used to establish a network baseline. The yellow category would include performance data indicating degradation in the network operations that requires investigating the situation to prevent potential network outages. The yellow category events can help NOC staff to start proactively planning in rectifying a problem in its infancy. The red category will include events, alarms, traps, or other performance data points that indicate a more serious network problem such as complete resource depletion, link outages, or a situation where SLAs are being violated. The events in the red category require immediate corrective action. Most of the network management software vendors, such as "JEL Console" on Netcool's software, provide a capability to color-code the events or traps.

Overall, this color categorization tremendously helps NOC operations in prioritizing their activities to efficiently manage the VoIP network. The following categorization of performance counters value ranges provides an example of establishing criteria for organizing performance data in an enterprise VoIP network using CUCM clusters for call processing. Some of the call-processing server-related information, such as CPU and memory utilization, is specific to the CUCM. The VoIP call quality metrics, such as delay and call success rate, apply to all VoIP networks. Similar criteria can be applied t.o SP VoIP networks.

Green: Network operation is efficient (no degradation):

- Greater than 99.999 percent Call Completion Rate

- Total CPU usage under 68 percent

- Dial Tone Delay under 251 ms

- Media cut-through/clipping under 500 ms

- Media latency/delay under 150 ms

- No calls failed because of depletion of bandwidth resources (CAC-related performance counters)

- CUCM virtual memory consumption less than 1.4 GB (on a smaller-size server with less VoIP endpoints, such as an MCS-7825)

- CUCM virtual memory consumption less than 2.1 GB (on a larger-size server with more VoIP endpoints and more memory, such as an MCS-7845)

Yellow: Network operation becomes less efficient (slight degradation):

- 98.0–99.99 percent Call Completion Rate, some calls dropped or denied

- Total CPU usage of 68–80 percent

- Dial Tone Delay of 251–400 ms

- Media cut-through/clipping of 500–800 ms

- Media latency/delay of 150–300 ms

- CUCM virtual memory consumption greater than 1.4 GB but less than 1.7 GB (on a smaller-size server with less VoIP endpoints, such as an MCS-7825)

- CUCM virtual memory consumption greater than 2.1 GB but less than 2.55 GB (on a larger-size server with more VoIP endpoints and more memory, such as an MCS-7845)

Red: Network operation is not efficient, failures are occurring (major degradation):

- Less than 98 percent Call Completion Rate

- New calls are not accepted or calls in progress drop

- Total CPU spiking for a measurable time period over 80 percent for periods longer than 5 seconds

- Dial Tone Delay greater than 400 ms

- Media cut-through/clipping greater than 800 ms

- Media latency/delay greater than 300 ms

- CUCM virtual memory consumption greater than 1.7 GB (on a smaller-size server with less VoIP endpoints, such as an MCS-7825)

- CUCM virtual memory consumption greater than 2.55 GB (on a larger-size server with more VoIP endpoints and more memory, such as an MCS-7845)

Enterprise Case Study—Analyzing Network Performance

This case study looks at a subset of performance counters on the CUCM to analyze system health and capacity. It provides general guidelines on interpreting individual areas and of the system such as CPU resources and call capacity by looking at a group of counters. Chapter 8 discusses how to correlate multiple performance counters to draw more holistic conclusions to optimize the performance of the entire VoIP network.

CPU Rate and Critical Processes

The CUCM can have as many as 50 or more processes running concurrently. However, CPU utilization of only the critical processes as well as overall CPU utilization are tracked. These processes include CCM, RISD, and CTIManager. All of these are directly responsible for processing calls and features. Figure 7-15 includes the average value of the CPU usage rate for each of the processes and the average CPU usage rate for the entire system.

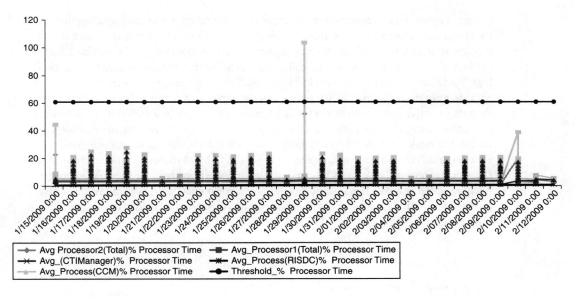

Figure 7-15 *Average CPU Rate for Critical Processes*

Referring to Figure 7-15, we can understand which process is consuming most of the CPU cycles. If the peak value is not large, most likely there isn't any problem. However, if the average value continues to have a large percentage of CPU utilization, more investigation/confirmation is needed. For dual CPU systems, the maximum value is 100 percent times the CPU count. For occasionally high CPU utilization and threshold crossing, NOC staff must pay special attention to the CCM application such as CCM (ccm.exe), as shown in Figure 7-15. A normal CUCM operation will typically show that all the processes show the same trend for CPU utilization that approximately aligns with the total CPU utilization. This trending data can render itself by overlapping lines over time, as apparent in Figure 7-15. But any exceptions should be noted by looking for a deviation in the trend shown by increasing gaps between the lines representing the monitored processes. Generally no action is required if this deviation lasts for a short amount of time. Persistent deviations should be investigated, and particularly threshold crossings need immediate action, as shown on January 29, 2009 in the graph in Figure 7-15.

Rate of Active Calls

Monitoring the rate of active calls provides an overview of the CUCM's capability to successfully process a call in a reasonable amount of time that does not impact the end user experience, such as delayed dial tone or delay in call establishment. CallsInProgress counts the number of times that the user picks up the receiver and starts dialing the digits. Using this value, we can estimate the average time it takes for the user to pick up the receiver to start talking. When the rate of Active Calls is low, it might be assumed that the server is slow to respond to the calls, and it is likely that the server is experiencing a high load. Calls requiring multiple elements, such as a DSP resource for transcoding, address resolution through an SIP proxy or gatekeeper, or a feature server, might cause the rate of active calls to go down for any number of reasons associated with these call-brokering elements. On weekends, when the number of calls is low, this value is not very useful. On working days, when the number of calls is high, this value makes a good reference. Figure 7-16 contains the average value for the CallsInProgress count per hour and the average rate of the Active Calls to compare and see how many of the calls in progress become active in a day. Because both values are approximately equal and the call active rate remains high during business hours, it can be considered a healthy trend, as shown in Figure 7-16.

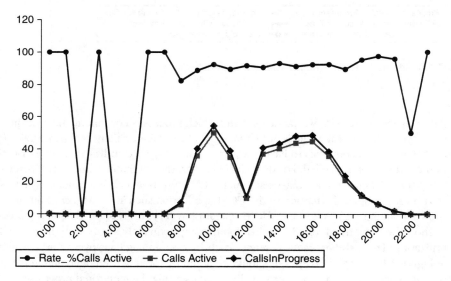

Figure 7-16   *Rate of Active Calls*

Tracking Trunk Utilization for PSTN Access

Tracking the trunk utilization for PSTN access provides important feedback about when add to capacity or move idle resources into trunk groups that are overflowing. This example analyzes outbound calls by tracking the number of attempted calls that were rejected because of being busy per unit of time (day), and outbound and inbound calls were tracked by monitoring the number of calls that were completed per unit of time (day). Looking at the graph in Figure 7-17, we can confirm how many of the calls have changed to the talking state and how many have been denied per each PRI span unit. A PRI span unit is typically a grouping of individual channels or DS0s into T1 or E1 trunks. In this example, the PRI trunks are controlled by the CUCM through MGCP.

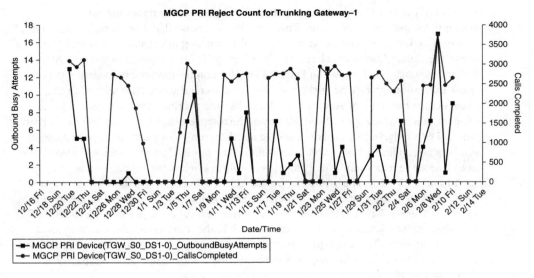

Figure 7-17   *MGCP PRI Call Reject Count Versus Completed Calls*

If the number of denied calls is large, we might want to think about adding new lines. Note that when line grouping is applied, there might be lines that have all their channels completely used up, whereas other lines still have free channels, but the OutboundBusyAttempts still counts the total busy attempts for the entire span. The graph in Figure 7-17 shows that the calls completed trend with calls that experience OutboundBusyAttempts. This can be interpreted as the MGCP-controlled PRI trunks are available when the call volume increases (by cross checking with call completed counters) and the capacity limit is reached; hence, the occurrence of OutboundBusyAttempts. However, the call blocking rate because of OutboundBusyAttempts never exceeds 0.5 percent when we focus on the data points corresponding to 2/7 Tue on the date/time axis. This is when the system experienced the highest number of OutboundBusyAttempts (about 17), when the number of calls completed was around 3700. If the call-blocking rate as agreed upon in the SLA is 0.5 percent or lower, this would require increasing the trunk capacity. The trend seen in Figure 7-17 between December 20 and December 31 shows an anomaly where the OutboundBusyAttempts is unusually high for almost a negligible number of completed calls. Because this time period coincides with a holiday period, it can be safely ignored.

Note that when line grouping is applied for tracking these counters, line grouping includes all the MGCP-controlled spans (T1 or E1 trunks) that serve the same route destinations such as local or long-distance. There might be trunks or spans in the line grouping that have all their channels used up, whereas other trunks still have free channels. This is because of the fact that the OutboundBusyAttempts counter tracks the total busy attempts for the entire line group. Therefore, keep in mind that when a high rate of rejection is observed, if we don't have more granular grouping information, we cannot be sure about the cause. Always check the graph in Figure 7-17 in conjunction with the MGCP PRI status usage graph shown in Figure 7-18. This graph shows the individual span or T1 trunk utilization by looking at the maximum and average active channels. With the exception of the week of August 13, it looks like the maximum trunk capacity of 23 channels is almost always utilized. However, the average usage is around five to six channels or DS0s only. In this situation, adding another trunk to absorb the capacity might not be a good idea. A better approach would be to observe other trunks and identify the ones where the maximum active channels are always below the maximum capacity (23 for T1 and 31 for E1). This trunk can be added in the Route Group and/or Route List in the CUCM dial plan that will correspond to the routing destination. For example, if the graphs in Figures 7-17 and 7-18 are representative of local calls, perhaps a less-utilized long-distance trunk can serve as a backup trunk in the route group dedicated to local route destinations. This concept in leveraging other trunks for a given route destination is discussed in more detail in Chapter 8 in the section "Maximizing Trunk Capacity and Avoiding Call Blocking."

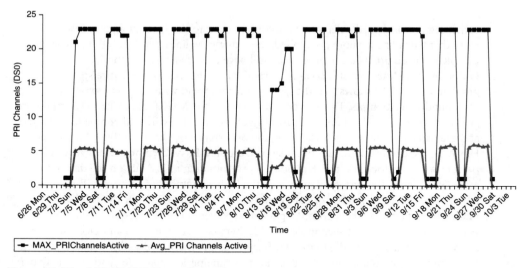

Figure 7-18   *MGCP PRI Usage Status*

Trend Analysis Best Practices

For accurate performance analysis, the interpretation of graphed data should follow these guidelines:

- Each time you make the new graphs, make sure that these graphs use the same time scale and polling frequency as the previous ones.

- Investigate if there are reports of trouble tickets or events such as a shutdown because of maintenance or power failure.

- Investigate to find out whether there is any relevant event in the event/trace logs for the call-processing servers.

- Look to see how the performance graphs differ from the same performance counters for different but similar network segments/customers/implementations. For example, by observing and comparing performance data trends, two CUCM clusters in different regions can provide valuable feedback when applying critical changes to the network, such as CAC parameter adjustments or DSP resource allocations. Investigate to see whether there is any noticeable difference between servers with the same functionality. For example, servers within the same CUCM cluster should exhibit the same performance trend for CPU and memory utilization. A varying difference can point to an unbalanced load on the servers in the same CUCM cluster.

Performance Analysis from Call Agent Perspective

The Call Agent or the CMS is an ideal networking component to collect the network KPIs. In most cases, it is impacted by VoIP network anomalies. Through the use case of the BTS 10200, which is a Cisco VoIP Softswitch solution, we cover the KPIs needed to assist in isolating faults and determine performance issues. This deployment scenario is shown in Figure 7-6, and the KPI concept described here can be applied to both SP and enterprise VoIP environments. The following sections cover the performance counters for tracking the call traffic, the PSTN leg, and the On-net leg. We also look into strategies to help optimize the Call Agent's performance. In the following sections, we also look into BTS 10200 call traffic, PSTN trunk, SS7 signaling, MGCP, and SIP-related performance counters, along with some techniques to alleviate VoIP infrastructure issues.

Performance Analysis for VoIP Call Traffic

The BTS 10200 has a rich set of call traffic performance counters. Following the KPI strategy discussed in the section "Proactive Monitoring Through Performance Counters," earlier in this chapter, we show different call traffic leg dashboards. The dashboards reflect the full day's traffic, thus capturing the busy hour along with clearly identifying any anomalies, if they exist.

We start off with the overall call traffic metrics. Table 7-2 lists the graphed BTS 10200 call-related counters that we discuss as use cases.

Table 7-2   *BTS 10200 Performance Counters*

Call Counter	Description
CALLP_ORIG_ATTMP	The number of originating call attempts of all types on the reporting call agent
CALLP_TERM_ATTMP	The number of terminating call attempts of all types on the reporting call agent
CALLP_ORIG_FAIL	The number of originating call attempts of all types that failed on the reporting call agent
CALLP_TERM_FAIL	The number of terminating call attempts of all types that failed on the reporting call agent
CALLP_SS7_ORIG_ATTMP	The number of originating SS7 call attempts on the reporting call agent
CALLP_SS7_TERM_ATTMP	The number of SS7 terminating call attempts on the reporting call agent
CALLP_SS7_ORIG_FAIL	The number of SS7 originating call attempts that failed on the reporting call agent
CALLP_SS7_TERM_FAIL	The number of SS7 terminating call attempts that failed on the reporting call agent
CALLP_SIP_ORIG_ATTMP	The number of originating SIP call attempts on the reporting call agent
CALLP_SIP_TERM_ATTMP	The number of SIP terminating call attempts on the reporting call agent
CALLP_SIP_ORIG_FAIL	The number of SIP originating call attempts that failed on the reporting call agent
CALLP_SIP_TERM_FAIL	The number of SIP terminating call attempts that failed on the reporting call agent
CALLP_MGCP_ORIG_ATTMP	The number of originating MGCP call attempts on the reporting call agent
CALLP_MGCP_TERM_ATTMP	The number of MGCP terminating call attempts on the reporting call agent
CALLP_MGCP_ORIG_FAIL	The number of MGCP originating call attempts that failed on the reporting call agent
CALLP_MGCP_TERM_FAIL	The number of MGCP terminating call attempts that failed on the reporting call agent

Figure 7-19 illustrates some of the call traffic counters. The overall call traffic flow for a particular voice switch can be determined by this dashboard.

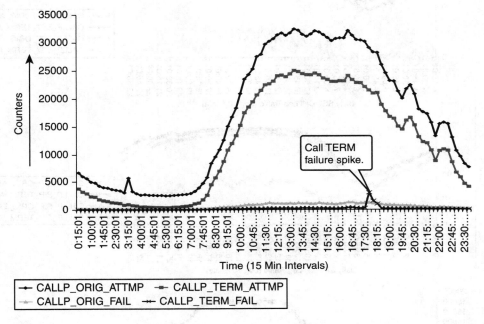

Figure 7-19 *VoIP Call Traffic Performance Counters*

Figure 7-19 reflects a full day's worth of call traffic counters for a VoIP switch. Notice the bell-shaped curve reflecting a busy hour that's stretched from morning until the evening time. The network operator can track the call performance through this dashboard. The busy-hour numbers reflect the peak call volume numbers.

Also notice the spike in the call termination failures around 5:30 p.m. The terminations need to be further investigated.

Figure 7-19 depicts that baselining the traffic pattern allows the anomalies to clearly become visible to the network operator.

The counters shown in Figure 7-19 are CALLP_ORIG_ATTMP, CALLP_TERM_ATTMP, CALLP_ORIG_FAIL, and CALLP_TERM_FAIL. Including this set of KPIs in the dashboard enables the operator to track the calls originating and terminating at switch along with the respective failures. The increase in failures would immediately be noticed and prompt the operator to take corrective action.

Figure 7-20 reflects a dashboard for all the call legs for a BTS 10200–based solution.

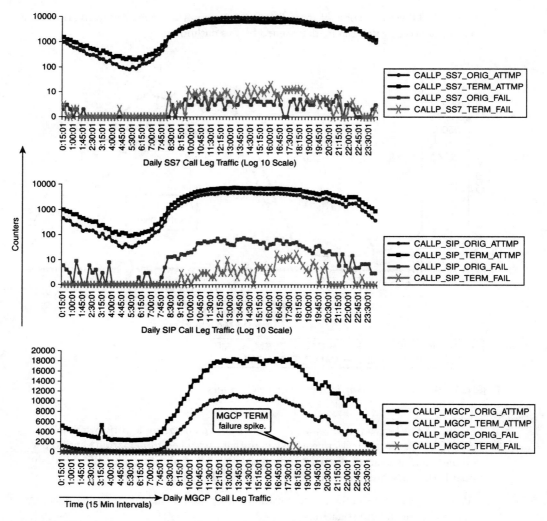

Figure 7-20   *VoIP Call Leg (SS7, SIP, and MGCP) KPI Dashboards*

The dashboard in Figure 7-20 follows the theme of the earlier case study in Figure 7-1. Here the various call legs are exposed with the origination and the termination perform-ance counters. The VoIP operator can easily narrow the anomaly to the MGCP call leg. This could be related to a particular far-end switch having resource issues, thus terminat-ing the call abnormally.

Performance Analysis for a PSTN Network (PSTN Trunk and SS7 Signaling)

Continuing with the Call Agent performance counter analysis, this section looks at the PSTN segment of the VoIP network. A few SS7 counters are discussed as KPIs in a use case. These counters are also documented in Table 7-3.

Table 7-3 *PSTN Bearer Trunk Usage and SS7 Traffic (ISUP) Performance Counters for a Trunk*

Performance Counters	Description
ISUP_IAM_TX	The total number of IAM messages sent on the reporting trunk group
ISUP_IAM_RX	The total number of IAM messages received on the reporting trunk group
ISUP_REL_TX	The total number of REL messages sent on the reporting trunk group
ISUP_REL_RX	The total number of REL messages received on the reporting trunk group
TRKGRP_INCOM_USAGE	Summation of the number of trunk circuits within the reporting trunk group marked as busy or in the maintenance state with terminating calls taken every 100 seconds during the interval
TRKGRP_OUTG_USAGE	Summation of the number of trunk circuits within the reporting trunk group marked as busy or in the maintenance state with originating calls taken every 100 seconds during the interval
TRKGRP_TOTAL_USAGE	Summation of the incoming usage and the outgoing usage counters for the reporting trunk group
TRKGRP_TOTAL_OVERFLOW	The number of outbound trunk call attempt failures because of all trunks within the reporting trunk group being in a busy state
TRKGRP_TOTAL_OOS_TRK	The number of transitions to the locally blocked state for all trunks within the reporting trunk group

Tracking the bearer trunk usage and SS7 signaling counters, the operator has a handle on monitoring the PSTN leg of the VoIP network. Figure 7-21 reflects a dashboard of KPIs captured in Table 7-3.

The dashboard reflects a busy-hour scenario for a particular trunk and the corresponding SS7 signaling exchange. Here a potential high number of call releases compared to call completion can mean capacity, trunk, or gateway problems.

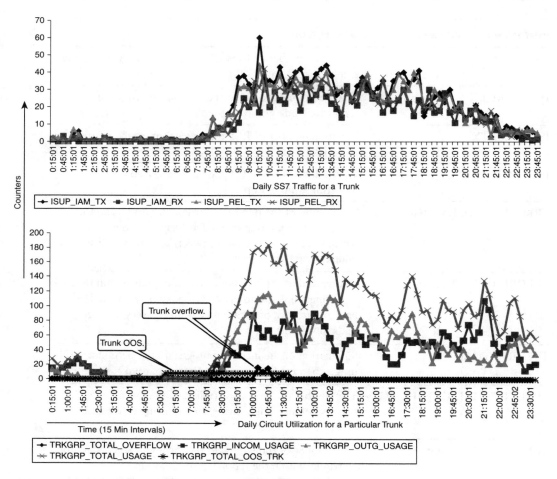

Figure 7-21   *Daily Bearer Trunk and SS7 Signaling KPI Dashboard*

The dashboard also helps in providing an understanding of which trouble areas area to focus on. The current state of the dashboard acts a baseline for the trunk usage and SS7 traffic. A increase in number of call releases would be a trigger point for the operator for investigating the cause.

A high IAM with increasing call release can also indicate a capacity issue on the PSTN bearer trunks.

The second graph in Figure 7-21 is for trunk KPIs. The captured performance counters are trunk total overflow, incoming trunk usage, outgoing trunk usage, total trunk usage, and total out-of-service trunks.

We can observe a pattern on the dashboard that most of the trunk seizers during the day are for outgoing trunks and during the evening the incoming trunk seizures are high. This could be a candidate for a time of day (ToD) policy application.

Also, we notice an overflow of trunks, which is a clear indication of capacity issues. A nonzero TRKGRP_TOTAL_OOS_TRK is also observed, and the value should be a trigger for investigation.

Performance Analysis for an SIP Network

This section highlights SIP signaling counters to shed some light on the SIP component of the VoIP network. Table 7-4 captures a subset of SIP performance counters. These counters act as a KPI dashboard to track the SIP health of the VoIP network.

Figure 7-22 reflects the SIP dashboard, which can act as a baseline. Here the total number of successful incoming and outgoing messages should be about equal to the total number of incoming and outgoing messages, respectively. Any difference between the two counts, especially a rapidly growing difference, might indicate a problem.

Table 7-4   *SIP Dashboard Performance Counters*

Performance Counters	Description
SIP_TOTAL_INCOMING_MSG	The number of SIP messages the reporting call agent attempted to receive
SIP_TOTAL_SUCCESS_INCOMING_MSG	The number of SIP messages the reporting call agent successfully received
SIP_TOTAL_OUTG_MSG_ATTMP	The number of SIP messages the reporting call agent attempted to send
SIP_TOTAL_SUCC_OUTG_MSG	The number of SIP messages the reporting call agent successfully sent

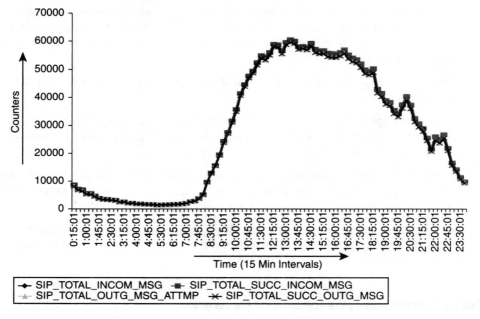

Figure 7-22   *Daily SIP Signaling KPI Dashboard*

Performance Tracking for a Session Border Controller (SBC)

This section looks at a KPI dashboard for an SBC. In a majority of VoIP deployments, the VoIP operator is terminating the SIP trunk from its CMS to an SBC that is connected to a PSTN provider's SBC. This enables the VoIP operator to focus its resources on the IP infrastructure. However, keeping on top of a healthy PSTN-bound SIP connectivity becomes the challenge. The KPIs that need to be tracked are at a minimum the ingress and egress SIP connections. Figure 7-23 shows an SBC dashboard.

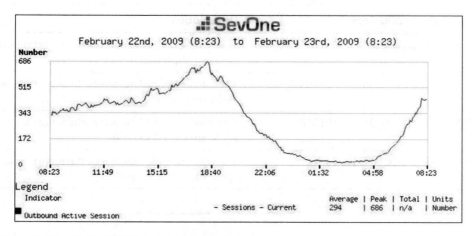

Figure 7-23   *Daily SBC Session Dashboard*

The dashboard shows the daily sessions going through the VoIP operator's SBC. Thus, any changes from the baseline behavior are clearly visible. This enables the VoIP operator to develop an SLA tracking system to hold the PSTN SP accountable for agreed-upon session throughput.

Performance Information Through the Call Detail Records (CDR)

The BTS reports QoS statistical data for the call in the CDRs. This data must be extracted as KPIs and transformed into a dashboard. The following list reflects the CDR-reported QoS statistics:

- OrigQosPacketsSent
- OrigQosPacketsRecd
- OrigQosOctetsSent
- OrigQosOctetsRecd
- OrigQosPacketsLost
- OrigQosJitter
- OrigQosAvgLatency
- TermQosPacketsSent
- TermQosPacketsRecd
- TermQosOctetsSent
- TermQosOctetsRecd
- TermQosPacketsLost
- TermQosJitter

The list reflects the received and sent packet throughput for the specific call, the latency, and the jitter statistics. A pseudo-real-time dashboard for these stats on a 15-min interval can keep the VoIP operator on top of VoIP call quality.

The following statistical data also needs to be extracted from the CDRs to track the call serviceability:

- Call termination code (cause code)
- Call type
- Duration
- Associated trunk
- Signaling duration

Thus, tracking and correlating the CDR information and tying it with overall KPI strategy mentioned in previous sections would enhance the VoIP operator's management experience, as the CDRs provide another set of key information for operating the VoIP network effectively.

Performance Enhancement Schemes and Their Effect on VoIP Network Monitoring

A VoIP network operator can take several actions to optimize the network's performance. We cover some of the common ones in the following sections.

Effect of DNS Caching

The DNS plays a key role in VoIP deployment. Typically, almost every IP address maps to a fully qualified domain name (FQDN). This enables the operator to assign multiple IPs to an FQDN. FQDN has its challenges; the FQDN's Time To Live (TTL) can expire or the current FQDN's IP address can become unreachable. In either case, a DNS lookup needs to be done. This is expensive in the VoIP system, as the FQDN-to-IP mapping needs to be fast and efficient to expedite the call setup. In the latter case, you are stuck with a DNS query, but for the first case, a few options can be applied:

- DNS caching can be introduced as a solution to alleviate the TTL issue in the CMS, which has UNIX as the operating system. The DNS caching facilitates the VoIP network devices, in particular the VoIP switch to become a caching server. Thus, in essence, the DNS query is made only if the IP address cannot be reached.

- A second option, which is available sometimes, is to rely on the VoIP switch's software itself. The software has a built-in capability to cache the FQDN-to-IP map, and it renews the IP only in case of a failure. This caching feature availability needs to be verified through the vendor.

- You can also use static entries to map the FQDN-to-IP addresses; this is configured through the file option under the UNIX /etc/nsswitch.conf file.

Server Load Balancing

Server load balancers are also introduced to front the DNS servers. Basically, a DNS server farm is created, and the VoIP switch sees only a primary and secondary DNS VIP. The load balancer not only provides an even distribution of load for DNS queries but also provides high availability because it takes the nonfunctional DNS servers out of rotation.

Firewall

Firewalls can become a VoIP operator's nightmare. There have been numerous times that the VoIP network has taken a drastic outage because of a bad access control list (ACL) push during a maintenance window. Countless hours are spent troubleshooting the problem. An effective KPI-based dashboard can isolate all the key legs. This includes the baselined DNS traffic and the link traffic dashboards related to the firewall and the relative VoIP network. The link drops are easily noticed before and after the ACL change, thus narrowing the root cause to a set of routers or switches where the ACL was pushed.

Optimizing the SBC

As more and more traffic is routed toward the SBCs, the VoIP SP should introduce traffic-optimizing schemes for the SBC. Two of the common ones are

- Introduction of load balancers (through a SIP Virtual IP [VIP]) to front an SBC server farm. This enables high availability for the SIP traffic along with the flexibility of dynamic growth through the additions of servers in the SBC server farm.

- Introduction of an SIP proxy (cluster) in the network that does the routing to the right SBC VIP based on the geographic location of the termination number.

In most VoIP deployments today, both scenarios are prevalent.

To summarize our strategy so far, we started out with explaining layered KPI tracking methodologies and tied them with detailed call use cases. Further, we looked into a practical BTS 10200 performance analysis example. This was covered through the statistics information available from the BTS 10200. Lastly, we covered some optimization strategies that affect the critical DNS infrastructure of the VoIP network.

The case studies provided here are to give the reader an idea of how to formulate an operations strategy. Some of the other important areas for tracking the services are flow through provisioning, billing, trouble tickets, SNMP statistics, voicemail stats, announcement stats, and Communications Assistance for Law Enforcement Act (CALEA).

A best practice is to periodically collect the KPIs. In most cases, the network devices report the stats in 5-, 15-, 30-, or 60-minute intervals. Also, a proper storage and retrieval mechanism for these statistics needs to be in place. Collectors and correlation software facilitates this activity. We have covered this process at length in Chapter 6.

Performance Analysis from a DOCSIS Network

Monitoring VoIP performance in a DOCSIS network requires collecting and analyzing information from various layers in the network, including the physical layer, the DOCSIS layer, the IP layer, and the application layer.

Correlating this information in a meaningful and easy-to-understand manner can provide valuable information to the voice SP on the health of the VoIP service. This section looks at the different parameters that need to be monitored, whereas the correlation of this information is explained in the section "Tools for Monitoring DOCSIS Networks—VoIP Dashboard," later in this chapter.

VoIP Endpoints

The VoIP endpoints can provide valuable information about the quality of the voice call. This information can be collected and analyzed on a periodic basis to gauge the health of the VoIP service.

For example, in a PacketCable environment, the MGCP/NCS endpoints report call statistics that include voice quality metrics such as delay, jitter, and latency. This information is reported by each endpoint when the call completes and the Softswitch tears down the call leg. Similar information is also available in the Softswitch logs that can provide such data on a per-call basis. The endpoints can provide additional Layer 1 (physical layer), Layer 2 (DOCSIS), and Layer 3 (IP) statistics for the call that the call agent might not be able to provide. This can help with in-depth troubleshooting if the problem is isolated to a specific endpoint and lies between the CMTS and the endpoint.

Figure 7-24 illustrates the information reported by VoIP endpoints in a PacketCable environment.

```
DLCX
IP SRC: 27.178.66.166/2727
IP DST: 72.151.194.22/2427

MGCP
   Command [DLCX] : Delete Connection
   Transaction ID = 50528418
   Endpoint       = aaln/18m0019479a1678.test.net
   Version        = MGCP 1.0 NCS 1.0
Parameters:
   Connection ID      [I] :  1137
   Call ID            [C] :  13db

Server: CMS-1
Timestamp: Mar 28, 2009 16:57:00.550464000 EDT
250

IP SRC: 72.151.194.22/2427
IP DST: 27.178.66.166/2727

MGCP
   Response Code   : [250] -- Connection was deleted.
   Transaction ID = 50528418
   Commentary     = Connection was deleted
Parameters:
   Connection Params   [P] :
      PS: Packets Sent      = 1573,
      OS: Octets Sent       = 251680,
      PR: Packets Received  = 2362,
      OR: Octets Received   = 377920,
      PL: Packets Lost      = 0,
      JI: Jitter            = 0,
      LA: Latency           = 18, PC/R
      PS: Packets Sent      = 2362, PC/R
      OS: Octets Sent       = 377920, PC/R
      PL: Packets Lost      = 0, PC/R
      JI: Jitter            = 0
```

Figure 7-24   *Call Statistics Reported by MGCP/NCS Endpoints*

The information in Figure 7-24 is reported by the endpoint when the call has completed and the connection is deleted by a Delete Connected (DLCX) message sent by the call agent to the endpoint. The endpoint responds with a message with code 250, which indicates to the call agent that the connection has been successfully deleted.

If the endpoint experienced quality issues for this voice call, the information reported previously (packet loss, delay, and jitter) gives details on what could be causing this problem. The VoIP SP can then look at other KPIs in the network (link utilization, CPU, and so on) to find the root cause of this problem.

Similarly, in the case where VoIP is deployed over a public infrastructure, the VoIP endpoints can provide call statistics containing data related to voice quality.

Figure 7-25 illustrates the information reported by an SIP-based VoIP endpoint.

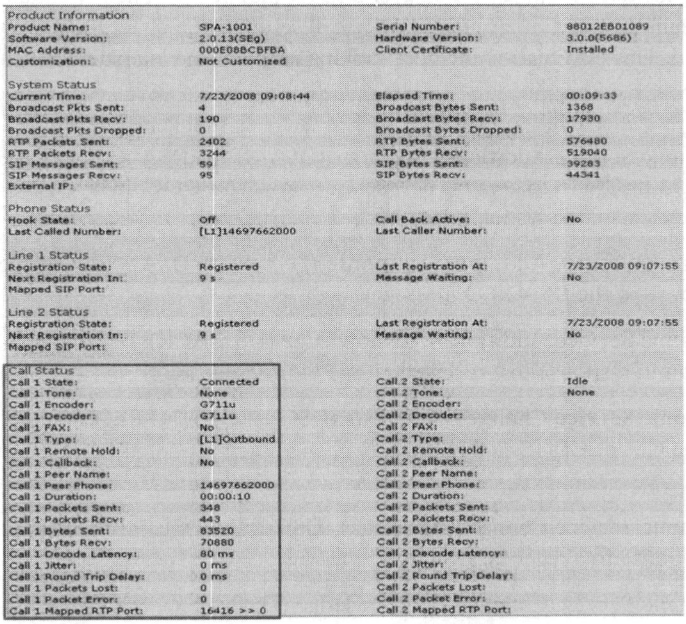

Product Information			
Product Name:	SPA-1001	Serial Number:	88012E801089
Software Version:	2.0.13(SEg)	Hardware Version:	3.0.0(5686)
MAC Address:	000E08BCBFBA	Client Certificate:	Installed
Customization:	Not Customized		
System Status			
Current Time:	7/23/2008 09:08:44	Elapsed Time:	00:09:33
Broadcast Pkts Sent:	4	Broadcast Bytes Sent:	1368
Broadcast Pkts Recv:	190	Broadcast Bytes Recv:	17930
Broadcast Pkts Dropped:	0	Broadcast Bytes Dropped:	0
RTP Packets Sent:	2402	RTP Bytes Sent:	576480
RTP Packets Recv:	3244	RTP Bytes Recv:	519040
SIP Messages Sent:	59	SIP Bytes Sent:	39283
SIP Messages Recv:	95	SIP Bytes Recv:	44341
External IP:			
Phone Status			
Hook State:	Off	Call Back Active:	No
Last Called Number:	[L1]14697662000	Last Caller Number:	
Line 1 Status			
Registration State:	Registered	Last Registration At:	7/23/2008 09:07:55
Next Registration In:	9 s	Message Waiting:	No
Mapped SIP Port:			
Line 2 Status			
Registration State:	Registered	Last Registration At:	7/23/2008 09:07:55
Next Registration In:	9 s	Message Waiting:	No
Mapped SIP Port:			
Call Status			
Call 1 State:	Connected	Call 2 State:	Idle
Call 1 Tone:	None	Call 2 Tone:	None
Call 1 Encoder:	G711u	Call 2 Encoder:	
Call 1 Decoder:	G711u	Call 2 Decoder:	
Call 1 FAX:	No	Call 2 FAX:	
Call 1 Type:	[L1]Outbound	Call 2 Type:	
Call 1 Remote Hold:	No	Call 2 Remote Hold:	
Call 1 Callback:	No	Call 2 Callback:	
Call 1 Peer Name:		Call 2 Peer Name:	
Call 1 Peer Phone:	14697662000	Call 2 Peer Phone:	
Call 1 Duration:	00:00:10	Call 2 Duration:	
Call 1 Packets Sent:	348	Call 2 Packets Sent:	
Call 1 Packets Recv:	443	Call 2 Packets Recv:	
Call 1 Bytes Sent:	83520	Call 2 Bytes Sent:	
Call 1 Bytes Recv:	70880	Call 2 Bytes Recv:	
Call 1 Decode Latency:	80 ms	Call 2 Decode Latency:	
Call 1 Jitter:	0 ms	Call 2 Jitter:	
Call 1 Round Trip Delay:	0 ms	Call 2 Round Trip Delay:	
Call 1 Packets Lost:	0	Call 2 Packets Lost:	
Call 1 Packet Error:	0	Call 2 Packet Error:	
Call 1 Mapped RTP Port:	16416 >> 0	Call 2 Mapped RTP Port:	

Figure 7-25 *Call Statistics Reported by SIP Endpoints*

The information illustrated in Figure 7-25 is reported by the SIP endpoint after a call is completed. It provides details about the call statistics (packets sent, packets received, bytes sent, bytes received, and so on) and key performance data (delay, jitter, and latency) about the call. This information can be polled from the endpoint (through SNMP or by accessing the endpoint through a web browser and displaying call statistics) and can be available on the Softswitch (by looking at the logs).

If the VoIP endpoint experiences voice quality issues, the VoIP SP can look at the call statistics and performance data to figure out whether the issue is caused by network resource overutilization (link utilization, CPU, memory, and so on) in its own infrastructure or whether the problem is related to resources on the public infrastructure (packet loss, delay, and jitter reported by the endpoint), for example, the last-mile connection provided by the Network Access Provider (NAP). If the problem is caused by the public infrastructure, the VoIP SP might not be able to implement a fix because it does not own the infrastructure. Possible solutions can include asking the end customer to upgrade the last-mile connection bandwidth to a higher service level or the VoIP SP signing a service-level agreement (SLA)–based contract with the NAP to provide better quality of service (QoS) to the VoIP SP's traffic.

DOCSIS/DQoS

In a cable network, looking at the Layer 2 (DOCSIS) performance counters can help identify problems that can either impact voice call setup or the quality of the call after it is established. This information includes looking at the following parameters at the DOCSIS layer:

- **Dynamic Service Addition—Requests (DSA-REQ):** DSA-REQ messages are sent by the CM or CMTS to create a new service flow. This DOCSIS layer message is used to create dynamic flows for voice calls.

- **Dynamic Service Addition—Response (DSA-RSP):** DSA-RSP messages must sent in response to the DSA-REQ messages.

- **Dynamic Service Addition—Acknowledge (DSA-ACK):** DSA-ACK messages must be sent in response to a received DSA-RSP message.

- **Dynamic Service Change—Requests (DSC-REQ):** DSC-REQ messages can be sent by the CM or CMTS to dynamically change the parameters of an existing service flow.

- **Dynamic Service Change—Response (DSC-RSP):** DSC-RSP messages must be sent in response to a received DSC-REQ message.

- **Dynamic Service Change—Acknowledge (DSC-ACK):** DSC-ACK messages must be sent in response to a received DSC-RSP message.

- **Dynamic Service Delete—Requests (DSD-REQ):** DSD-REQ messages can be sent by the CM or CMTS to delete a single existing upstream or downstream service flow.

- **Dynamic Service Delete—Response (DSD-RSP):** DSD-RSP messages must be sent in response to a received DSD-REQ message.

- **Dynamic Service Delete—Acknowledge (DSD-ACK):** DSD-ACK messages must be sent in response to a received DSD-RSP message.

As mentioned previously, the DSA, DSC, and DSD messages are exchanged in a three-way handshake between the CMTS and CM or MTA. By tracking these messages, the voice SP can easily identify problematic trends. For example, if the number of DSA-REQ, DSA-RSP, and DSA-ACK messages do not track closely, there might be a resource issue on the CMTS that can result in call setup failures. A DSA failure can also result in a DQoS messaging failure on the CMTS. This can trigger some alarms in network-monitoring stations or probes that are tracking the number of voice call failures in the network.

Similarly, performance data from PacketCable DQoS counters can provide a good view of the health of the VoIP service by tracking the number of successful versus number of failed messages.

Table 7-5 displays the commonly used PacketCable gate messages.

Table 7-5   *PacketCable Gate Messages*

Gate Command	Direction
Gate-Alloc	CMS -> CMTS
Gate-Alloc-Ack	CMTS -> CMS
Gate-Alloc-Err	CMTS -> CMS
Gate-Set	CMS -> CMTS
Gate-Set-Ack	CMTS -> CMS
Gate-Set-Err	CMTS -> CMS
Gate-Info	CMS -> CMTS
Gate-Info-Ack	CMTS -> CMS
Gate-Info-Err	CMTS -> CMS
Gate-Delete	CMS -> CMTS
Gate-Delete-Ack	CMTS -> CMS
Gate-Delete-Err	CMTS -> CMS
Gate-Open	CMTS -> CMS
Gate-Close	CMTS -> CMS

The Gate-Alloc messages are sent by the Gate Controller (GC) to the CMTS at the time of the call setup to request resource allocation for the call. The Gate-Set messages are sent by the GC to the CMTS to initialize or modify the operational parameters of the gates (a gate is a logical entity/resource on the CMTS that is assigned at the time of call setup and is deleted when the call is completed). Similarly, when the GC wants to find out the current parameter settings of a gate, it sends the CMTS a Gate-Info message. When the GC wants to delete the gate on the CMTS, it sends a Gate-Delete message. This is not typical because the gates are normally deleted by the CMTS upon receipt of the DSD-REQ from the MTA or a Gate-Close message from the CMS.

If we look at the example illustrated in Figure 7-26, we can see that by tracking the Gate-Set messages and the number of Gate-Set-Ack and Gate-Set-Err counters, the voice SP can identify problematic trends such as resource allocation issues on the CMTS. These allocation issues can be caused by overutilization or because of certain parameters requested for the call that the CMTS cannot support (for example, invalid or missing parameters or subscriber exceeding his or her gate limit and so on).

Figure 7-26 illustrates the PacketCable Gate-Set information reported by the CMS.

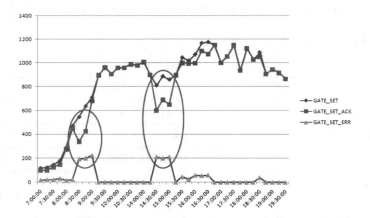

Figure 7-26   *Gate-Set Messages Reported by the CMS*

These failures (Gate-Set-Err) can directly correlate to the number of call setup failures on the CMTS being monitored. The two areas of the graph marked from 08:15:00 to 09:15:00 and from 14:15:00 to 15:15:00 report higher-than-normal Gate-Set-Err messages. During the same intervals, the number of Gate-Set-Acks drops sharply, indicating a problem with call setup or some other anomaly affecting the VoIP service. By looking at the other KPI metrics (DOCSIS layer messaging, CMTS resource utilization—maximum number of gates on the CMTS, subscriber reaching his or her gate limit, and so on), the voice SP can narrow the root cause of this problem.

CPU Impact/Link Utilization

Another important parameter to monitor in the transient network is the CPU utilization on network devices as well as link utilization. If the CPU resources are maxed out, either because of an instantaneous or a sustained event, it can impact the VoIP service. New call setups can fail because the CPU might not be able to process these requests, or existing calls might experience voice quality issues because the CPU might be too busy to process the VoIP packets. As a result, voice packets can get dropped in the network.

Figure 7-27 illustrates a momentary spike in CPU utilization because of a burst of traffic on this device.

Because of this spike, the network device can drop packets. If the packets being dropped include VoIP signaling, it can cause call setup failures.

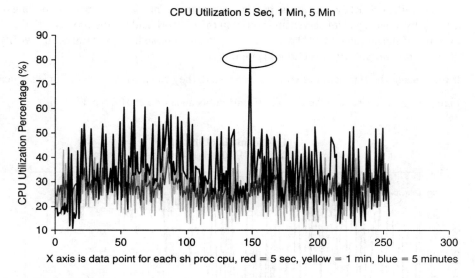

Figure 7-27 *Momentary CPU Spike Caused by a Traffic Burst*

Similarly, if there is a sustained CPU event on the device, it can cause voice quality issues as packets might either get delayed or dropped because of the CPU cycles being maxed out.

Figure 7-28 illustrates a sustained spike in CPU utilization because of a rogue server in the network polling this device.

Figure 7-28 *Sustained CPU Spike Caused by Rouge Server Polling Activity*

Because of this sustained CPU event, the VoIP service can be degraded for an extended period of time. The VoIP SP would need to find the root cause of this sustained event by looking at the other key indicators (traffic rate under the interfaces, increase in polling activity by a management server, and so on) in the network and fix the problem. For example, if the sustained CPU spike is caused by an increase in SNMP traffic, the VoIP SP can rate-limit SNMP traffic to mitigate this issue.

If the links in the transient network are congested, they can also impact the voice service.

Figure 7-29 illustrates a spike in link utilization because of a traffic burst.

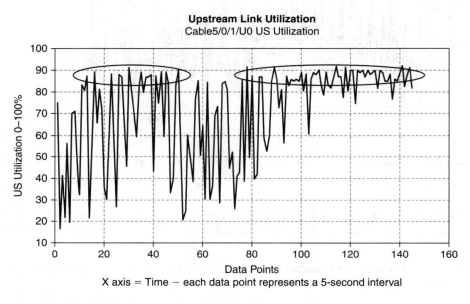

Figure 7-29   *Link Utilization Spike Because of Traffic Burst*

The graph shows two different periods where the upstream link utilization goes beyond 90 percent because of a surge in upstream traffic from the end users. In this situation, the link can be congested and packets (including VoIP signaling data if sent as best effort) can be dropped. If the dropped packets include VoIP traffic, this could adversely affect the VoIP service.

A partial summary of the different parameters that need to be monitored in a DOCSIS VoIP network are listed in Table 7-6. This table is not meant to be an exhaustive list of everything that needs to be monitored but illustrates some of the parameters that can be monitored to track voice quality in a DOCSIS network. Other parameters might need to be monitored based on an SP environment, for example, if the SP uses SIP instead of MGCP/NCS for VoIP call signaling.

As mentioned previously, it is useful to monitor the DSA, DSC, and DSD counters (included in the docsQosDynamicServiceStatsTable table) for CMTS cable interfaces. These counters help identify any call setup failures and resource allocation failures on the

CMTS. Also, the PacketCable DQoS (Gate-Alloc, Gate-Set, and so on) counters should be monitored to track and identify issues that can impact the VoIP service.

Table 7-6 *SNMP Polling Recommendations for a DOCSIS VoIP Network*

Object Name	Object Description	OID
DOCS-IF-MIB		
docsIfSigQCorrecteds	Codewords received on this channel with correctable errors	1.3.6.1.2.1.10.127.1.1.4.1.3
docsIfSigQUncorrectables	Codewords received on this channel with uncorrectable errors	1.3.6.1.2.1.10.127.1.1.4.1.4
docsIfSigQSignalNoise	At the CMTS, describes the average signal/noise ratio of the upstream channel	1.3.6.1.2.1.10.127.1.1.4.1.5
DOCS-QOS-MIB		
docsQosDynamic ServiceStatsTable	Statistics associated with the dynamic service flows for an MTA or similar device	1.3.6.1.2.1.10.127.7.1.6
CISCO-PROCESS-MIB		
cpmCPUTotal5sec	The overall CPU busy percentage in the last 5-second period	1.3.6.1.4.1.9.9.109.1.1.1.1.3
cpmCPUTotal1min	The overall CPU busy percentage in the last 1-minute period	1.3.6.1.4.1.9.9.109.1.1.1.1.4
cpmCPUTotal5min	The overall CPU busy percentage in the last 5-minute period	1.3.6.1.4.1.9.9.109.1.1.1.1.5
CISCO-DOCS-EXT-MIB		
cdxIfUpChannelAvgUtil	The average percentage of upstream channel utilization at the CMTS	1.3.6.1.4.1.9.9.116.1.4.1.1.7

Trace Log Monitoring on Softswitch and Network Devices

Trace log monitoring is critical in tracking the health of the VoIP network. A key set of KPIs need to be extracted from logs and monitored through dashboards. Syslog retrieval and analysis are covered in Chapter 6. This section focuses more on security and other aspects of trace logs such as disk storage.

Analyzing and Correlating Syslog Messages

The messages appearing in the syslogs need to be extracted on a periodic basis. They need to be analyzed for the following key buckets:

- By source vendor
- By functional groups
- By protocols
- By services

Furthermore, the syslogs need to be correlated through an NTP server for time discrepancies. This is to minimize the drift in timestamps (clock). Also, the issues seen are often of different time zones, where the network device can be reporting time in its local time zone, but for the VoIP operator, the event needs to be reported in a consistent nondrift time to keep things in perspective and to make fault isolation possible.

FQDN also plays a key role in syslog analysis and correlation. Reporting of only the IP addresses for VoIP network elements makes the fault isolation aspect very tough. The syslog needs to captures events with an FQDN rather than an IP address. FQDNs are more descriptive about the VoIP network element, thus making it easy to locate and isolate the issue.

The syslogs also derive alarm reporting of some key metrics when certain thresholds are crossed. This topic is covered in Chapter 6.

We describe a simple use case that ties back to the example shown in Figure 7-5. In this graph, a spike can be noticed in the MGCP RSIP messages, which is one of the KPIs being monitored for the particular VoIP switch. The next set of logical steps should be to look into the VoIP switch trace logs. In this particular case, looking into the BTS trace logs, it was noticed that there was a single MTA stuck in a state of constant RSIP, creating an RSIP storm.

However, the fallout of this problem is that the actual Call Agent performance was affected, and the users were starting to report issues. The problem was tied back to reported trouble tickets and an alarm (Alarm = Calls Failing at the CA, MGCP 500 error codes, Device = mga device instance Id).

The following RSIP trace snippet can be used as a sample to create pattern filters in the syslog servers. The RSIP messages can then be further correlated for a particular time duration to determine abnormal frequency and in turn trigger an alarm on an event.

```
. .      MGA   00-00 .     |rsip 479 aaln/2@va-brix-1.hrndevtest.cisco.com MGCP
  1.0 NCS 1.0
|
. .      MGA   00-00 .     |RM: forced
|
. .      MGA   00-00 .     ||snd_rcv.c:260
```

```
5 15:29:36.787 MGA  01-02 -----------|ReceiveThread: Message is a command of
  type=1|mcs_threads.c:268

5 15:29:36.787 MGA  01-02 -----------|Hash Bucket Empty for Recent Response
  (Dest IP [167773859] Port [2427] Trans ID [479] ).|mcs_recent_resp_idx.c:648

5 15:29:36.787 MGA  01-02 -----------|Could not find Recent Response for IP
  [167773859] Port [2427] Trans ID [479]|mcs_recent_resp_idx.c:391

4 15:29:36.787 MGA  01-02 -----------|mga_mgcp_netw_msg: Recvd RSIP for
  ep=aaln/2@va-brix-1.hrndevtest.cisco.com nInstance=2|mga_sig_mgcp.c:194

5 15:29:36.787 MGA  01-02 -----------|tmm_update: handle: 26, counter_id 21,
  operation 0, value 1 succeeded|tmm_api.c:1306

4 15:29:36.787 MGA  01-02 -----------|mga_process_rsip_for_threadalloc:
  ep=aaln/2@va-brix-1.hrndevtest.cisco.com rd=0|mga_term_init.c:3109

3 15:29:36.787 MGA  01-02 -----------|RSIP aggregation
  type=RSIP_AGGREGATION_TERM_LEVEL rm=1 rd=0|mga_term_init.c:3268

3 15:29:36.787 MGA  01-02 -----------|RSIP recvd for term<—:ep=aaln/2@va-brix-
  1.hrndevtest.cisco.com admin=INS oper=0x100 rd=0 rm=1|mga_term_init.c:345
```

Log Files Management

This section discusses an approach for log management for VoIP switches and network devices. Security is a crucial subject for the integrity of the network because the management servers are designed to extract detailed information from all the devices. As such, their access needs to be limited in a way that it is not compromising the security of the network while allowing the authorized users and systems to access the network for monitoring and management. Similarly, the management data, including syslog, performance counters, Call Detail Records, and trace/debug logs, needs to be managed so that they can be used for reactive troubleshooting and proactive monitoring and trend analysis without impacting the devices in the VoIP network.

Security

If you want to configure a unique SNMP community string for traps, but you want to prevent SNMP polling access with this string, the configuration must include an access list. In this example, the community string is named "comaccess," and the access list is numbered 10:

```
snmp-server community comaccess ro 10
snmp-server host 172.20.2.160 comaccess
access-list 10 deny any
```

This example sends the SNMP traps to the host specified by the name myhost.cisco.com. The community string is defined as comaccess:

```
snmp-server enable traps
snmp-server host myhost.cisco.com comaccess snmp
```

Using Infrastructure Access Control Lists (ACL) at the Network Boundary

Although it is often difficult to block traffic transiting your network, it is possible to identify traffic that should never be allowed to traverse your infrastructure devices and block that traffic at the border of your network. Implementing ACLs is a network security best practice and should be considered as a long-term addition to good network security and as a workaround for this specific vulnerability. The ACL example that follows should be included as part of the deployed infrastructure access list that protects all devices with IP addresses in the infrastructure IP address range:

```
! — Permit SNMP (UDP port 161) packets from trusted hosts
! — destined to infrastructure addresses.
!
access-list 150 permit udp TRUSTED_HOSTS MASK INFRASTRUCTURE_ADDRESSES MASK eq 161
!
! — Deny SNMP (UDP port 161) packets from all other sources
! — destined to infrastructure addresses.
!
access-list 150 deny udp any INFRASTRUCTURE_ADDRESSES MASK eq 161
!
! — Permit/deny all other Layer 3 and Layer 4 traffic in
! — accordance with existing security policies and
! — configurations.
!
! — Permit all other traffic to transit the device.
!
access-list 150 permit ip any any
!
! — Apply iACL to interfaces in the ingress direction.
!
interface GigabitEthernet0/0
 ip access-group 150 in
!
```

Note If managed devices are behind a firewall, the firewall must be configured to allow management traffic to pass through. Network management tools such as Unified Operations Manager can have limited support in a network that uses Network Address Translation (NAT). The management server must have IP and SNMP connectivity to the NAT IP addresses for the devices behind the NAT. A single Unified OM server cannot manage duplicate IP addresses across NAT domains. If overlapping IP address ranges exist, a separate Unified OM server must be deployed for each NAT domain.

These are the ports used by the network management tools and can be selectively opened for the specific IP addresses belonging to them.

Table 7-7 lists some common ports used by network management tools.

Table 7-7   *Network Management Server Port Utilization Requirements*

Protocol	Port	Service
UDP	161	SNMP polling
UDP	162	SNMP traps
TCP	80	HTTP (not preferred)
TCP	443	HTTPS (preferred)
TCP	1741	Tool's HTTP server for GUI
UDP	514	Syslog
TCP	8080	Determining status of Softswitch web service
TCP	8443	SSL port

Storage Location (Local Versus Remote) and Archiving Logs

It is recommended to offload the log files and CDRs from the call-processing servers to a remote storage. The primary responsibility of Softswitches is to maintain the endpoints and process calls. As a result, the Softswitch constantly generates trace files and call records. To manage the local disk storage, file offload should be scheduled during off-peak hours to a remote location. If the log files are kept on a local server, they will deplete the disk space and will impact the system performance, possibly to an extent where it might crash. Usually, the call-processing application manages the disk space by recycling the files. This might cause the CDRs to be lost if they are not periodically offloaded from the Softswitch. Logs files should not be parsed on the local servers unless it is imperative to troubleshoot a problem in real time. Also, CDRs should not be parsed on the local servers. Instead, after they are offloaded to a storage server, they can be analyzed to generate reports.

Log files and CDR files are often quite large, and their transfer using File Transfer Protocol (FTP) or Secure FTP (SFTP) might fail because of timeout or another issue. The network management system must track the proper file transition by subscribing to appropriate alarms or traps such as CiscoDRFSftpFailure to track disaster recovery file transfer from the Unified CM to a remote server. The disaster recovery file will be quite large in size because it contains the complete configuration of the call-processing system, OS settings, and any other settings to restore a server from the ground up. Therefore, it is important to ensure that this file gets created successfully and archived properly to ensure its availability when the situation demands disaster recovery. It is recommended that the network administrator perform a periodic audit to ensure the integrity of the remote storage and the link utilization during the bulk file transfers.

There are often compliance requirements for storing the performance data and call records in addition to troubleshooting and trend analysis for capacity planning. SPs often have to comply with the Communications Assistance for Law Enforcement Act (CALEA) in the United States, where the law enforcement agencies request Call Detail Records when needed as part of investigation. The health-care industry requires compliance with certain regulations such as the Health Insurance Portability and Accountability Act (HIPAA). Similarly, in the financial industry, the Securities and Exchange Commission (SEC), also requires communication data archival for certain periods requiring trade confirmations over the phone. The SP industry generally operates in a regulated environment. As such, it has compliance requirements that include SLA conformance and require data safekeeping for certain time periods. CALEA, HIPAA, and the SEC are related to United States. Other countries might have similar regulatory compliance requirements requiring CDRs and performance data archiving.

Tools and Scripts

This section discusses some of the tools that can be used to proactively monitor the VoIP network and provide the network operator the ability to track and identify problematic areas.

Tools for Monitoring an Enterprise VoIP Network

This section discusses some of the applicable Cisco offerings for enterprise VoIP network management tools. The details about SNMP MIB sets and performance counters, as discussed earlier in this chapter, can be used with network management tools from any vendor if it offers standards-based collection methods such as SNMP, SOAP, HTTPS, and FTP.

Cisco Unified Operations Manager (CUOM)

The CUOM provides a unified view of the entire VoIP network infrastructure and the call control devices in the Unified Communications framework, including the following:

- Cisco Unified Communications Manager (Unified CM)
- Cisco Unified IP Phones
- Cisco Unified Communications Manager Express (Unified CME)
- Cisco voice gateways
- Router and switches
- Cisco Unity and Unity Connection
- Cisco Unity Express

- Cisco Unified Contact Center Enterprise (Unified CCE), Unified Contact Center Express (Unified CCX), and Unified Customer Voice Portal (Unified CVP)

- Cisco Unified Presence

- Cisco Emergency Responder

- Cisco Unified MeetingPlace (Unified MP) and Unified MeetingPlace Express (Unified MPE)

The CUOM consolidates the current operational status of each element of the Cisco Unified Communications network to provide diagnostic capabilities for faster problem isolation and resolution.

The CUOM uses SNMP to monitor the operational status of managed devices and supports three versions: authentication, encryption, and message integrity. SNMP v3 can be used if security is desired for traffic management. The CUOM needs only SNMP read access to collect network device information. In addition to SNMP, the CUOM interfaces with other devices in the network in the following ways:

- AVVID XML layer (AXL) to manage the Unified CM. AXL is implemented as a Simple Object Access Protocol (SOAP) over HTTPS web service

- Skinny Client Control Protocol (SCCP) and Session Initiation Protocol (SIP) to Cisco Unified IP Phones for synthetic tests

- Internet Control Message Protocol (ICMP) or Ping Sweep for IP Phones, Cisco IOS routers and switches, and other voice and nonvoice devices

- Windows Management Instrumentation (WMI) for Windows-based PCs and servers

The Unified OM polls the managed devices for operational status information at every predefined interval and can be subscribed to receive traps and alarms from these devices. The amount of management traffic varies if the managed devices are in a monitored or partially monitored state and if any synthetic tests are performed.

The CUOM is also capable of generating synthetic calls and performing tests in batches to proactively test VoIP endpoints to isolate a problem or simulate scenarios to diagnose possible fault sources in the network.

The following diagram displays a real-time autorefresh dashboard that provides status information about all the Unified Communications clusters and the elements of the clusters in the given network. This dashboard, containing the service level view, is designed so that it can be set up and left running, providing an ongoing monitoring tool that signals you when something needs attention. When a fault occurs in the network, the Operations Manager generates an event or events that are rolled up into an alert. If the alert occurs on an element, it is immediately shown on the Service Level View, as depicted in Figure 7-30.

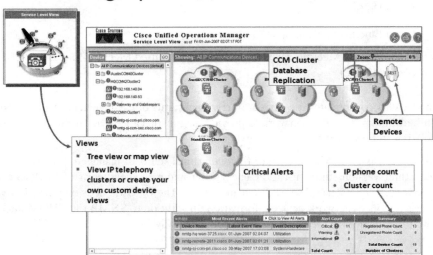

Figure 7-30   *Unified OM Dashboard with Service Level View*

Cisco Unified Service Manager

Cisco Unified Service Monitor (CUSM) monitors and supplements the CUOM by focusing on the voice quality of calls by analyzing CDRs and call diagnostic records extracted from the CUCM and the Cisco 1040 Sensor to monitor and gather voice quality statistics on real calls rather than simulated calls in the network. The CUSM compares the collected voice quality statistics against a predefined Mean Opinion Score (MOS) threshold. If the voice quality falls below the threshold, the Unified SM sends an SNMP trap message to the CUOM to indicate that a potential issue has been identified. This early detection of call degradation enables network administrators to take proactive steps to improve the call quality.

The Cisco 1040 Sensor is a hardware device that predicts a subjective quality rating that an average listener might experience on VoIP calls. It can measure various quality impairment metrics that are included in the IP header of RTP streams, such as packet loss, delay, jitter, and concealment ratio. This computed quality rating is converted to an MOS value equivalent such as a K-factor. The MOS value is included in syslog messages that are sent to the CUSM every 60 seconds; thus the Cisco 1040 Sensor monitors the voice quality almost on a real-time basis.

The Cisco 1040 Sensor has two Fast Ethernet interfaces, one of which is used to manage the sensor itself and the other that is connected to the Switch Port Analyzer (SPAN) port on the Cisco Catalyst switch to monitor the actual RTP streams. A Cisco 1040 Sensor can monitor 100 simultaneous RTP streams. If there are more RTP streams than the Cisco 1040 Sensor can handle, the Cisco 1040 Sensor will perform a sampling of those RTP streams, and the resulting MOS value will be diluted. A pair of Cisco 1040 Sensors must be deployed at both sides of the WAN cloud to monitor the voice quality of calls across the WAN, as illustrated in Figure 7-31.

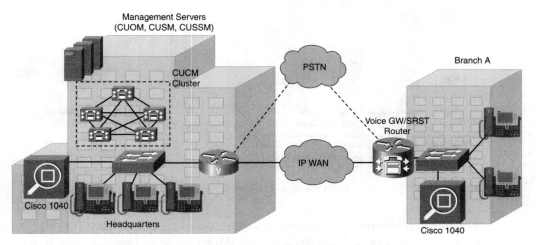

Figure 7-31 *Unified SM and Cisco 1040 Deployment*

The Cisco 1040 Sensor is more likely to experience degradation of voice quality because of packet loss, delay, and jitter inherent in the WAN. Therefore, the RTP streams (from campus to branch) should be monitored by the Cisco 1040 Sensor at the branch site. By the same token, the sensor in the central site should monitor the incoming RTP streams in that segment across the WAN. These RTP streams provide important voice quality statistics, and their associated K-factor values as recorded by the probes (MOS score equivalent) should be analyzed on a weekly basis. The analysis includes sorting out calls with a K-factor below 3.5 and correlating them to specific network segments, as explained earlier in the section "Complex Correlations."

Cisco Unified Service Statistics Manager

Cisco Unified Service Statistics Manager (Unified SSM) can integrate with only one Unified OM but can integrate with multiple Unified SMs. The Unified SSM extracts call statistics data from the Unified OM and Unified SM databases. The data-extraction process is performed by the Unified SSM agent. This data is used to perform advanced call statistics analysis and generates reports for executives, operations, and capacity-planning functions. The Unified SSM provides both out-of-the-box reports and customizable reports that provide visibility into key metrics such as call volume, service availability, call quality, resource utilization, and capacity across the Cisco Unified Communications system. The Unified SSM then stores the extracted data in its own SQL database. Figure 7-32 illustrates a sample report showing weekly call volume by call type.

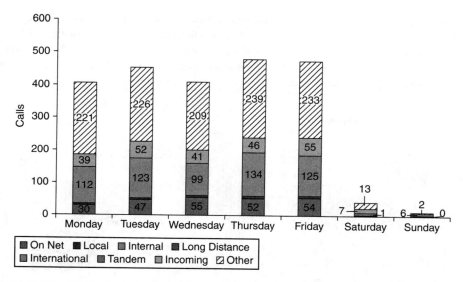

Figure 7-32 *Unified SSM-Generated Weekly Call Volume Report by Call Type*

There are two different data collection approaches within the Unified SSM. The first approach is called a *raw data collection*. With this approach, the Unified SSM instructs the Unified SSM agent to retrieve all call statistics data directly from the Unified OM and Unified SM databases. All retrieved data is then saved in the Unified SSM's database for up to 30 days. The advantage of this approach is that it provides the Unified SSM with a comprehensive data source to perform detailed analysis and report generation.

The second approach is called *monitor-based data collection*. With this approach, the Unified SSM instructs the Unified SSM agent to transfer the processed call statistics data only. The advantage of this approach is fewer traffic loads over the network, and the processed data can be stored in the Unified SSM database for up to three months. To process the original call statistics data in the Unified OM and Unified SM databases, a specific monitor instance must be created in the Unified SSM administration console, and that monitor instance must be associated with the appropriate Unified SSM agent. The monitor instance extracts only the data based on predefined attributes. For example, for Call Volume Monitor, the attributes include number of completed calls On-net, number of failed calls On-net, average duration per call On-net, and so on. Each monitor instance has a unique list of predefined attributes. The monitor instance then polls and extracts the data every 15 minutes, and the Unified SSM agent aggregates the processed data from its associated monitor instance(s) and sends it to the Unified SSM every 30 minutes.

If the Unified OM and Unified SM are deployed separately from the Unified SSM, the Unified SSM agent must be installed on every instance of the Unified OM and Unified SM. The executable installation file of the Unified SSM agent can be downloaded from the Unified SSM Web administration page and installed locally on the Unified OM and

Unified SM. With Unified SSM agents distributed on the Unified OM and Unified SM, the Unified SSM is able to control and manage the data-extraction process. The Unified SSM connects all distributed Unified SSM agents through TCP on port 12124, and Unified SSM agents send call statistics data back to the Unified SSM through TCP on port 12126.

Tools for Monitoring Service Provider VoIP Networks

There are several test probes available in the industry that not only passively monitor the network and report on network conditions but also actively test VoIP traffic paths in the network to identify anomalies that can affect the VoIP service.

Although it is difficult to list all the possible tools available for VoIP monitoring, this book discusses some tools for illustration purposes only. These tools include the Cisco IP SLA feature (refer to the IP SLA case study in Appendix A), IXIA's IxRave solution, Agilent's Network Analysis and Troubleshooting solutions, and so on.

> **Note** The authors do not endorse any of the tools listed in this section. Certain tools/applications have been discussed for illustration purposes only. It is up to the enterprise or SP to test and choose the best-suited tool for its network based on its set of requirements.

IXIA's IxRave Solution

The IxRave software provides a combination of end-to-end, access, and core network testing capabilities to provide network monitoring and fault isolation ability to enterprise as well as SP customers.

The IxRave solution can provide the following services:

- **Fault isolation:** IxRave isolates network segments, allowing each segment to be tested and diagnosed; this can help reduce the mean time to diagnose and repair problems in the network.

- **Turn-up validation:** IxRave can verify that new services have been correctly provisioned and can satisfy SLA requirements.

- **Predeployment network assessment:** IxRave takes measurements to ensure that the network performs correctly as new updates, devices, and services are deployed. This can help minimize disruptions introduced by updates in the network.

- **Ongoing active monitoring:** IxRave utilizes scheduled performance tests to ensure network and customer quality of experience (QoE) and SLA levels. Customer problems and complaints are avoided by proactive testing.

By testing over the actual customer path and performing a sequence of tests, IxRave permits quick isolation of problem sources without an expensive truck roll. Test software can be downloaded to the customer premises endpoint from the voice SP's support website. Other endpoints in the network can be hardware probes, embedded endpoints, or Ixia chassis. An end-to-end test is first run to verify that a problem exists; this is supplemented by tests for the local loop, backhaul, and core segments. After a network baseline is established, subsequent tests are run to compare and identify problems in the network. IxRave monitors network thresholds and alerts the management staff when defined thresholds are exceeded; this helps proactively isolate and prevent problems before they become service impacting.

IxRave also integrates active monitoring and IP testing with existing OSS and NMS systems. This limits the number of databases required to store customer or network element information. IxRave can access information from existing databases, avoiding support and synchronization of multiple databases and eliminating the need for new elements in the provisioning chain and for staff training.

IxRave can send SLA and fault information to existing service management systems. SNMP traps generated based on test thresholds can be captured by existing alarm-monitoring systems. IxRave can be integrated with existing passive monitoring solutions and network element alarms. Using received SNMP information, network management systems frequently trigger thousands of alarms, but the carrier's operations staff often does not know what effect these events have on customers' quality of experience. However, certain triggers cause IxRave to automatically run end-to-end tests that validate the customer's QoE. In this way, IxRave can complement existing monitoring solutions with the addition of test and measurement metrics specific to a customer's SLA.

IxRave Case Study—Voice Assurance for Cable Networks

The IxRave system provides one central view of data needed to proactively assure voice quality in cable networks. This is accomplished by various applications testing, gathering, and displaying data in a user-friendly graphical display using good, marginal, or bad indicators.

Here are the details of how this systems works:

- The system proactively tests voice quality to every customer's MTA in the network using network loopback or network continuity.

- During the test call, a voice wave file is sent and received through a loopback within the MTA, and then the received wave file is analyzed using a standard Perceptual Evaluation of Speech Quality (PESQ) algorithm. The user establishes the voice-quality PESQ threshold for the network.

- The use of the customer's telephone is not affected during the execution of these tests, as described in PacketCable specification EC-MGCP-I10-040402 Section 4.3.

- Users can initiate a test call, get a voice-quality PESQ score, obtain SNMP data from network elements, and look at the CDRs for the test call from one web-based interface.

- Testing can be initiated by a web-enabled device in the field to test the MTA for voice quality at time of install or after repair has been completed to baseline the voice quality for future reference.

- The system automatically places results for customer calls and test calls on a map-based application, with drill-down capability to view hubs, nodes, and MTAs. The map can include a layer for the CAD maps of the physical cable plant. The map application provides a graphical relationship of the physical plant and network topology that might be experiencing poor quality.

Figure 7-33 illustrates how the system plots the physical map of the cable plant and the customer status.

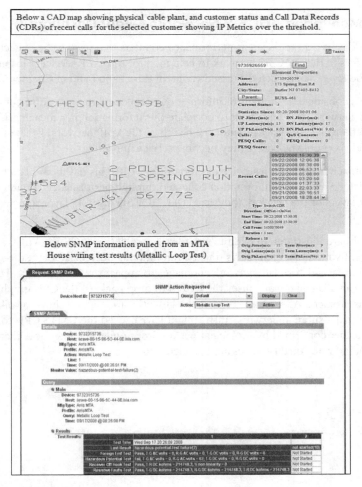

Figure 7-33   *Physical Map of the Cable Plant and Customer Status*

- The system automatically analyzes CDRs from the network Softswitch to determine whether a customer might have experienced poor voice quality on a real call. CDR IP metric results that do not meet thresholds administered by the user are placed on the MAP for visual network health indications based on where poor call quality is occurring. Calls are then made to these customers to verify voice quality by the active test system.

- The system has an SNMP application to gather MIB data from cable modems, CMTSs, and other network devices with color-coded results.

- This system requires one server to be connected to Media Gateways for placing of test calls to customer MTAs that are loopback capable without the need for the deployment of remote probes.

Tools for Monitoring DOCSIS Networks—VoIP Dashboard

We have discussed the different performance counters that need to be collected from various devices in the DOCSIS network to help monitor the VoIP service. This information covers the layered approach discussed earlier in this chapter and in Chapter 6.

Because a considerable amount of information is collected to cover different aspects of the VoIP network, it is important to correlate this information in a meaningful way. This can help the voice SP easily monitor the health of the VoIP network, isolate problematic areas, and fix problems before they become service impacting.

One approach is to create a service dashboard for VoIP. This dashboard can provide a consolidated view of the information collected through performance counters and cover different layers of the network (physical layer, DOCSIS layer, IP layer, application layer, and so on).

The dashboard illustrated in Figure 7-34 gives a complete view of the VoIP network performance. By monitoring the KPIs tracked through this VoIP dashboard, the voice SP can easily identify and correlate network anomalies to impact the VoIP service. For example, if there is a CPU spike that causes the CMTS to drop packets, the CPU graph can pinpoint this problem. At the same time, if the link utilization graph shows a spike, it can be assumed that a surge in network traffic might have caused the CMTS CPU to spike. The can be further confirmed by looking at packet drops under the interfaces that customers experiencing voice quality issues are connected to.

Figure 7-34 summarizes the different VoIP signaling flows in SP environments.

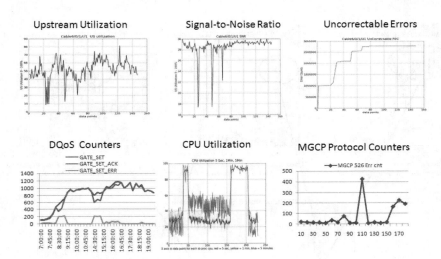

Figure 7-34   *VoIP Dashboard for DOCSIS Networks*

The voice SP can then drill down into specific areas to find the root cause of the problem. This approach can help reduce network problem resolution time, improve productivity of operational staff, and increase network availability by quickly narrowing the scope of the problem.

In Figure 7-35, the drill-down view clearly highlights the processes that are contributing to the spikes in CPU utilization. For example, the two CPU spikes in the Overall CPU Utilization graph between data points 40 and 50 and between data points 150 and 200 are caused by the SNMP process. When the overall CPU utilization jumps close to 100 percent, the SNMP process takes up to 60 percent of the CPU cycles. In this case, the voice SP can analyze the SNMP data being polled from this device and figure out the root cause of the CPU spikes. If there is a rogue server in the network periodically polling this device, the voice SP can either shut down this server or simply block this server by applying an access control list (ACL) on the ingress network interface (uplink) of the device.

Figure 7-35 illustrates a drill-down view of the CPU utilization process.

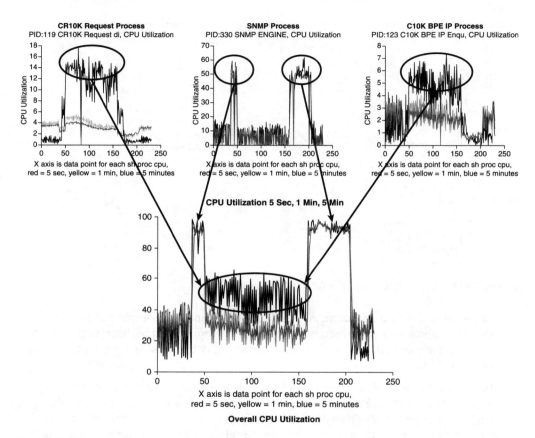

Figure 7-35   *CPU Utilization Drill-Down View*

After a baseline of the network is established, the VoIP dashboard can help identify problems because of changes in the network, such as a software upgrade, new feature deployment, and so on, by looking at changes in data patterns. This dashboard can also be used for trend analysis and capacity planning, which is discussed in detail in Chapter 8.

Tools for Monitoring VoIP Network Health Through Protocols

Earlier we saw the protocols flowing through the network in Figure 7-7. We want to mention a set of potential industry tools that can be used to track the protocol KPIs depicted in Figures 7-2 through 7-7. Figure 7-36 shows the tools by protocol-monitoring categories.

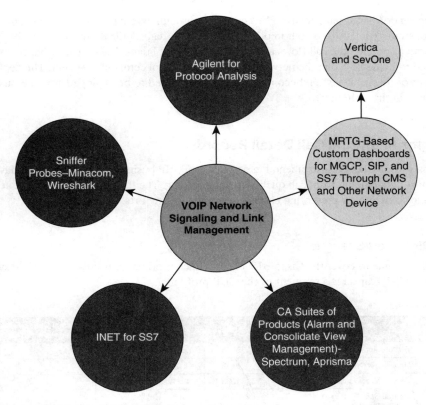

Figure 7-36   *VoIP Signaling Protocol Tools*

The idea behind the tools is to effectively do fault isolation and do performance analysis for various VoIP segments. The key aspect to be highlighted here is the development of custom dashboards. These dashboards can be developed through Multi Router Traffic Grapher (MRTG) graphs or any other graphing tool and can be instrumental in deriving high resiliency on the network by rapid fault isolation. SevOne is a tool that is often seen in the industry to capture and group these MRTG-like dashboards. The dashboards can be fed data collected from the various tools mentioned in Figure 7-36. These include SS7 monitoring and testing (INET), packet capture (Wire shark, Minacom), SNMP Manager (Alarm and Correlation, CA products), Protocol tool (Agilent), and other custom dashboard tools.

The vendor-based data needs to be correlated through external sources, so typically a tool strategy needs to be put in place to collect this data. Agilent products have been found to provide the call flows and individual protocol data. The caveat is that the probes need to be deployed in the network to feed into the collectors like Agilent. The deployment of these probes can become difficult as they need to be deployed at numerous places in the VoIP network.

Tools for Analyzing Call Detail Records

Analysis of CDRs is instrumental in deriving the VoIP business. The CDRs these days report data ranging from call quality to call profiling. The following sections look at some scenarios for presenting effective custom CDR reports.

SP CDR Report Scenario

We continue to cover the Cisco BTS 10200, which is primarily used as an SP VoIP switch. Figure 7-37 captures a segment of the CDR profiling.

Cause Code	Definition	Total Calls	% of Total	Call Completion	Abnormal Call Termination	Subscriber Out of Order
0	Feature Activation/Deactactivation	5038	1.61%	X		
1	Unallocated/Unassigned DN	3152	1.0%	X		
13	Service Was Denied	1563	0.5%	X		
16	Normal Call Clearing	273421	87.48%	X		
17	User Busy	10272	3.29%	X		
18	No User Responding	9	0.0%	X		
19	No Answer, User Alerted	37	0.0%	X		
21	Call Was Rejected	167	0.1%	X		
28	Invalid/Incomplete Number	5250	1.68%	X		
31	Normal, Unspecified	670	0.2%	X		
47	Resource Unavailable	119	0.03%	X		
102	Recovery on Timer Expiry	128	0.04%	X		
3	No Route to the Destination	79	0.02%		X	
34	No Circuit Available	3515	1.12%		X	
111	Network Out of Order	1581	0.5%		X	
41	Temporary Failure	3053	0.98%		X	
42	Switching Equip. Congestion	88	0.03%		X	
127	Interworking, Unspecified	3480	1.11%		X	
22	Terminating Number Was Changed	75	0.02%			X
27	Destination Out of Order	861	0.28%			X

(annotations: 95.93%, 3.77%, 0.3%)

Figure 7-37 *VoIP Call Completion/Failure Characterization for Case Study*

Call Detail Records are generated by the VoIP switch and are primarily used for billing purposes and call type profiling. They also contain call behavior information that helps in diagnosing systematic problems in the VoIP network.

A Call Termination Cause Code is one of the key attributes of CDRs that helps us profile VoIP call traffic health.

Figure 7-37 clearly shows that the call traffic is profiled by cause code. We notice the call cause codes broken down for a switch, and we can also further narrow the root cause of the call anomalies, thus eventually fix the issue.

A CDR sample can be taken in pseudo-real-time frequency. A good sampling frequency is every 15 minutes.

A dashboard can be created to track the call termination cause codes. We describe some custom dashboards in the following section.

Customizing CDR Reporting for Effective Monitoring

Vertica is an example of a tool that can help in providing a correlated and customized view into the CDRs. The data is collected from different VoIP switches, augmented through scripts to produce flat file records, and then periodically uploaded into this tool for reporting.

The tool allows developing a category base grouping of statistics and then provides graphical reports among other types of reports. In Figure 7-38, we see the grouping available for various CDR statistics. Also, we see a breakdown of Call Terminations Cause Codes. In this particular scenario, the cause code represents bearer resources not present versus the number of calls per hour.

Dashboard Views for the VoIP Network

As an example in Figure 7-39, we summarize a collapsed view of some of the dashboards presented so far.

A Service dashboard is presented for an Off-net call with an SIP leg. Figure 7-39 reflects the dashboard.

Following the dashboard strategy in Figure 7-39, the VoIP operator can be on top of the SIP health in the network. The views presented in this figure have been covered in earlier sections but have been presented here as a collapsed service dashboard for providing an SIP view.

Figure 7-38 highlights some of the features and views available through this tool.

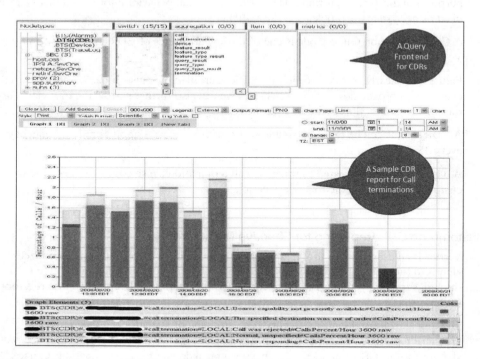

Figure 7-38   *Custom CDR Reporting Tool and Dashboard*

Software Maintenance

Software maintenance is a key component of the overall network management strategy. It is equally important for all types of networks, not just VoIP networks. In this section, software maintenance refers specifically to operating systems that are deployed on network devices such as routers, switches, and so on. It consists of the following major areas:

- **Software release management:** Software release management analyzes currently deployed software for consistency across devices of the same platform and function.

- **Software lifecycle management:** Software lifecycle management analyzes the process for managing the software revisions.

- **Software resiliency:** Software resiliency focuses on currently deployed device operating software and on the processes used to identify, select, certify, implement, and operate software in the network. It provides key metrics on how well release management and lifecycle management processes are performing.

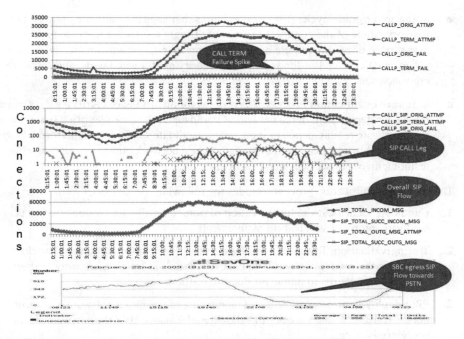

Figure 7-39 *Custom Service Dashboards for an Off-net SIP Call Flow*

Software Release Management

Software release management focuses on the consistent deployment of approved versions of software in the network. Typically, the SP has a software quality assurance (Q/A) team that performs software qualification testing according to defined specifications (requirements defined by the SP's engineering and operations teams) before the software is approved for deployment in the production network.

Key performance indicators for software release management are the number of software versions deployed consistently among various platform configurations and the overall health regarding deferred software (not suitable for deployment because of defects found in the software) deployed in the network.

Software Lifecycle Management

Software lifecycle management focuses on the processes used to plan, design, implement, and operate software in the network.

This is a documented process that determines how software is managed within the network and should include the following components:

■ **Software version control:** A strategy for deploying a common software image across identical devices in the network based on software feature or hardware requirements. The goal of the exercise is to optimize software consistency and network availability, and to reduce operating costs.

■ **Software upgrade cycle:** A set of conditions that trigger a software upgrade in the network. Some of the conditions that might trigger a software upgrade include a new service or feature requirement, new hardware requirement, service-impacting software defects, end-of-life software, a security alert–related issue, and software consolidation.

■ **Software qualification process:** The process for qualifying or certifying a software release for deployment in the network. Candidate releases are identified based on feature/hardware requirements. In some cases, candidate selections are performed twice or more if showstoppers (issues that can cause a network outage or a device reboot resulting in loss of service) are identified.

The software qualification process includes the following five steps:

1. **Candidate identification:** Identification of potential software versions that meet hardware, protocol, and feature requirements.

2. **Testing/validation:** Testing of candidate software in a lab environment that closely resembles the production network, using such tools as traffic generators, sniffers, and WAN simulators to mimic production traffic.

3. **Pilot deployment:** Deployment and monitoring of the target software in a small noncritical area of the network to validate performance after the software has been successfully tested.

4. **Limited deployment:** Deployment of target software to other smaller areas of the production network using the normal deployment process, after completing a successful pilot.

5. **Full-scale deployment:** Mass deployment of the target software using the normal upgrade process after being validated through the limited deployment phase.

Table 7-8 provides an example of software version control.

Table 7-8 *Software Version Control*

Item	Description	Hardware Platforms	Features	IOS Version	Qualification Status
1	Core routers	CRS-1, GSR	QoS, BGP	3.6.1 (XR)	Under test
2	Core/distribution switches	6500	Sup 720	12.2(18)SRC3	Qualified 4/1/09
3	Access switches	2960	DHCP snooping	12.4.14-T	Qualified 12/1/08
4	Aggregation/ edge routers	7300, 7206VXR	OSPF, MPLS	12.4(10)T	Qualified 11/15/08
5	Broadband aggregator	uBR10K	uRPF, modem steering	12.2 (33)SCB	Under test

Periodic review of security alerts, vendor software maintenance status, upgrade cycle, and the certification process should be performed to ensure that the goals and requirements of the organization are being met. These reviews should include the number of software upgrades during a given period, the different versions of software trains deployed, the number of software-related network problems, and so on.

Software Resiliency

Software resiliency focuses on currently deployed device-operating software and on the processes used to identify, select, certify, implement, and operate software in the network. Key performance indicators for network resiliency can be used for quality initiatives or as a measure of success. Key performance indicators for software resiliency include the following:

- **Number of software versions deployed in the network:** This is an indicator of how well the software management process is working. It can help identify a breakdown in software management process and software revision controls if there are devices in the network running deferred or unapproved software versions.

- **Percentage of devices conforming to the approved software version:** An indicator of software consistency in the network. Having software consistency in the network is important because it can help prevent network outages and degradation or loss of service if only approved software is deployed.

- **Amount of production downtime or availability impact because of software issues or quality:** An indicator of business impact because of software instability.

- **Amount of proactive and reactive support time required to maintain software versions:** An indicator of software management support requirements. Useful for budgeting software support costs.

Periodic Auditing of a VoIP Network

A periodic audit of the network carrying media traffic and call-processing entities helps with the proactive identification of systemic or component-level issues, thereby improving availability, reliability, and Mean time to Recovery (MTTR) for VoIP applications. An audit also helps establish a baseline that aids in determining capacity requirements, establishing monitoring thresholds, and in some instances, identifying opportunities to grow the business by identifying network resource utilization levels. The example in Figure 7-40 shows the call volume during a regular business day as collected by a CDR processing engine and normalized for the entire month. The CDR processing engine is part of an auditing tool such as CUSSM, UCAT, and so on. This can serve as a reference and provide context to the performance data gathered from the call control/processing devices such as CPU loads, memory utilization, and disk I/O.

Figure 7-40 illustrates the call volume during a regular business day.

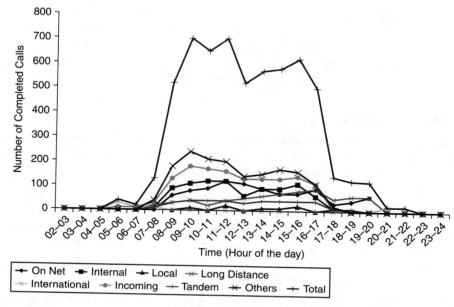

Figure 7-40   *Call Volume During a Business Day*

In large networks, it is a continuous process, where parts of the network are audited on a monthly, quarterly, or biannual basis. It is recommended to have more frequent audits done on the network and the call control elements especially if their configurations are constantly in flux by virtue of move, add, changes, and deletion (MACD) of endpoint or network resources.

Part of the audit includes network discovery to create an inventory of all the devices in it. In this process, the audit tool finds software and hardware version information along with configuration and performance data. This information provides guidance and a road map for maintaining the network devices by deploying stable software releases and proactively catching any issues that are not reported by the vendors of the network equipment. Cisco offers auditing tools through Cisco Advanced Services and through its partners. One such tool is the Unified Communications Audit Tool (UCAT). The example in Figure 7-41 shows details of network elements, their software or firmware versions, and their risk exposure. It also identifies hardware that reaches its end-of-life or end-of-support cycle as announced by Cisco. This information lets the network operator plan an upgrade strategy in advance.

This information can be collected by an audit tool such as Cisco Network Collector (CNC), Network Profiler (NP), or UCAT that can inventory the devices in the SP or enterprise network with hardware and software details.

In Figure 7-41, UCAT inventories all the call-processing devices in the network including Unified CM and Cisco Unity, including their respective versions and bug exposure. It also captures a number of IP phones currently deployed and their respective boot firmware images. This tool further filters the bug list by component. In this example, it looks like the CP-7970 and CP-7971 phones that are being deployed are exposed to three bugs, where two of them are not applicable because in this network the phones are controlled using SCCP instead of SIP. However, they are exposed to one of the bugs, which is CSCef68573. This bug identifies a problem where the 7970 phone marks the DSCP value in the frame as 0. The default value should be set as 0x68 so that the signaling packets originating from these IP phones can be assigned to proper queues in the QoS scheme as configured in the network. Instead, they are placed in the queues that are going to receive "best-effort" treatment, which leaves these signaling packets vulnerable to excessive delay or drop.

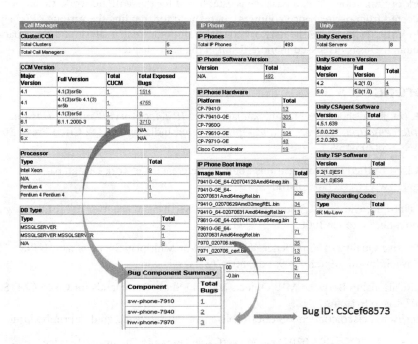

Figure 7-41 *Software Inventory Audit and Stability Risk Analysis*

The information received during the audit of a pilot deployment enabled the network administrator to proactively implement a workaround by configuring the switch port to not trust the phone and rewrite the DSCP value correctly. Also, the next release of the Unified CM that fixes this issue can be planned accordingly. However, in the absence of this discovery, the CP-7970/71 IP phone users could possibly have experienced erratic phone behavior as a result of signaling protocol transmission degradation, as discussed in Chapter 6.

Summary

Proactive monitoring of the VoIP network using the KPIs mentioned in this chapter can help enterprise and voice SPs better manage and operate their network and provide high-quality and reliable service to their customers.

Collection, categorization, and correlation of performance counter data from various devices is a good way of gauging the health of the VoIP network and the service. This approach is also an effective way for problem isolation and proactive identification of problematic areas in the network to avoid impact to the VoIP service.

The data collected from the VoIP network can also be used for trend analysis and network optimization. After a baseline is established for the VoIP network, the information collected from performance counters can be used for trending the various KPIs. This can help VoIP SPs optimize the network for voice through trend analysis and for capacity planning. This is discussed in more detail in the next chapter.

References

1. Kaza, Ramesh and Asadullah, Salman. *Cisco IP Telephony: Planning, Design, Implementation, Operation, and Optimization.* Indianapolis, IN: Cisco Press; 2005. ISBN: 1-58705-157-4.

2. Cisco Unified Communications Solution Reference Network Design 6.0, http://www.cisco.com/go/srnd.

3. Dantu, R., Ghosal, D., and Schulzrinne, H. "Securing Voice over IP," IEEE Network. September/October 2006.

4. "Active Quality of Experience Verification for Multiplay Services with IxRave," IXIA IxRave Solution, March 2009, Ixia, http://www.ixia.com.

5. "Relationship Between MIB Objects and CLI Show Commands for Cisco CMTS," 2007, Cisco Systems, http://www.cisco.com/en/US/docs/cable/cmts/mib/reference/guide/ubrmibb.html.

6. "PKT-SP-DQOS-I03-020116—PacketCable Dynamic Quality-of-Service Specification," January 2000, CableLabs, http://www.cablelabs.com/packetcable/downloads/specs/PKT-SP-DQOS-I12-I03-020116.pdf.

7. "PKT-SP-DQOS-I12-050812—PacketCable Dynamic Quality-of-Service Specification," August 12, 2005, CableLabs, http://www.cablelabs.com/packetcable/downloads/specs/PKT-SP-DQOS-I12-I05-050812.pdf.

8. "Understanding Call Detail Records," January 2009, Cisco http://www.cisco.com/en/US/docs/voice_ip_comm/cucmbe/service/6_0_1/car/carcdrdef.html#wp1061813.

9. Madani, Habib, BRKNMS-2831, "Voice Health Management for Service Providers," Cisco Networkers 2007 at Anaheim California, June 2007, Cisco.

10. "Operations and Maintenance Guide," Cisco BTS 10200 Softswitch Release 5.0, February 2010, Cisco, http://www.cisco.com/en/US/partner/docs/voice_ip_comm/bts/5.0/maintenance/guide/OMGuide50.html.

11. "Cisco Unified Communications SRND," February 2010, Cisco, http://www.cisco.com/en/US/solutions/ns340/ns414/ns742/ns818/landing_uc_mgr.html.

12. "Enterprise Quality of Service SRND," November 2005, Cisco, http://www.cisco.com/en/US/docs/solutions/Enterprise/WAN_and_MAN/QoS_SRND/QoS-SRND-Book.html.

Trend Analysis and Optimization

Being proactive about Voice over IP (VoIP) network management is key to reducing reactive issues. This requires analyzing the VoIP network traffic and calling behavior and resource utilization periodically. To perform trend analysis, the network administrator should not only collect key data over a period of time but should also correlate different parameters. At the same time, looking at the rather large amount of data for different parameters is not pragmatic for daily operations.

An aircraft's cockpit has many dials on its instrument panel. Most of them are for measuring the position of the aircraft (for example, altitude and pitch) and the performance of the plane, including speed and engine thrust. The instrument panel has an array of telemetry equipment to monitor weather patterns and receive navigation information. Figure 8-1 shows the cockpit of NASA's space shuttle. The source of the image shown in Figure 8-1 is the National Aeronautics and Space Administration (NASA) at http://spaceflight.nasa.gov/gallery/images/shuttle/sts-101/html/jsc2000e10522.html.

The pilot constantly keeps an eye on the main panel highlighted by the superimposed square in the picture. This dashboard, in the form of a digital display, shows the most critical information, including altitude, pitch, and speed. This dashboard also includes radar showing telemetric data about weather patterns in the path and the presence of other aircrafts. As needed, any changes can be correlated to the adjacent panels shown by the arrows. For example, at cruising altitude, nautical speed and engine thrust should correlate. Otherwise, tailwind or headwind speed should provide the explanation. Similarly, the radar can show change in the weather pattern or anticipate air turbulence. Before the pilot can make a decision to change course, he or she might need to consult other parameters such as fuel availability to support the change in the most optimal flight path to maintain a smooth flight.

The VoIP network administrator's job is analogous to that of an aircraft pilot. The network administrator is constantly monitoring the key metrics, establishing a pattern, correlating them to draw conclusions about problem root causes as they happen or begin to develop, and proactively planning for network resource optimization or capacity increase before any service-level agreements (SLA) are violated. This chapter discusses how to perform trend analysis through key metrics and using a dashboard as the profiling tool. A concept of network utilization and efficiency is introduced with supporting examples to show how to measure the utilization of the network and call-processing resources to improve their efficiency. This chapter also discusses how to track traffic patterns and other metrics that can be correlated with other data to pinpoint a potential problem and also provide feedback and guidance in optimizing the VoIP design and the underlying transport network.

Figure 8-1   *Space Shuttle Cockpit*

Trend Analysis Through Key Metrics

We have talked about the key metrics and how they can be used to isolate and troubleshoot network-related issues. When using these metrics to perform trend analysis, we have to consider the following aspects:

- The identification of key VoIP metrics
- Effective collection of these metrics
- Reporting, correlation, storage, and visualization of these metrics

The effective trending, when performed on a routine basis, can result in optimal running of the VoIP network. So far in this book, we covered a set of metrics for trending along the lines of

- Call signaling

- Voice bearer traffic

- VoIP features (for example, voicemail, announcement server, or directory service)

- System load metrics (CPU, memory, disk, load, link utilization)

- On-net and Off-net traffic

Each of these categories needs to be tracked for peak usage, average usage, weekday usage, weekend usage, service degradation, and errors for egress and ingress paths. The collection should be at least on a 15-minute basis. Typical performance counter statistics are stored on the network elements on a 15-minute cumulative basis. Because the metrics are cumulative, we do not miss out on a real-time view but rather get a pseudo-real-time picture.

The collected metrics need to be efficiently stored in a database so that they can be data-mined, and then correlated reports can be generated on a real-time basis for these metrics.

The trending effort should enable the network operator to

- Optimize the current network by trending overutilization, underutilization, and effective utilization

- Better predict the network growth

- Identify the busy hour and hot spots (events that cause the usage to spike)

- Identify possible security attacks

- Identify potential network/product anomalies

The following sections highlight examples of using a dashboard and describe how it can be used as a profiling tool.

Dashboard as a Profiling Tool

A collection of key metrics are the building blocks for profiling through the dashboard. Some of the key dashboard characteristics must reflect the following:

- An aggregate view of the VoIP network

- VoIP business impact:

    - Call traffic profiling

    - Small business service impact

- All the VoIP network- and business-related components

- Nested or detailed drill-down view of the network and correlated views

- Changes to be accommodated in the network

- A real-time view of the VoIP network

A service provider (SP) can support many small businesses. Dashboards provide a mechanism for the business customer and the SP to track an agreed-upon SLA.

A VoIP network is an organic entity because components come in and out of service. These state changes need to be actively reflected in the dashboards for them to be dependable profiling tools.

These state changes need to be tracked by a trained eye to identify optimization opportunities and capacity issues. We illustrate the impact of these changes through various scenarios for the SP and enterprise networks and describe how the results can be analyzed.

The dashboards can be built on these key metrics as they are collected from the devices in the VoIP network. They can then be correlated into the aggregated graphical or tabular views on the dashboards to reflect the various components of the VoIP network. An example of such abstracted views is peak call success percentage, which needs to be derived. Similarly, all aggregate views need to be derived.

The following sections build on these concepts and describe the various VoIP network metric categories and how changes can be introduced into the network to perform optimization.

Network Utilization and Efficiency

Optimizing a production network carrying media traffic requires monitoring its resources on a continuous basis. The resources monitored include link utilization, CPU load and memory utilization of its nodes in the critical path, performance of the routing and signaling protocols, endpoint registration status, and resources such as digital signal processors (DSP) and other application servers. Monitoring individual resources provides feedback, which needs to be characterized and correlated so that decisions can be made for performance tuning, capacity planning, new application introduction, and network technology refresh cycle planning. Network resource characterization includes capacity, utilization, and efficiency.

Capacity is defined as the maximum volume of traffic carried by a network. *Utilization* is the ratio of the carried traffic to the network capacity. *Efficiency* is the ratio of the carried traffic to the capacity utilized for carrying that traffic. The following three scenarios regarding network utilization and efficiency are possible:

- **Scenario 1:** Network capacity is used efficiently, *and* there is unused capacity available for carrying additional traffic load.

- **Scenario 2:** Network capacity is being used efficiently, *but* there is no extra capacity left for additional traffic load.

- **Scenario 3:** Network capacity is not used efficiently, *and* there is still unused capacity available for carrying additional traffic load.

The goal of the VoIP network architect or designer should be to design networks that can run efficiently with room for future growth, as mentioned in Scenario 1. The network operator should constantly strive for maximum efficiency and optimize the utilization to its full capacity before adding more resources in the network (Scenario 2). Scenario 3 might require design changes to optimize the network's efficiency.

Safeguarding Network Resources from Potential Security Threats

Converged networks lead to converged threats. A VoIP network is exposed to both data attacks and voice attacks as data and voice are converged. As discussed earlier, media traffic needs to be delivered as quickly and as smoothly as possible with minimum delay, packet loss, and jitter. For this to happen, Real-time Transport Protocol (RTP) packets carrying voice traffic get higher priority in the transport IP network to leverage priority and low-latency queues in an unrestrictive manner. This exposes the transport network to security threats such as distributed denial of service (DDoS) attacks, where the hacker might hijack the priority queues by generating packets with the same profile and priority as the voice traffic.

This can be observed by comparing call completion rate trending data with traffic flows specific to voice-signaling or media queues. If the amount of traffic using the voice-signaling or media queues is increasing without the corresponding change in call attempts or call completion rate, this can indicate a DDoS attack, as shown in Figure 8-2.

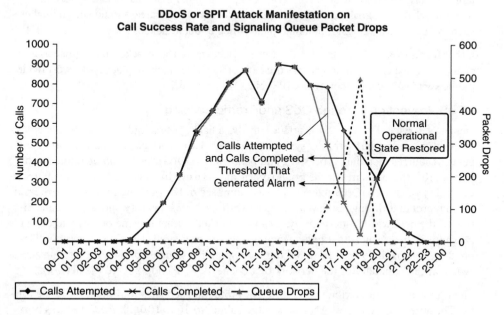

Figure 8-2   *Comparing Traffic Flows and Call Attempts/Completion Rate*

Figure 8-2 shows the pattern of attempted calls and completed calls during a day. Until about 4:00 p.m., the call completion rate drops compared to calls attempted. When the differential between calls attempted and completed reaches a threshold of 5 percent, an alarm is generated. This graph shows a superimposed traffic pattern of the signaling queue on the gateway with Internet access. At the same time (4:00 p.m.), the queue starts to drop packets.

It is apparent that this network—including the links, QoS scheme, and call-processing devices—can handle at least 900 calls. But when under DDoS attacks that flood the priority queues and other queues reserved for signaling traffic, the call completion rate drops, even when the call attempt rate subsides. This is a clear correlation that is further confirmed by deeper analysis of system logs on Cisco Unified Communications Manager (CUCM) that show normal activity and normal CPU utilization. The network administrator deploys security measures to block the access to the voice-signaling queue and selectively restores it by deploying an access control list (ACL) on the compromised gateway. This results in the restoration of normal operation around 7:30 p.m., where the queue drops have stopped and the call completion rate matches the call attempt rate. However, because the call attempt rate is going down because of the end of the business day, it must be monitored for next 48 hours.

Similarly, signaling traffic is also vulnerable to attack, including Spam for Internet Telephony (SPIT). SPIT typically leverages a Session Initiation Protocol (SIP) proxy impersonation to send unsolicited messages in bulk to IP telephony endpoints. VoIP phishing (vishing) involves caller ID spoofing and then call rerouting to dummy IVR systems for further exploitation of the VoIP call processing resources. Because both of these attack VoIP signaling, the traffic pattern for signaling traffic should see anomalies such as queue drops for the signaling queue when under attack, as shown in Figure 8-2.

Other forms of attack can involve theft of service, where the attacker might pose as a genuine service endpoint and access the network services including access to worldwide public switched telephone network (PSTN).

DDoS Prevention Through QoS and Traffic Policing

Denial of service (DoS) or DDoS attacks involve a hacker using multiple PCs by installing virus or worms on the operating system, which then launch attacks on the network, choking its resources including processing and bandwidth. DDoS attacks can be mitigated by taking proactive measures at the boundary of the network by having a conditionally trusted quality of service (QoS) scheme that also imposes traffic policing to reclassify the traffic and reroute it through a scavenger queue with the lowest priority and smallest allocation of the bandwidth. Alternatively, it can be blocked. Either scheme or a hybrid of the two approaches ensures that low-latency queues are protected. The flow diagram in Figure 8-3 explains the overall flow and the corresponding sample configuration for an access switch at the boundary of the network that connects VoIP endpoints to the network.

The following configuration shows different protection schemes to protect the voice traffic. The access list Voice-VLAN restricts the subnet to 10.1.110.0/24. All the access ports are set to trust class of service (CoS) markings only from a Cisco IP phone using the **mls qos trust device cisco-phone** command. In addition to this, the policy-map IPPHONE+PC is set to drop voice traffic exceeding 1 Mbps, which should cover conference calls using wide-band codecs. Also, the data traffic policing in this example is set to remark traffic to dscp 8 (equivalent to CoS 1) if it exceeds 6 Mbps.

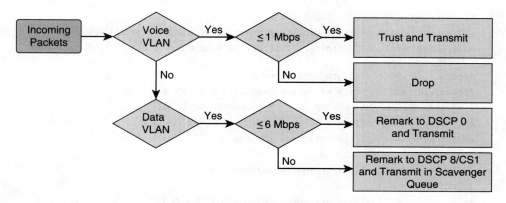

Figure 8-3 *Conditional Trust and Traffic Policing on Access Switches*

```
AccessLayer-SW1(config)# class-map Voice-VLAN
AccessLayer-SW1(config-cmap)# match access-group name VVLAN-ANY
AccessLayer-SW1(config-cmap)# class-map Data-VLAN
AccessLayer-SW1(config-cmap)# match access-group name DVLAN-ANY
AccessLayer-SW1(config-cmap)# exit
AccessLayer-SW1(config)#
AccessLayer-SW1(config)# policy-map IPPHONE+PC
AccessLayer-SW1(config-pmap)# class Vioce-VLAN
AccessLayer-SW1(config-pmap-c)# police 1000000 8192 exceed-action drop        !
  Excess VoIP dropped
AccessLayer-SW1(config-pmap-c)# class Data-VLAN
AccessLayer-SW1(config-pmap-c)# set ip dscp 0
AccessLayer-SW1(config-pmap-c)# police 6000000 8192 exceed-action dscp 8     !
  Excess Data remarked
AccessLayer-SW1(config-pmap-c)# exit
AccessLayer-SW1(config-pmap)# exit
AccessLayer-SW1(config)#
AccessLayer-SW1(config)#
AccessLayer-SW1(config)# interface range FastEthernet0/1 - 48
AccessLayer-SW1(config-if)# switchport access vlan 10
AccessLayer-SW1(config-if)# switchport voice vlan 110
AccessLayer-SW1(config-if)# mls qos trust device cisco-phone     ! Conditional
  trust
AccessLayer-SW1(config-if)# mls qos trust cos                    ! Trust CoS from IP
  Phone
AccessLayer-SW1(config-if)# service-policy input IPPHONE+PC           ! Policing
  policy
AccessLayer-SW1(config-if)# exit
AccessLayer-SW1(config)#
AccessLayer-SW1(config)# ip access-list standard Voice-VLAN
AccessLayer-SW1(config-std-nacl)# permit 10.1.110.0 0.0.0.255        ! Voice-VLAN
  subnet
```

```
AccessLayer-SW1(config-std-nacl)#
AccessLayer-SW1(config-std-nacl)# ip access-list standard Data-VLAN
AccessLayer-SW1(config-std-nacl)# permit 10.1.10.0 0.0.0.255        ! Data-VLAN
  subnet
AccessLayer-SW1(config-std-nacl)# end
AccessLayer-SW1(config)# wrr-queue bandwidth 5 25 70 0
! Q1 gets 5% BW, Q2 gets 25% BW, Q3 gets 70%, Q4 is PQ
AccessLayer-SW1(config)#
AccessLayer-SW1(config)# wrr-queue cos-map 1 1
! Maps Scavenger/Bulk to Q1
AccessLayer-SW1(config)# wrr-queue cos-map 2 0
! Maps Best Effort to Q2
AccessLayer-SW1(config)# wrr-queue cos-map 3 2 3 4 6 7
! Maps CoS 2,3,4,6,7 to Q3
AccessLayer-SW1(config)# wrr-queue cos-map 4 5
! Maps VoIP to Q4 (PQ)
AccessLayer-SW1(config)#
```

Because DDoS attacks often exploit the operating system vulnerabilities, it is recommended to install virus scanners and adware blockers and keep the OS up to date by applying vendor-recommended patches.

Securing Signaling Protocols

Signaling protocols can be secured by using Secure SIP (defined in RFC 3261) and using Transport Layer Security (TLS) to prevent proxy impersonation, signaling attacks, SPIT, call hijacking, and theft of service. Authentication of endpoints and IP trunks further bolsters the signaling security. TLS can use certificates for manual authentication and SIP digest.

Encryption of signaling protocols further bolsters the security in addition to authentication, which provides a means for access control and identity confirmation. For example, SIP messages can reveal sensitive information if intercepted, which can include both parties' SIP Uniform Resource Identifiers (URI), IP addresses, and port numbers associated with media, allowing eavesdropping. SIP messages can also contain users' geographic information and current activity level that can lead to vishing and SPIT expeditions by hackers. TLS can offer point-to-point encryption, but Secure Multipurpose Internet Mail Extensions (S/MIME) (as defined by RFC 2633) allow end-to-end encryption. However, these security measures do have call-processing overhead attached to them in terms of bandwidth consumption and CPU utilization for encryption, especially when we have economies of scale such as at call agents.

Firewalls and Address Hiding

VoIP firewall traversal also provides additional security, although it can cause scalability concerns if firewall is in the critical path of major egress points for VoIP traffic. A Cisco Firewall offers special features for the stateful inspection of VoIP signaling protocols using the **ip inspect <SIP/H323/MGCP>** command. Network Address Translation (NAT), Simple Traversal of UDP through NAT (STUN), or Traversal of UDP through Relay NAT

(TURN) under Interactive Connectivity Establishment (ICE) framework hides the private enterprise network and critical service nodes or servers in the service provider network by making them more secure and accessible at the same time. The Session Border Controller (SBC) or the border element (BE) can provide another level of network address masking while performing other tasks such as call filtering, call normalization (codec interoperability and fast-start and slow-start call setup methods), and bandwidth management. SBE/BE simplifies firewall interoperability by allowing only one conduit to be opened for access to the aggregation point, that is, the SBC or the BE. This is illustrated in Figure 8-4.

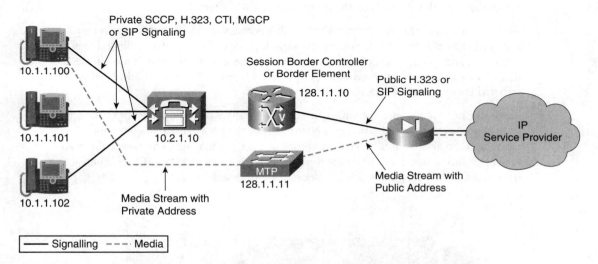

Figure 8-4   *Using a Firewall with NAT to Hide IP Address Space*

In Figure 8-4, the Cisco Unified Communications Manager (CUCM) and all the endpoints, including IP phones and gateways, have private IP addresses. The SBC/BE and the media termination point (MTP) have public addresses. MTP and SBC functionalities can be offered in one physical device. Also, there can be several MTPs and/or SBCs for redundancy and scalability. All the media from the service provider's network are sent through the MTP. There is no direct connectivity between the phones, Unified CM, and the outside world.

Theft of service, involving using network resources to place long-distance calls that incur high toll charges or exploiting the resources by intertrunk transfers, is covered later in this chapter.

Dashboard for Trunk Resources Utilization

A VoIP service provider's PSTN bearer trunk utilization is growing rapidly with the fast growth of its customer base. A dashboard to efficiently monitor the Trunk Carrier Identification Codes (CIC) is important. As an example, an SP network that deploys more than 16 VoIP switches (BTS 10200) is on target to increase that number to 40 by the end of 2010.

Each of these switches typically manages and controls on average three to four Trunking Gateways, which in turn terminates 100+ trunks from the telco carrier. These trunks are T1 or E1 trunks that consist of 24 or 32 DS0s/CICs, respectively. Each CIC represents a physical circuit or a voice channel between the two switches, and a collection of CICs in turn represents the voice capacity. Hence, monitoring of the state of these voice channels is highly critical. If these CICs are not available to carry voice traffic, the service provider's voice network is critically impaired as Off-net or PSTN calls do not go through. Again, this depends on the percentage of CICs affected.

Another important reason to monitor these CICs is because the SP routes certain categories of Off-net calls, such as emergency, long-distance calling, toll-free calling directory assistance, and so on, to a specific sets of trunks. If these particular trunks are affected, that entire category of service is down, unless the desired traffic is route-advanced or rerouted through a redundant path.

So far, the need to monitor the CICs is described. Ideally a periodic snapshot of the CICs in an abnormal state would give a dashboard view of the health of an SP's network. This dashboard can also be used to keep inventory of the CICs that are to be turned up or are in maintenance mode. Refer to Figure 8-5 for a command-line interface (CLI) output of the trunk CIC information, which can derive the dashboard.

```
1018  92   ADMIN_INS      TERM_DOWN               LBLK   IDLE   TERM_FAULT DES_LOC_F
1018  93   ADMIN_INS      TERM_DOWN               LBLK   IDLE   TERM_FAULT DES_LOC_F
1018  94   ADMIN_INS      TERM_DOWN               LBLK   IDLE   TERM_FAULT DES_LOC_F
1018  95   ADMIN_INS      TERM_DOWN               LBLK   IDLE   TERM_FAULT DES_LOC_F
1018  96   ADMIN_INS      TERM_DOWN               LBLK   IDLE   TERM_FAULT DES_LOC_F
1021  96   ADMIN_INS      TERM_ACTIVE_CTRANS_BUSY ACTV   TRNS_OBSY  NON_FAULTY
1024  95   ADMIN_INS      TERM_ACTIVE_CTRANS_BUSY ACTV   TRNS_OBSY  NON_FAULTY
1031  23   ADMIN_INS      TERM_ACTIVE_CTRANS_BUSY ACTV   TRNS_IBSY  NON_FAULTY
1050   1   ADMIN_INS      TERM_DOWN               LBLK   IDLE   TERM_FAULT DES_LOC_F
1050   2   ADMIN_INS      TERM_DOWN               LBLK   IDLE   TERM_FAULT DES_LOC_FC
1050   3   ADMIN_INS      TERM_DOWN               LBLK   IDLE   TERM_FAULT DES_LOC_FC
1050   4   ADMIN_INS      TERM_DOWN               LBLK   IDLE   TERM_FAULT DES_LOC_FC
1050   5   ADMIN_INS      TERM_DOWN               LBLK   IDLE   TERM_FAULT DES_LOC_FC
1050   6   ADMIN_INS      TERM_DOWN               LBLK   IDLE   TERM_FAULT DES_LOC_FC
1050   7   ADMIN_INS      TERM_DOWN               LBLK   IDLE   TERM_FAULT DES_LOC_FC
1050   8   ADMIN_INS      TERM_DOWN               LBLK   IDLE   TERM_FAULT DES_LOC_FC
1050   9   ADMIN_INS      TERM_DOWN               LBLK   IDLE   TERM_FAULT DES_LOC_FC
1050  10   ADMIN_INS      TERM_DOWN               LBLK   IDLE   TERM_FAULT DES_LOC_F
1050  11   ADMIN_INS      TERM_DOWN               LBLK   IDLE   TERM_FAULT DES_LOC_F
1050  12   ADMIN_INS      TERM_DOWN               LBLK   IDLE   TERM_FAULT DES_LOC_F
1050  13   ADMIN_INS      TERM_DOWN               LBLK   IDLE   TERM_FAULT DES_LOC_F
```

Figure 8-5   *Faulty CIC Dashboard*

Thus, faulty resources need to be tracked and trended; the CIC example shows the cost impact of not performing this exercise.

Feedback for Change Control

It is important for the voice SP to monitor and track changes in the network, and any impact to network resources and services as a result of the change. There should be a feedback mechanism that ties changes in the network to the trouble-ticketing systems, for example, an increase in the number of customer calls after a software upgrade on a network device. If the call center continues to experience high call volume, the voice SP would need to verify whether anything else changed in the network that could contribute to the high call volume. If nothing else changed except for the software upgrade, the voice SP would either need to take steps to mitigate the problem because of the software upgrade or back out the change and revert to the old software image. This information needs to be fed back to the change management process.

Any changes in the network should go through a change control process. The changes need to be tested and verified in a test environment before they are implemented in a production environment. The VoIP dashboard can be used for trending key performance indicators such as CPU utilization, interface statistics (link utilization, errors, drops, and so on), memory utilization, and so on. This gives the voice SP a comprehensive view of network performance and helps identify capacity issues and plan for upgrades. Capacity planning and upgrade strategies are discussed later in this chapter in the section "Resource Optimization and Capacity Planning."

Let's look at an example of how a change in the network can impact services such as VoIP. In this case, the voice SP upgrades the software on an edge router (this could be a cable/DSL/ETTH aggregator) that connects thousands of subscribers to its network.

After the software upgrade, the voice SP notices an increase in customer calls complaining about voice call failures (not able to place/receive calls, loss of dial tone, and so on). When looking at the VoIP dashboard (as described in Chapter 7, "Performance Analysis and Fault Isolation"), the voice SP notices an increase in CPU utilization and spikes in link utilization on the LAN/WAN interface.

Using the VoIP dashboard view, the voice SP starts to drill down on different performance indicators and looks at trending data to understand the impact of the software upgrade on the edge router, other network resources, and services such as VoIP. The voice SP first verifies the impact on call signaling, which can directly influence call completion rate and cause voice call setup/teardown failures. By looking at the call-signaling trending data, the voice SP can quickly identify problematic trends in the network.

Figure 8-6 illustrates the increase in call-signaling failures (Media Gateway Control Protocol [MGCP] errors) on the edge router before and after the software upgrade.

Looking at Figure 8-6, we can clearly see an increase in MGCP errors around 3/1/09. The number of MGCP errors jumped from 55 on 2/28/09 to 1895 on 3/1/09. This date coincides with the date when the software was upgraded on the edge router. By looking at the call-signaling trending data, we can see an obvious correlation between the software upgrade on the edge router and the increase in call-signaling failures. The spike in signaling failures also coincides with the high number of customer calls in the voice SP's trouble-ticketing system. This correlation establishes a clear link between the software upgrade on the edge router and the spike in call failures in the network.

Next, you need to find the root cause of the call-signaling failures. By looking at the CPU utilization trending data, the voice SP notices that the increase in CPU utilization coincides with the software upgrade on the edge router. Figure 8-7 illustrates the increase in CPU utilization on the edge router before and after the software upgrade.

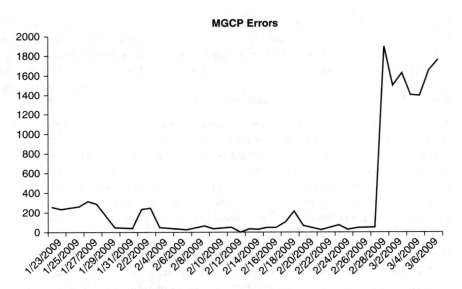

Figure 8-6   *Call-Signaling Trending Data on the Edge Router*

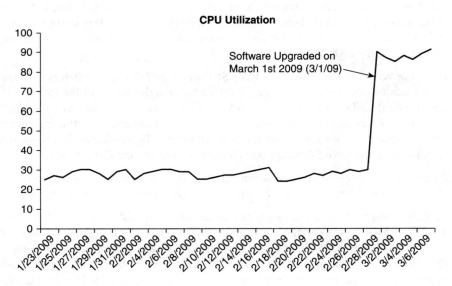

Figure 8-7   *Edge Router CPU Utilization Trending Data*

As depicted in Figure 8-7, the CPU on the edge router jumped from the 25–30 percent range (baseline) to the 85–90 percent range right after the software upgrade. To figure out the trigger for this jump in CPU utilization, the voice SP needs to look at other performance indicators to see what else was impacted after the software upgrade.

As mentioned in Chapter 7, the voice SP can drill down further on the CPU utilization data to find out which processes are contributing to the increase in CPU utilization. In this case, the high CPU utilization was being caused by an increase in Simple Network Management Protocol (SNMP) traffic and a new feature that was introduced in the network after the software upgrade.

Because SNMP traffic is typically processed by the route processor (process switched or processed in software) and not the forwarding engine (switched in hardware), it can drastically impact the CPU utilization. In this case, the SNMP process consumed up to 60 percent CPU cycles after the software upgrade. The baseline CPU utilization before the software upgrade was around 30 percent. Because of the increase in CPU cycles caused by SNMP traffic, the overall CPU utilization jumped to around 90 percent.

The CPU utilization of 90 percent could not solely contribute to the high number of call-signaling failures, so there can be other factors contributing to these failures. The next thing to verify is any change in SNMP traffic patterns in the network. This information can be obtained by capturing sniffer traces off the WAN uplink of the edge router or by retrieving data off a deep packet inspection (DPI) device connected to the WAN uplink interface of the edge router.

The information from the sniffer capture can reveal useful statistics, such as the source of the increased SNMP traffic (maybe a rogue server in the network aggressively polling information from the edge router), type of SNMP data being polled (Management Information Base—MIB, Object ID—OID), and so on. In this case, the voice SP notices certain SNMP information (VoIP call quality metrics) being aggressively polled from the edge router. This additional data could easily impact the link utilization of the WAN interface of the edge router. The next thing the voice SP looks at is the WAN link utilization trending data to see whether the WAN link was being saturated, causing packet drops under the interface.

By looking at the WAN uplink utilization data, the voice SP notices an increase in input and output utilization right after the software upgrade on the edge router. Because the software upgrade was performed on 3/1/09, Figure 8-8 illustrates an increase of about 40 percent in WAN link utilization on the edge router.

In Figure 8-8, we can see that input utilization increased from an average of 40 percent to 90 percent, whereas the output utilization increased from an average of 30 percent to 70 percent after the software upgrade.

Validating IP Trunk Introduction

A new technology introduction into a VoIP network needs validation. Any new technology has its own caveats along with the promised improvements and new features. Consider a situation where an enterprise has decided to expand its call capacity by introducing IP trunks using SIP. This enables it to leverage its Multi-Protocol Label Switching (MPLS) network and the bundled offer of the SIP trunk as part of the Virtual Private Network (VPN) offering, allowing it to use the same network resources for data, voice, and video. This is typically associated with lower toll charges for calls placed on IP trunks in a bundled offer by a service provider. Bandwidth utilized by IP trunks is directly proportional to the number of calls these trunks process, although the call-processing capacity of IP trunks can be increased without adding new hardware such as T1/E1 network modules, DSPs, and slot availability in gateways.

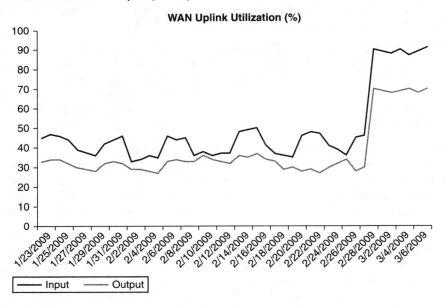

Figure 8-8   *Edge Router WAN Link Utilization Trending Data*

At the same time, when the SIP-based IP trunk is introduced in the network, the administrator should confirm its absorption by monitoring its performance using key performance indicators (KPI) such as call completion rate (in lieu of total call attempts) and call quality metrics.

The graph in Figure 8-9 depicts the call success rate for an IP trunk in the context of total calls attempted using this trunk.

The diagram includes several KPIs plotted on the same graph, including call success rate versus total calls attempted and failure rate of fax and point of sale (PoS) calls (for credit card transactions) for the same trunk.

The VoIP network administrator observed the call success rate for the SIP-based IP trunk slipping during business hours and dropping to 66 percent at its peak. The VoIP operations dashboard received an alarm when the call success rate dropped below 95 percent consistently for 30 minutes. The trend in call success drop did follow the overall call volume, but it did not match it in the same proportion consistently. Also, there were no indications of "channel not available" incidents on the operational dashboard, which happens because of lack of bandwidth availability to place calls. The immediate suspicion went to the call type using less common codecs, such as fax or point of sale calls. The administrator was able to filter these calls based on the calling party or the originating number because they have been assigned according to a specific pattern. The results were superimposed to correlate the data points. It became apparent that the call completion (success) rate for the SIP trunk is inversely proportional to the call failure rate for fax and PoS calls consistently throughout the day. After the suspicion was confirmed, the network operations personnel raised a trouble ticket with the service provider for further investigation and troubleshooting specific types of calls. This elimination process and pinpointing the issue reduced the problem resolution time. It also enabled the Network Operations Center (NOC) personnel to implement a temporary workaround to route the fax and PoS calls through a backup time-division multiplexing (TDM) trunk. This external call-routing design is discussed in the section "Maximizing Trunk Capacity and Avoiding Call Blocking," later in this chapter.

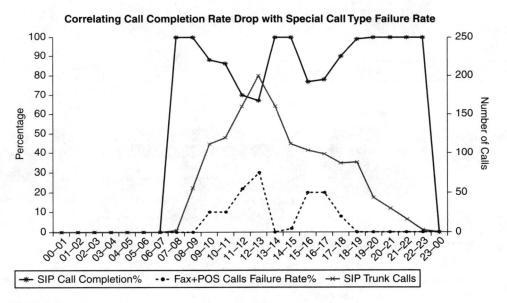

Figure 8-9 *Correlating SIP Call Failures to Fax and PoS Calls*

These examples illustrate how the VoIP dashboard can play a key role in identifying problems in the network and help in finding the root cause. The VoIP dashboard can also be used to identify areas of the network that might need to be optimized and can be used for capacity planning by looking at KPIs and monitoring trends.

Profiling in an SP VoIP Network

We build on the concepts covered earlier in the chapter and go over a dashboard use case for an SP VoIP. Dashboards can become a powerful tool for the SP, because they allow the SP to identify trends in the network. These identified trends can drive optimization- and capacity-related activities. Figure 8-10 depicts an SP scenario. The graphs show signaling protocol flows, the corresponding stats, and trunk connectivity in an SP VoIP network.

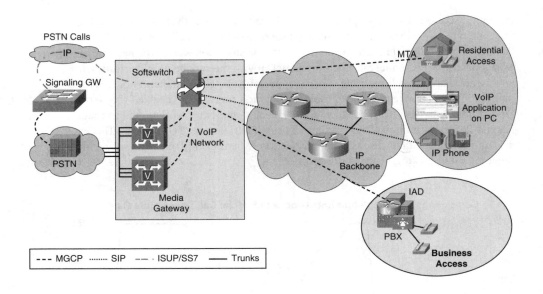

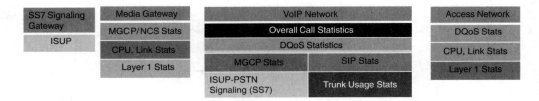

Figure 8-10 *Profiling in an SP VoIP Network*

The signaling flows and trunk connectivity described in the figure are represented by the following functional components:

■ An MGCP and ISDN User Part (ISUP)–based PSTN interfacing section

■ A central VoIP switch, which is a switching point for MGCP, SIP, dynamic quality of service (DQoS), ISUP, and trunk allocation

■ An MGCP- and SIP-based residential access

■ An MGCP-based business access

The corresponding metrics need to be collected from the VoIP switch and all the other component elements. SPs usually introduce out-of-band tools to collect statistics from the network elements; this way they correlate the stats back to the switch for reliability and accountability.

We discussed the layered concept in Chapter 7. Here we build on that concept to tie it to our dashboard use case. Metrics from network elements can be collected from the physical layer (Layer 1 in the OSI model)—for example, radio frequency (RF) metrics on the Cable Modem Termination System (CMTS)—or T1 slips and bit error rate on the Media Gateway side. These metrics help identify and rule out media layer issues.

Tracking the network element–related CPU and link utilization metrics would help trend load and network link utilization characteristics. Further, the VoIP switch is a central point of the VoIP network, because it sees all switching-related protocols and most resource allocation flows, thus facilitating the collection of the flow metrics.

The metric layers help isolate the issue affecting the overall call success behavior. Like peeling an onion, you can identify trends based on the dashboards. The accumulative dashboard reflects trends at an aggregate level, which helps lead to specific drill-down views.

Tracking the signaling flow metrics and trunk utilization through a series of dashboards can help the SP to quickly identify the trends. This is illustrated in Figure 8-11.

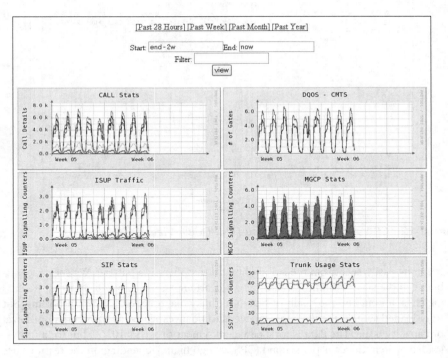

Figure 8-11 *SP VoIP Dashboard*

Figure 8-11 reflects an aggregate dashboard built from the metrics collected from the VoIP network. We use the stats collected from the switch for our discussion. Trends can easily be seen from the various dashboards in the figure. We can observe bell-shaped graphs with peaks representing the busy hour. The counters are not elaborated in this dashboard because it is introduced to show a consolidated view of different call traffic and VoIP subnetwork statistics.

The call metrics would act as the main driver; a dip in the calls or a plateau of the peaks would be a trigger to look into specific dashboards.

To isolate specific trends, we now drill down and expand on the specific dashboards for call metrics, DQoS metrics, ISUP metrics, trunk usage, and SIP metrics.

Figure 8-12 reflects the call metric dashboard.

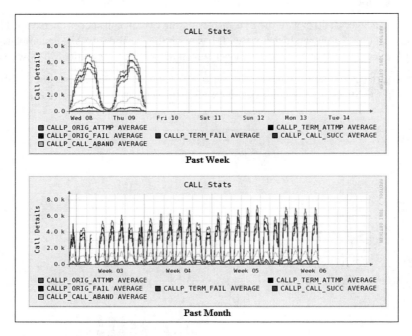

Figure 8-12 *SP VoIP Dashboard—Call Drill-Down View*

It can be easily seen from the call dashboard that there is a pattern. We see a clear gap between the success and the call origination attempts; this gap indicates that we are losing calls. It could be the result of hang-ups, busy dial tone, or network problems. But on a fully loaded switch, the discrepancy needs to be investigated as a potential optimization or capacity issue. The data is for a new deployment of a BTS 10200, but for a fully loaded switch, the calls per second (CPS) can go high. Depending on the features used and the switch hardware, the CPS can go over 100 CPS.

The DQoS drill-down view is shown in Figure 8-13; it is representing the access network metrics for our SP use-case scenario.

Network problems related to the CMTS are visible through these DQoS gate counters. A trend showing the difference in Gate Set attempts and Gate Set successes would be a clear indication of a CMTS resource allocation issue. The ISUP metric drill-down view is shown in Figure 8-14.

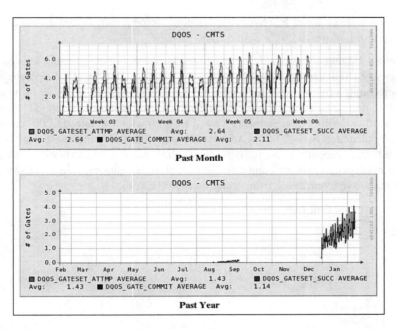

Figure 8-13   *SP VoIP Dashboard—DQoS Drill-Down View*

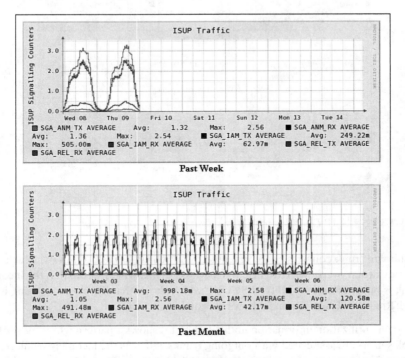

Figure 8-14   *SP VoIP Dashboard—ISUP Drill-Down View*

This view is reflective of PSTN-bound signaling protocol metrics and gives a good representation of the overall PSTN-bound calls.

Figure 8-14 shows a higher number of call releases than call completion. This could mean capacity or trunk gateway resource problems. Trending data can also help determine what to focus on. If a trend is seen for a high number of call releases, it should be an alert for the SP to perform an investigation in the PSTN Signaling Transfer Point (STP) link setup. The monthly dashboard view shows a trend for a growing call rate, and this can help in capacity planning. A High Initial Address Message (IAM) with increasing call release can also indicate a capacity issue. In some cases, a high number of IAMs with a lower number of releases can indicate hung calls. Any trunk issues causing failure in IAMs can also be extracted by contrasting trunk-usage and ISUP graphs.

The trunk usage drill-down view is presented in Figure 8-15.

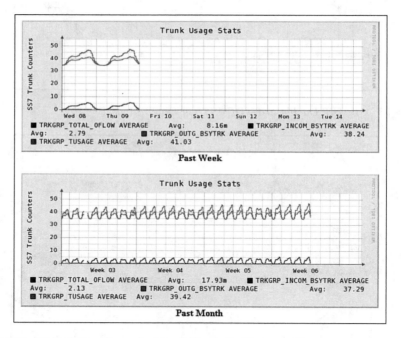

Figure 8-15   *SP VoIP Dashboard—Trunk Usage Drill-Down View*

Dashboard trunk counters are represented by TRKGRP_TOTAL_OFLOW, which shows the trunk total overflow; TRKGRP_INCOM_BSYTRK, which shows that the incoming trunk is busy; TRKGRP_OUTG_BSYTRK, which shows that the outgoing trunk is busy; and TRKGRP_TUSAGE, which shows the total trunk usage.

The trunk utilization information in the dashboard shows a pattern where most of the trunk seizers are associated with outgoing trunks, whereas the incoming trunk seizures are low. The overflow of trunks is also low. When the overflow of trunks and the total trunk utilization are high during the same time interval, there is a potential capacity issue.

Figure 8-16 describes the SIP drill-down view.

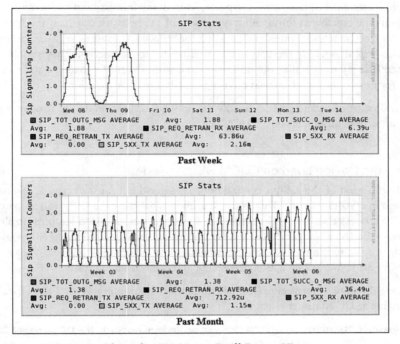

Figure 8-16   *SP VoIP Dashboard—SIP Usage Drill-Down View*

The dashboard SIP counters are represented by SIP Outgoing messages, SIP Outgoing Successes, and 5xx Errors. Problem trends can be an increase in retransmits or an increase in 5xx errors. Trends showing an increase in SIP invite transmits (TX) with a lower number of receives (RX) can also indicate a resource issue on the SIP endpoint.

We have described an SP voice switch-centric use case. The key here is the value of dashboards for key metrics to help identify trends in the network. We also elaborated on how these dashboards can help identify specific trends for capacity, optimization opportunities, and other potential issues.

Profiling in an Enterprise VoIP Network

Profiling KPIs in enterprise networks requires an understanding of the business practices and the associated communication patterns. For example, a Unified Communications system in a manufacturing setup might see increased activity during regular business hours. However, if there are multiple shifts, the communication behavior is the same throughout the day. Similarly, a retail shop sees varying business activity during the day based on the pattern of traffic into the store. The retail setup might see spikes in traffic during the busiest time of the year, such as the week before Christmas or Black Friday, the day after Thanksgiving Day in the United States that is considered to be the busiest shopping day of the year. This provides context for analyzing the Busy-Hour Call Attempts (BHCA), call-blocking statistics, CPU utilization, and link utilization.

In general, CPU and link utilization should correspond with the call traffic. Call blocking should be minimal to none and is expected during the busiest hour of the day. Call quality should be consistent throughout, regardless of the network traffic, if the QoS design is done correctly. Yet, any threshold crossing for utilization levels or service availability metrics should motivate the network administrators to take action. They should use this data to first try to optimize the utilization to absorb the traffic load and to proactively plan for growth before threshold crossings result into SLA violations.

The following two examples show how the network utilization data is used to improve the capacity management on a CUCM.

Balancing the Device Load on CUCM Cluster Nodes

CPU utilization of the call-processing nodes is a good indicator of critical resources utilization in a VoIP network. This example is about analyzing the CPU utilization of the CUCM process that is mainly responsible for processing calls and features on all the nodes in the UC Manager cluster. The UC Manager employs a cluster technology to scale its capacity to register a large number of endpoints including IP phones, trunks, and other media resources such as conferencing and transcoding DSPs.

Figure 8-17 shows the average CPU utilization in 1-hour intervals for 2 days for all the nodes in the UC Manager clusters.

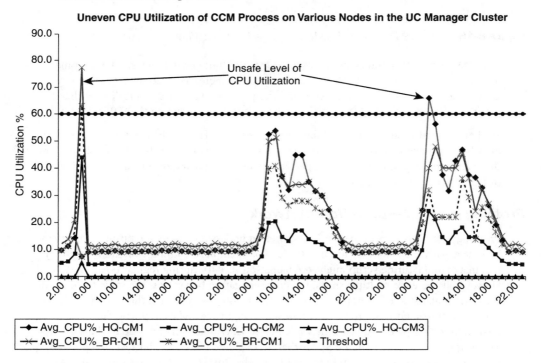

Figure 8-17 *Average CPU Utilization Measured on Unified CM Cluster Nodes over 2 Days*

This figure shows that all the nodes have different levels of CPU utilization, but these nodes approximately follow the same pattern. The pattern shows an elevated CPU utilization during the busier hours of the typical workday between 9:00 a.m. and 4:00 p.m. The threshold for safe average CPU utilization is established at 60 percent for the version of UC Manager and the platforms used in the cluster. The nodes are also referred to as subscribers. It is also observed that HQ-CM1 and HQ-BR1 not only have a higher CPU utilization than the other nodes in the cluster but also exceeded the 60 percent threshold for safe operation twice in the sampled data for 48 hours.

The pattern of utilization is expected because it matches the BHCAs, so we need to analyze the load on these servers. It turns out that the load on the five nodes in the cluster is not evenly distributed when the cluster is audited. Figure 8-18 shows the distribution of IP phones, trunks, and media resources on all the nodes in the cluster.

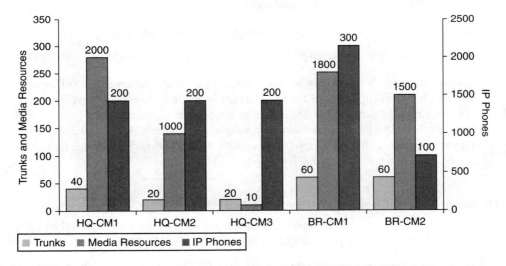

Figure 8-18 *Uneven Device Loads on the Nodes in the Unified CM Cluster*

This cluster serves an enterprise that has about half of its workforce concentrated on a campus with some future growth plans, and the other half distributed in various branch offices. From the graph shown in Figure 8-18, we can draw following conclusions:

■ The campus site is served by HQ-CM1, HQ-CM2, and HQ-CM3.

■ The branches are served by BR-CM1 and BR-CM2.

■ The branches served by BR-CM1 and BR-CM2 might be the same (or close), but they can differ in size. This is evident by the same number of trunks registered with each node dedicated for branches but a varying number of phones and media resources.

■ This also points to a different IP phone–to–trunk ratio in branches that we discuss in the section "Resource Optimization and Capacity Planning," later in this chapter.

- Because the media resources are common and collocated for the entire campus, the system administrator configured them up front by evenly distributing them on HQ-CM1, HQ-CM2, and HQ-CM3.

- The IP phones are not distributed evenly on the three campus nodes. It appears that the strategy for their assignment was to fill the first node to its capacity and then move to the second and the third node.

- The trunk distribution for the campus is also uneven, where there are twice as many trunks (40) on HQ-CM1 than on HQ-CM2 and HQ-CM3 (20 each).

- The uneven load on the five nodes does correspond to their respective CPU utilizations, as shown in Figure 8-17.

The remediation plan for this situation requires that

- HQ-CM1, HQ-CM2, and HQ-CM3 should all have 30 trunks each.

- Half of the 2000 registered phones on HQ-CM1 should be transferred to HQ-CM3.

Similarly, the branch nodes BR-CM1 and BR-CM2 should take advantage of the virtual resource management aspect of the UC Manager software by moving 100 media resources from BR-CM1 to BR-CM2 while making sure that these resources are not split between the nodes for the same branch for ease of management. An even phone distribution would mean 1650 phones split between BR-CM1 and BR-CM2, although they should not be split for the same branch.

Maximizing Trunk Capacity and Avoiding Call Blocking

The CUCM provides multiple layers of route redundancy. Figure 8-19 depicts the dial plan logic specific to the external call-routing logic for the CUCM.

A call placed by an endpoint registered with the CUCM, typically an IP phone, filters through class of service implementation configured through partitions and calling search spaces, and matches a route pattern based on the dialed digits. The route pattern can perform digit manipulation, for example, stripping the access code. This route pattern is associated with a route list that contains a prioritized list of route groups. The associated route groups have the option to perform digit manipulation by either prefixing digits or stripping leading digits or masking digits of dialed digits. The route groups point to actual devices that can be gateways, gatekeepers, or IP trunks that ultimately egress the call outside of the enterprise's VoIP network.

The network administrator of a retail chain observes a spike in call blocking on a VoIP management dashboard on December 18, 2009, during the busiest time of the day prior to Christmas. He correlates this with the trunks serving the busiest market determined by the highest BHCA counter. The trunk utilization is plotted in Figure 8-20. The call-blocking rate is also superimposed on the same graph, along with a maximum threshold of the individual trunks (equivalent to 23 channels in ISDN T1 in U.S. markets) for illustration purposes.

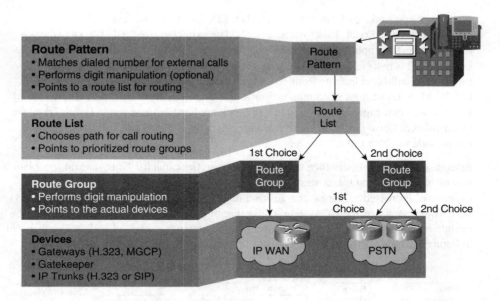

Figure 8-19   *External Call Routing in the CUCM*

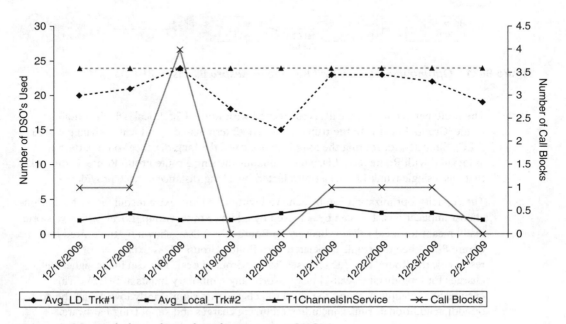

Figure 8-20   *Unbalanced Trunk Utilization on a CUCM*

There are two trunks dedicated to this market: LD_Trk#1, mainly reserved for long-distance calls, and Local_Trk#2 for local calls. The expected use for LD_Trk#1 is long-distance calls to suppliers and point of sale (PoS) transactions to credit card companies. Local_Trk#2 is reserved for local calls and terminates incoming calls. At the same time the spike in call-blocking rate (four blocked calls) was observed on December 18, 2009, LD_Trk#1 demonstrated maximum utilization. It can be inferred that the BHCAs exceeded the maximum capacity, resulting in call blocking. For the same week, Local_Trk#2 was underutilized. Local_Trk#2 is capable of carrying long-distance and PoS calls but might not provide the best toll rates compared to LD_Trk#1.

Adding additional long-distance trunks can increase the capacity. Note that this problem was manifested during the busiest season of the year. The additional capacity is at the expense of increased recurring charges that might go to waste for the most part of the year. The capacity of the UC system, consisting of local and long-distance trunks, can be optimized using the CUCM route path redundancy feature explained earlier. The diagram in Figure 8-21 shows the external call-routing configuration on the CUCM.

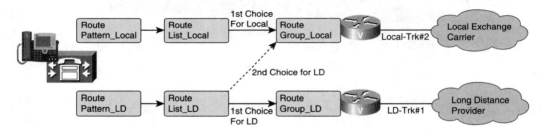

Figure 8-21   *Optimizing External Call Routing to Absorb Bursts in BHCAs*

The route pattern for local calls is associated with Route List_Local, which contains Route_Group_Local with the trunk Local_Trk#2 terminated by a local exchange carrier (LEC). Similarly, for routing the long-distance calls, the long-distance route pattern is associated with Route_List_LD, which contains the single route group Route_Group_LD that has a single trunk LD_Trk#1 terminated by a long-distance service provider.

The capacity optimization can be achieved without adding extra resources with additional cost implications. This can be achieved by adding Route_Group_Local, which was originally meant for local calls exclusively, to Route_List_LD, as shown by the dotted line in Figure 8-7. When the trunk associated with Route_Group_LD is exhausted of its resources, the Route_List_LD chooses Route_Group_Local. This can incur higher toll charges for long-distance calls, but it absorbs any temporary bursts in BHCAs. This ensures SLA compliance, and the extra toll charges are significantly lower than the cost of adding additional trunks, including recurring charges and supporting hardware.

Call Detail Record–Based Trend Analysis

CDRs contain information related to a call including the exact time of the call, call duration, call origination and destination information, features used during the call, and billing rate information. The CDRs are often coupled with call diagnostic information, either in the same data packet or coupled with a corresponding data packet by virtue of a session identifier that is unique for every record.

The primary purpose for CDRs is to generate billing information in service provider networks and in multitenant environments in enterprise networks; they provide an in-depth picture of the VoIP network usage. This information can be used for the following:

- Benchmarking the network utilization and using it in the feedback loop for network resource monitoring and detecting or correlating anomalies by establishing a threshold or a trending reference

- Service-level agreement (SLA) conformance by looking into the call diagnostic reports associated with the CDRs, providing details about call quality and call completion statistics

- Network growth planning by looking at an increase or a decrease in call patterns and the use of features that provide feedback about when and where to add capacity and minimize toll charges

- Detecting toll fraud or theft of service by analyzing anomalous trends

Benchmarking

To benchmark the network utilization trends by analyzing CDRs, it is critical to archive these records for at least 2 years. In some cases, regulatory bodies can mandate this. Call-processing devices such as the Unified CM or the BTS 10200 call agent maintain these records on their hard drives. Typically, a partition is carved out so that the overflow of these records does not disturb other critical data kept on the hard drive, such as subscriber configuration and application settings. Because the hard drive space is limited, these records need to be offloaded and archived for trend analysis and, in some cases, for smaller periodic analysis. The call-processing nodes should be configured to offload these CDRs and other call diagnostic records for the last 24 hours to a separate billing and mediation server with a larger storage capacity.

Figure 8-22 shows hourly call behavior.

This must be a recurring pattern for all the business days in a week, with a standard deviation of 5 percent to qualify as a benchmarking data point.

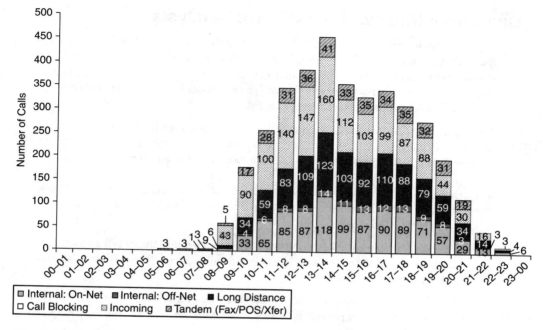

Figure 8-22   *Call Pattern During a Day*

Verifying VoIP Network Resources Capacity

The capacity ceiling of the VoIP network elements needs to be tracked. This capacity is almost always less than the advertised numbers on the product data sheets. A dashboard tracking these correlated views can help to identify the dense network elements in terms of availability. The dashboard would also reflect an inventory of elements that are ripe for optimization or capacity upgrade. Some of the general factors affecting this trending behavior can include the following:

- Underpowered system hardware (for example, CPU, memory, and link)

- Multiple feature combination affecting the performance and in turn the call capacity

- Software limitations

- Nonoptimized system architecture

The system hardware overutilization can be the most common issue for network elements to reach ceiling performance; the corresponding components need to be tracked through the dashboards. Thus CPU, memory, system load, and link usage need to be tracked.

The performance of a VoIP switch can degrade when it performs multiple functions simultaneously, such as providing PSTN access by way of TDM trunk termination, SIP trunk termination, and tandem call routing. A switch with a large cached database memory footprint can help in some cases, because this would reduce the route lookups and dial plan resolution if this information is cached in the memory instead of a database on a disk, which in turn causes the CPU to wait for disk I/O.

End-of-life product and particular software releases can cause the ceiling phenomenon. Also, software upgrades might be needed to accommodate bug fixes that are causing the system to not handle the high field call load.

System architecture decisions also affect the VoIP performance. One such critical component is the Domain Name System (DNS) infrastructure for a VoIP network. Many VoIP switch vendors recommend a local DNS cache to be turned up to minimize remote lookups.

The discussion so far has shed light on some of the pitfalls that can cause the VoIP network ceiling effect. We now look at Figure 8-23, where we give an example of monitoring the CPS against the advertised CPS on a VoIP switch.

The advertised CPS basically is the product CPS for the switch, but the number is often not reached, and the actual switch CPS plateaus before that. Figure 8-23 is based on a real-world switch scenario. Here the peak CPS was reached because of the various combinations of features used. The switch handles PSTN, cable-based access, and an SIP-based soft trunk call offload model. This takes CPU cycles for extra processing, thus the saturation of the peak CPS.

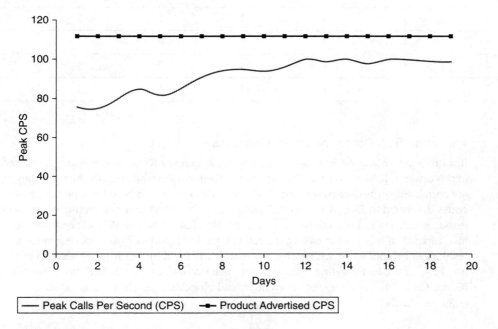

Figure 8-23 *Advertised CPS Versus Actual CPS*

SLA Conformance

An SLA for a VoIP network is not limited to the availability of dial tone. It includes the availability of all the services, such as mobility features, conferencing capabilities, and voicemail. All the placed calls, if dialed correctly and allowed by class of service, must complete successfully. Additionally, call quality should meet or exceed a certain limit. Generally, MOS scores or equivalent K-factors are used for measuring call quality.

Monitoring for Service Availability

Table 8-1 provides a monthly summary of service availability taken from UC Manager Call Detail Records.

Table 8-1 *Service Availability Based on Cause Codes*

Week	Cause Code 34	Cause Code 38	Cause Code 124	Cause Code 125	Cause Code 47
	Channel Not Available	Network Out of Order	Conference Full	Out of Bandwidth	Resource Unavailable, Unspecified
1	1	1	2	0	1
2	3	4	2	2	2
3	4	4	4	4	2
4	4	2	5	5	5
5	1	1	1	1	1

Validating SLA Conformance or Violations

It is often not enough to rely solely on single parameters or KPIs. As previously discussed in this chapter, KPIs are best understood when there is an established baseline and they are correlated with other parameters. If the SLA conformance is based just on the cause codes described in Table 8-1, the results might not present an accurate picture of network resource utilization. For example, Cause Code 34, Channel Not Available, needs further investigation. If it is trending along the call completion failures or call-blocking rate, the results can be confirmed. Otherwise, the matter will need further investigation. In Figure 8-24, we are investigating higher-than-normal occurrences of Channel Not Available (Cause Code 34). The first cross reference is call-blocking rate. Then the associated trunks are analyzed.

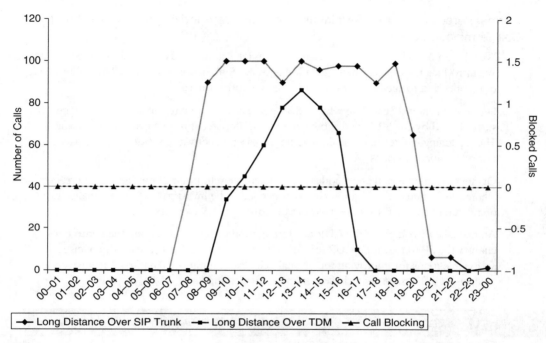

Figure 8-24 *Validating Call Capacity SLAs for Long-Distance Calls*

The dial plan in this situation was designed according to best practices, focusing on redundancy and resource maximization using the "sharing" concept described earlier in "Maximizing Trunk Capacity and Avoiding Call Blocking" section. In this design, an SIP trunk was configured as the primary resource to carry the maximum of 100 simultaneous long-distance calls. The route group with this SIP trunk along with another route group containing three E1 TDM trunks (with a net capacity of 92 calls) were part of the route list associated with a route pattern for long-distance calls.

The call-blocking rate and the utilization of the trunks associated with the long-distance route list are merged in Figure 8-9 for illustration purposes. Also, instead of looking at the three E1 TDM trunks individually, the graph shows the aggregated utilization. This graph represents the data collected during the busiest day in week 3 and 4, where we see the highest occurrences of Channel Not Available. The graph shows an SIP trunk utilized to its full capacity several times between 8:00 a.m. and 8:00 p.m., yet the call-blocking rate remains 0. This is a potential source for incrementing the Channel Not Available KPI/cause code value. During the same time frame, we see that the TDM trunks are supporting the long-distance call load after the SIP trunk is used to its full capacity. Hence, there is no SLA violation for service availability.

This stresses the fact that KPI analysis needs verification by way of correlation with other indicators.

Normal Versus Abnormal Termination Profiling: Categorizing and Correlating the Call Termination Code

We gave an overview of CDR profiling in Chapter 7 in the tools for CDR analysis section. We introduce the concept here again in Figure 8-25 to elaborate on the CDR call termination codes and to see how they are used to categorize the calls.

Figure 8-25 shows a use case where 95.93 percent of the call termination codes were considered valid by the SP. It also shows roughly 87.48 percent to be normal call clearing. The remaining call termination code delta of 8.45 percent was considered to be user errors or capacity issues.

Of the remaining termination codes, 0.3 percent seem to point to configuration errors, whereas 3.77 percent were anomalies that needed to be investigated. The anomalies can be related to the PSTN service provider or VoIP network capacity issues.

An exercise of compiling the CDR call termination codes and breaking them into buckets enables the SP to trend the CDR information. This trending further leads to concise actionable items for improving the normal call clearing number. The first 12 cause codes mentioned in Figure 8-25 represent normal call clearing.

Cause Code	Definition	Total Calls	% of Total	Call Completion	Abnormal Call Termination	Subscriber Out of Order
0	Feature Activation/Deactivation	5038	1.61%	X		
1	Unallocated/Unassigned DN	3152	1.0%	X		
13	Service Was Denied	1563	0.5%	X		
16	Normal Call Clearing	273421	87.48%	X		
17	User Busy	10272	3.29%	X		
18	No User Responding	9	0.0%	X		
19	No Answer, User Alerted	37	0.0%	X		
21	Call Was Rejected	167	0.1%	X		
28	Invalid/Incomplete Number	5250	1.68%	X		
31	Normal, Unspecified	670	0.2%	X		
47	Resource Unavailable	119	0.03%	X		
102	Recovery on Timer Expiry	128	0.04%	X		
3	No Route to the Destination	79	0.02%		X	
34	No Circuit Available	3515	1.12%		X	
111	Network Out of Order	1581	0.5%		X	
41	Temporary Failure	3053	0.98%		X	
42	Switching Equip. Congestion	88	0.03%		X	
127	Interworking, Unspecified	3480	1.11%		X	
22	Terminating Number Was Changed	75	0.02%			X
27	Destination Out of Order	861	0.28%			X

Figure 8-25 *Normal Versus Abnormal Call Termination*

Monitoring for Service Quality

Similarly, Table 8-2 shows the K-factor, which is derived from the statistical analysis of packet counts, concealment ratios, and concealment second counters. The K-factor ranges from 1 (poor) to 5 (excellent). As discussed in Chapter 6, "Managing VoIP Networks," this is essentially a computational model for depicting call quality that is otherwise measured subjectively through MOS scores. In Table 8-2, the results are aggregated for one month based on the gateway the call egresses or ingresses the VoIP network.

Table 8-2 *Monthly Average K-factor Scores Distributed by VoIP Gateways*

Gateway (DestDevice-Name)	Avg. K-factor 1 to 1.9 (Number of Occurences)	Avg. K-factor 2 to 2.9 (Number of Occurences)	Avg. K-factor 3 to 3.9 (Number of Occurences)	Avg. K-factor 4 to 5 (Number of Occurences)
10.200.14.4	3	5	16	905,238
10.200.14.6	1	1	0	6056

K-factor analysis is a good predictor for the network operator of frame loss but only after the problem becomes significant. It is a good idea to investigate K-factor scores below 3.0 to isolate potential problems. Table 8-3 shows call diagnostic records for a call with a low K-factor.

Table 8-3 *Sample of Call Diagnostic Records for a Low-K-Factor Score Call*

Date	Orig. Time	Term. Time	Duration (ms)	Origination	Destination	Orig Codec	Dest Codec	Orig Device	Dest Device	Avg. K Factor (MOS LQK)
19 Jan 2009	9:35:42 a.m.	9:35:42 a.m.	83	808293 6200	4793	G711-Ulaw	G711-Ulaw	10.200. 19.112	10.200. 14.4	2.8

Verifying Toll Savings (On-net Versus Off-net Profiling)

A network administrator should make every attempt in the dial plan design and call admission control (CAC) to manage bandwidth so that at least all the internal calls are routed through enterprise- or service provider–owned networks. However, this not only requires QoS to guarantee preferential treatment for media traffic over the IP network, but also the overall available bandwidth needs to be managed. It is a design best practice to back up any internal call-routing scheme with an automated alternate routing (AAR) feature. AAR is a mechanism in the CUCM that allows the internal calls to complete through another trunk or PSTN if the allocated bandwidth is exhausted. Also, the bandwidth allocation parameters should be conservative initially to preserve call quality. CDRs can provide critical feedback to determine the true bandwidth demand so that the allocations can be adjusted or more bandwidth be added. Figure 8-26 shows internal call patterns. For illustration purposes, the CAC metric showing the out-of-bandwidth incidents is superimposed on the figure.

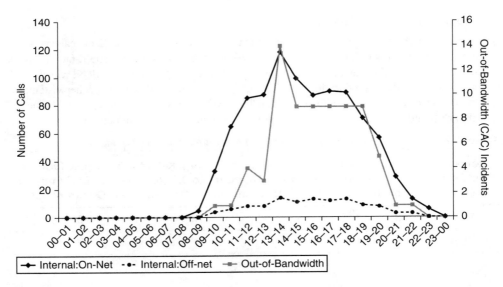

Figure 8-26   *Out-of-Bandwidth Incidents and Off-net Calls Relationship*

The first observation should be whether the call pattern matches the overall trend for the BHCAs shown in the "Benchmarking" section, earlier in this chapter. Notice that when the BHCA for internal calls peaks around 1:00 p.m., there is a corresponding spike in out-of-bandwidth incidents. Also, the internal calls are being offloaded to alternative resources (hence considered Off-net) in the same amount as out-of-bandwidth incidents. It is a good practice to monitor the call block rate for the same duration for internal calls to ensure SLA compliance. The network administrator now knows exactly how much bandwidth should be added or how much to adjust CAC parameters. An alternative method is to include a codec that consumes less bandwidth (for example, using G.729 instead of G.711), provided that it meets the call quality SLAs. This option does not require adding more bandwidth to keep the toll charges for Off-net calls to minimum.

Note   The time taken to isolate the fault is a key factor for the network operator. The KPIs tracked to isolate the VoIP network faults need to drive a reduction in this fault isolation time. The steps include going through an exercise to identify these KPIs as recommended in this chapter, correlate, and create visual representation of these KPIs. These KPI trends should be tracked by trained staff capable of taking timely corrective action. This methodology, when mature, allows for developing a resilient network.

Detecting Toll Frauds

This is primarily a concern in enterprise networks. Toll fraud in SP networks is mainly associated with hijacking of services, which involves vishing and SPIT, as discussed earlier in the section "Safeguarding Network Resources from Potential Security Threats."

The CDRs should be analyzed for the following anomalies:

- Excessive call attempts to (an international) number.

- Calls to unexpected or restricted destinations.

- Call activity during holidays or outside normal business hours.

- Repeated calls of short duration.

- A sudden increase in certain prefix number usage.

- Repeated call forwarding to external numbers and unexpected trunk-to-trunk activity. For example, watch out for inbound calls on local access trunks that are transferred outbound using long-distance trunks.

- Excessive calls to and from voicemail. This would require a voicemail system audit in addition to the CUCM audit and CDR analysis. An example of this fraud includes dialing into the voicemail system through a toll-free number and being forwarded out to a long-distance number.

Resource Optimization and Capacity Planning

As mentioned in the previous sections, the VoIP dashboard can be a powerful tool for the network operator to look at resource utilization and to identify areas of the network that might need to be optimized. It can also provide feedback to the network administrator when more capacity needs to be added based on trending the network utilization metrics. It can also provide information to help the network administrator optimize current resources before adding capacity, as discussed earlier in this chapter.

Network Resource Utilization and Optimization

In the section "Profiling in an SP VoIP Network," earlier in this chapter, we showed you a use case for an SP VoIP environment and an aggregate view of the dashboard for the VoIP Softswitch. The aggregate view helps in identifying problematic trends in the network and in isolating the area of the network contributing to the problem, for example, resource overutilization (CPU, link, memory, and so on) on edge routers in the access networks. Let's look at a specific example. If the aggregate view of the VoIP dashboard, as illustrated in Figure 8-11, shows a high number of call failures or a high number of DQoS failures, we can drill down into those specific views to find out the root cause of these failures. In the case of DQoS failures, we can look at the CMTS dashboard drill-down view to see whether certain CMTSs are getting overutilized while other CMTSs might be underutilized. This can point to a network resource optimization issue.

Figure 8-27 shows CPU utilization on four different CMTSs at the time of these DQoS failures.

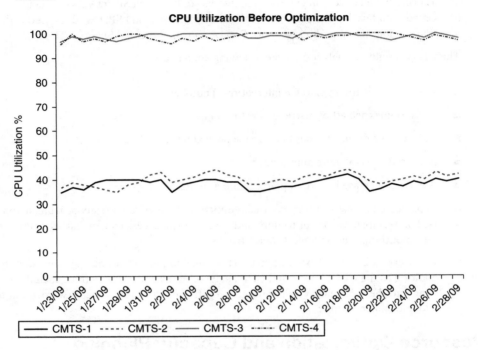

Figure 8-27   *CPU Utilization on Multiple CMTSs*

As illustrated in Figure 8-27, CMTS-1 and CMTS-2 are running around 40 percent peak CPU utilization, whereas CMTS-3 and CMTS-4 are running around 99 percent peak CPU utilization. So the DQoS failures, which might result in call failures, can be attributed to high CPU utilization on CMTS-3 and CMTS-4.

In this scenario, the network operator can optimize the access network and offload some calls from CMTS-3 and CMTS-4 to CMTS-1 and CMTS-2 to distribute the call volume more evenly. Optimizing the access network in this scenario might not be a trivial task because it might require the network operator to perform radio frequency (RF) node splits on the cable plant to move VoIP subscribers from one CMTS to another. It can also require changes on back-end provisioning, monitoring, and billing servers because the IP address of the VoIP subscribers can change when they are moved from one CMTS to another CMTS.

Figure 8-28 shows the results of this network optimization exercise and the impact on CPU utilization of the four CMTSs.

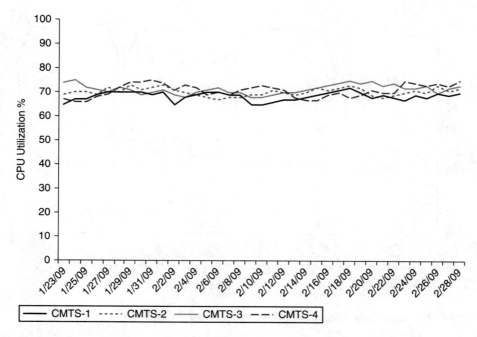

Figure 8-28 *CPU Utilization on CMTSs After Optimization*

As you can see from Figure 8-28, the CPU utilization on CMTS-3 and CMTS-4 dropped from 99 percent peak to around 70 percent peak utilization. The CPU utilization on CMTS-1 and CMTS-2 jumped from around 40 percent average to 70 percent peak utilization. So this might not be a long-term fix for this issue if the VoIP subscribers continue to grow in this area of the network. The network operator would have to weigh the benefits of this optimization exercise against the amount of effort/cost needed to perform this task.

If optimizing the access network in this scenario provides only a short-term solution because all CMTSs might experience high CPU utilization after a small period of time (3–6 months), network optimization might not be a fruitful exercise in this case. In this case, a better option for the network operator would be to upgrade the CPU on CMTS-3 and CMTS-4 to a more powerful processor or upgrade the chassis to a higher-capacity box that can handle the predicted VoIP subscriber growth for the next 12–18 months.

Let's look at another example. In this scenario, the aggregate view of the VoIP dashboard shows call failures and possible overutilization of trunks on some Trunking Gateways (TGW) in the PSTN-facing part of the network.

Figure 8-29 illustrates the scenario with call failures and trunk overutilization on TGWs.

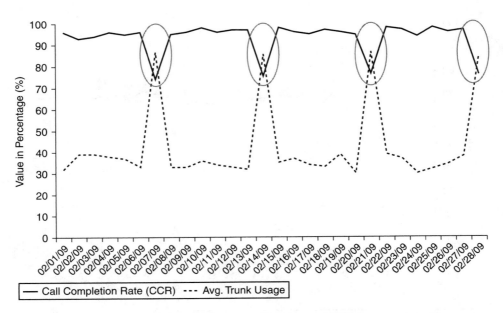

Figure 8-29   *Call Failures and Trunk Overutilization on TGWs*

As shown in Figure 8-29, there is a noticeable drop in successful call completion rate (CCR) and a clear indication of trunk overutilization on some TGWs. You can see that the CCR drops from around 99 percent to around 75 percent every time the trunk utilization jumps from 40 percent to around 85 percent. Because we are looking at the average trunk utilization view, there can be peaks of up to 100 percent on some trunks that can cause these call completion failures.

Overlaying the Call Stats data shown in Figure 8-11 as part of the VoIP dashboard that came from the Softswitch and comparing it to the Avg. Trunk Usage data coming from the trunking gateway as shown in Figure 8-29 can create this view. The view can also be tracked as a separate dashboard by itself. When we drill down further into the Avg. Trunk Usage data in the dashboard, we notice that some TGWs show all trunks as busy, whereas other TGWs have trunks in the idle state. If the Softswitch tries to set up calls on the TGWs that do not have any idle trunks, the calls fail.

Figure 8-30 shows average trunk utilization on four TGWs for the entire month of February 2009. You can see that TGW-1 and TGW-3 show a spike in trunk utilization on every Saturday of the month (February 7, 14, 21, and 28). TGW-2 and TGW-4 show a trunk utilization around 45 percent for the same days. This indicates that on Saturdays, the average trunk utilization on TGW-1 and TGW-3 is around 80 percent, which includes the maximum and minimum for the day. There might be times during the day where trunk utilization might have reached 100 percent, resulting in call failures if the Softswitch tried to set up calls on these trunks.

Figure 8-30 shows trunk usage on TGWs at the time of the call failures.

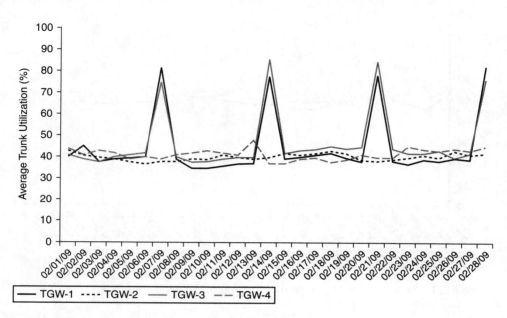

Figure 8-30   *Trunk Usage at the Time of Call Failures*

To resolve this issue, the network operator can optimize the trunk usage on these TGWs so that all TGWs get utilized evenly. The network operator can reconfigure the dial plan on the Softswitch to set up calls on the underutilized trunks and take advantage of the available trunk capacity. The trunk groups can also be optimized to use CICs in a round-robin fashion or use the least recently used (LRU) algorithm to set up calls on the TGWs.

Figure 8-31 shows trunk usage on the TGWs after the optimization.

As you can see from Figure 8-31, trunk utilization on all four TGWs is around 50 percent after the network operator made changes on the Softswitch to modify the dial plan and optimize trunk usage on these TGWs. This optimization exercise is an easy and cost-effective solution to the problem. Instead of adding more TGWs to the network and incurring additional cost, the network operator can solve this issue through optimization of resources on these TGWs.

Along the same lines, SIP traffic can be optimized through a load-balancing strategy; this topic is covered in the "Optimizing SBC" section in Chapter 7.

Deciding to optimize the network is a tricky issue for the network operator and requires some serious thought. If the cost of optimization outweighs the benefits, adding more capacity to the network might be a better option in the long run. Capacity planning and upgrade strategies are discussed in more detail in the next section.

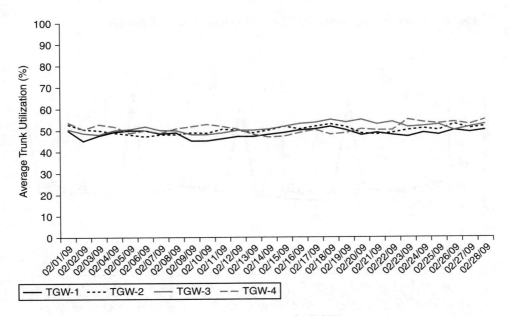

Figure 8-31   *Trunk Usage After Optimization on the TGWs*

Capacity Planning and Upgrade Strategies

As mentioned in the previous section, the VoIP dashboard can be used to identify the areas of the network that might need optimization for the VoIP service to scale. It can also help predict network growth by looking at trending data of the KPIs and can help the network operator put together a sound strategy for upgrading network resource- and capacity-planning exercises.

By looking at the trending data of KPIs, the network operator can predict when devices in the network might run out of capacity and cause outages or result in call failures.

Figure 8-32 illustrates an example of how trend analysis can help predict network growth.

As you can see from Figure 8-32, if the trending line follows a linear growth, the network hits capacity limits in the next 4–6 months. The network operator can use this information to put together an upgrade strategy, make the decision to optimize network resources, or add more capacity and order the equipment in advance before these capacity issues result in a network outage or cause call failures.

A technique used to predict capacity based on trending information is illustrated in Figure 8-32. Here, Microsoft Excel is used to plot and extrapolate trunk group average usage data. A Microsoft Excel spreadsheet allows the data to be plotted and estimates a function that best represents the scattered data. The function also has an R-squared value that represents the accuracy of the function. An R-squared value of 1 or close to 1 is an exact fit to the data. R-squared can be used for estimating the deviation.

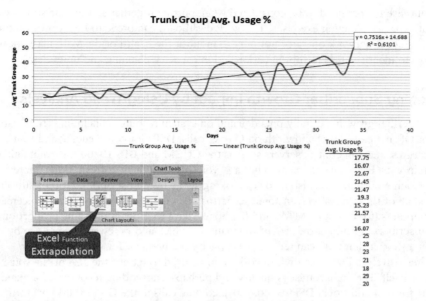

Figure 8-32 *Using the VoIP Dashboard to Predict Network Growth*

The graph shows the following extrapolated line:

y = 0.7516x + 14.688

The R-squared value is 0.6101. This extrapolation can be performed in Excel by selecting the f(x) option in the Chart Layouts section under Chart Tools. Thus, to predict the capacity 60 days out, the *x* value in the function is replaced by 60 to yield the corresponding average trunk usage value of approximately 59.

The following are some factors affecting the network growth and upgrade:

■ Predicting network growth might not be as simple an exercise as illustrated by the graph in Figure 8-32; it might not follow a linear growth pattern. The following other factors can influence growth in the network:

   ■ Change in business plans (mergers and acquisitions)

   ■ New feature/service introduction

   ■ Competitive pressure

■ Other factors can influence the upgrade strategy, including a technology refresh cycle (end-of-life, end-of-sale equipment)

■ Equipment lead time (equipment availability and orderability)

■ Facility readiness (rack space, power considerations, cabling, Internet connectivity, and so on)

All these factors need to be considered when putting together an upgrade strategy and for capacity-planning exercises. The VoIP dashboard can prove to be a powerful tool in helping the network operator not only to monitor the health of the VoIP network but also to identify areas for optimization and in capacity planning and upgrades.

Managing Subscriber Growth Impact by Using Trend Analysis

Subscriber growth naturally puts demand on system resources, including trunk capacity for Off-net calls, link bandwidth on On-net calls, DSPs for conferencing and transcoding features, and call control servers such as the CUCM and BTS 10200 for maintaining the endpoints. Features such as mobility and voicemail require a corresponding increase. Resource optimization also involves looking at the oversubscription ratio of subscribers and all of these resources. In some scenarios, the certain resources such as voicemail and endpoints (for example, MTAs and IP phones) have to grow in the same proportion as the subscribers. Shrinking the size of the mailbox, which allows fewer messages to be stored for a shorter duration, can reduce the capacity requirement for voicemail servers. Alternatively, a lower-bandwidth codec can be used to reduce the size of the voice message itself. The compromise is quality and perhaps transcoding resources, for example, in the form of additional DSPs to convert incoming calls in the G.711 (90-kbps) format to the G.729 (20-kbps) format. We discuss how to absorb the subscriber growth impact on call-processing entities using the CUCM example and managing network bandwidth using codecs.

Note   This discussion is not a substitute for a discussion on replacing traffic engineering, which was discussed in Chapter 6. We also do not offer design guidelines.

UC Manager Cluster Capacity

In an enterprise VoIP network, all the resources, including the conferencing resources, transcoders, gateways, voicemail servers, and other feature servers, work in concert with the CUCM as the nerve center of the network. Similarly, a call agent in a service provider environment plays the same central role, for example, a BTS 10200. Naturally, the CUCM is directly impacted by the subscriber growth. The Unified Communications Solution Reference Network Design (SRND) discusses this in exact details. The SRND also provides a capacity calculator known as the Unified Communications Sizing Tool (UCST) that can very accurately generate a list of resources needed according to the requirements. For more information on Cisco Unified Communications SRND, visit http://www.cisco.com/en/US/solutions/ns340/ns414/ns742/ns818/landing_uc_mgr.html.

Figure 8-33 presents a comparison of two redundancy options. One is cost-efficient, whereas the other is performance-efficient.

Both Unified CM clustering options are designed to serve 5000 phones in this example. The 2:1 redundancy option is definitely cost-efficient, but it is hard to manage from a configuration perspective because it can be seen from the CallManager grouping. The odd-numbered grouping makes it difficult to align resources other than phones with varying ratios such as trunks and DSPs for conferencing. Other management aspects including system maintenance and upgrade activities can briefly cause limited outages where certain features might not be available during the maintenance window. The 1:1 redundancy option is more efficient from a configuration, failover, and load distribution point of view. A 2:1 redundancy option must be chosen carefully after analyzing the CPU and memory utilization trends along with BHCA trends. These trends should be interpolated for expected subscriber growth using the growth function discussed in the section "Capacity Planning and Upgrade Strategies," earlier in this chapter.

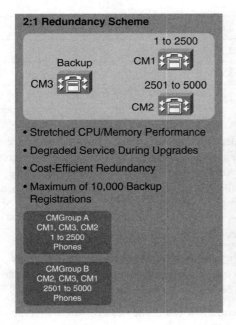 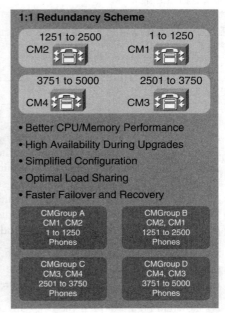

Figure 8-33 *Cost-Efficient Versus Performance-Efficient Redundancy Designs*

Network Bandwidth and Transcoding DSPs

Bandwidth becomes even more of a premium resource for media traffic when priority queues are carved out in the QoS scheme along with traffic shaping in the network. It might not always be possible to increase network bandwidth along with subscriber growth (not considered or expected at the time of initial design). Compression codecs such as G.729 offer solutions by reducing the amount of traffic for the same number of calls, hence requiring less bandwidth. However, this comes at the cost of reduced call quality and additional delay. Table 8-4 explains this relationship.

Table 8-4 *MOS and Delay Values for Various Compression Methods*

Compression Method	Bit Rate (kbps)	MOS	Compression Delay (ms)
G.711 PCM	64	4.1	0.75
G.726 ADPCM	32	3.85	1
G.728 LD-CELP	16	3.61	3 to 5
G.729 CS-ACELP	8	3.92	10
G.729 x 2 Encodings	8	3.27	10
G.729 x 3 Encodings	8	2.68	10
G.729a CS-ACELP	8	3.7	10
G.723.1 MP-MLQ	6.3	3.9	30
G.723.1 ACELP	5.3	3.65	30

Keep in mind that using a compression codec might require transcoding the calls back and forth to a clear channel or G.711 at the gateway, where the calls can egress or ingress using TDM trunks. This needs resources in the form of DSPs or software-transcoding resources in the call agent or the CUCM subscribers, which can tax its capacity. Also, a compression codec selection would require that CDRs are analyzed for call quality or K-factor, as mentioned earlier in this chapter in the SLA conformance section. Other potential problems using compression codecs are discussed in Chapter 7.

Similarly, DSPs are required to facilitate other features such as conferencing. There are several options of placing DSPs in a large VoIP network with multiple call sites and network ingress/egress points for calls. Figure 8-34 shows two of these options.

As shown in Figure 8-34, if the DSPs are centrally located, they can be safely oversubscribed by virtue of optimal use as a pooled resource. However, it can be taxing for the WAN links because the calls from distributed branches have to traverse them to reach the central location. It might also add to network delay, which can exacerbate echo and affect call quality. Also, the conferencing feature is vulnerable to any WAN outage. Conversely, the distributed resources keep the calls within the local branch, hence preserving the WAN bandwidth. However, the hardware cost would be significant. Also, the distributed resources are often associated with waste and less optimal use by virtue of undersubscription or failure to absorb unexpected bursts in demand. Trend analysis can be key in making this critical decision. If there are no queue drops observed for media queues on WAN links, minimum to no WAN outages (observed as alarms or SNMP traps), and no Cause Code 124s observed in CDRs (related to Conference Full), the central location of DSPs for conferencing is an acceptable (and cheaper) option. Otherwise, the SLA might dictate a distributed design given that the recurring cost of WAN bandwidth outweighs the cost of the hardware required for the conferencing feature in every branch. The key point is that the trend analysis and correlation of certain KPIs can provide feedback in making bandwidth cost–versus–hardware cost expenditure decisions.

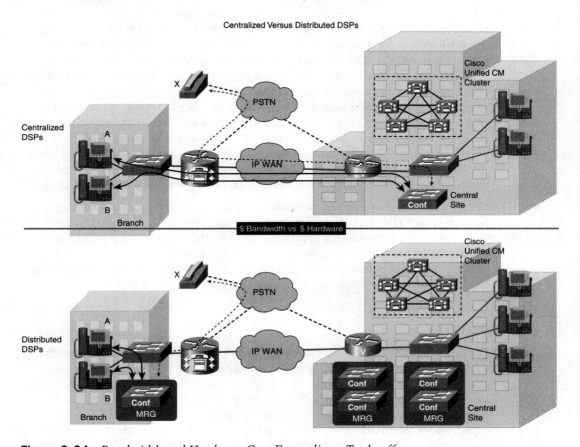

Figure 8-34 *Bandwidth and Hardware Cost Expenditure Trade-offs*

Considerations for Adding Trunk Capacity

The oversubscription ratio on voice networks is measured through a standard metric called a Line to Trunk Concentration Ratio (LCR), also referred to as an oversubscription ratio. Typically, an LCR might be as low as 4.5 (commonly depicted as 4.5:1), indicating that 4.5 lines compete for each trunk in the switch at the commissioning of the voice network. However, the actual ratio should be calculated using traffic engineering formulas that take into consideration the expected call volume and call-blocking rate using Erlang tables. This ratio can be much lower for busy call centers. Over a period of time, the customer calling behavior is validated through CDRs and other KPIs (BHCAs, call success rate, and call blocking rate), providing feedback to network operators to adjust the LCR ratio. This enables them to adjust trunk groups or route lists in SP and enterprise networks, respectively. This was discussed in the context of resource optimization earlier in this chapter. Subscriber growth can also be interpolated using trend analysis, as discussed in the "Capacity Planning and Upgrade Strategies" section, earlier in this chapter.

Network operators have the option of adding trunks in a central location, choosing between IP and TDM technologies, or expanding the scope of their network by virtue of interconnect alliances and using route servers/proxies that facilitate VoIP peering and business-to-business (B2B) communication. When using an IP-based technology for adding trunk capacity growth—for example, an SIP trunk—factors such as network overhead, interoperability, and hardware requirements must be considered.

Summary

The network management data, including KPIs and CDRs, can provide valuable feedback to network operators. The trending of these data points helps them optimize resource utilization, validate SLA conformance, manage changes by verifying trouble-free change absorption, perform capacity planning, and better plan upgrades. This can reduce the operational expenses (OPEX) and even save on capital expenditures (CAPEX) when capacity is increased only after maximizing the existing resources in the VoIP network. The network management dashboard plays a key role in achieving these goals by providing a structured method of visualizing and then correlating these metrics. Hence, an investment in establishing these operational best practices can yield immediate as well as long-term benefits.

References

1. Dismukes, Kim, "JSC2000-E-10522," STS-101 Shuttle Mission Imagery, March 2000, NASA, http://spaceflight.nasa.gov/gallery/images/shuttle/sts-101/html/jsc2000e10522.html.

2. Madani, Habib, BRKNMS-2831, "Voice Health Management for Service Providers," Cisco Networkers 2007 at Anaheim California, June 2007, Cisco.

3. Dimitriou, S., Tsioliaridou, A., and Tsaoussidis, V. "Introducing size-oriented dropping policies as QoS-supportive functions," *IEEE Transactions on Network and Service*

Management, Volume 7, Issue 1, 2010, IEEE, Digital Object Identifier: 10.1109/TNSM.2010.I9P0313.

4. "Cisco Unified Communications SRND," Feburary 2010, Cisco, http://www.cisco.com/en/US/solutions/ns340/ns414/ns742/ns818/landing_uc_mgr.html.

5. "Enterprise Quality of Service SRND," November 2005, Cisco, http://www.cisco.com/en/US/docs/solutions/Enterprise/WAN_and_MAN/QoS_SRND/QoS-SRND-Book.html.

6. Paquet, Catherine, *Implementing Cisco IOS Network Security (IINS)*, Indianapolis, IN: Cisco Press; 2009. ISBN-10: 1-58705-815-4.

7. Allspaw, John, *The Art of Capacity Planning: Scaling Web Resources*, Sebastopol, CA: O'Reilly Media; 2008. ISBN-10: 0-59651-857-9.

Scripts and Tools for Monitoring and Troubleshooting VoIP Networks

This appendix presents scripts that collect metrics data and parse out the key performance indicators (KPI). These are sample scripts written in Perl, Expect, and Awk that can be used by VoIP networker operators to automate some of the common tasks of data collection, data parsing, and data correlation to get KPIs. The following sections contain sample scripts and tools that allow the operator to automate tasks, such as the following:

- Polling Domain Name System (DNS) servers for availability and name resolution

- Performing a Session Initiation Protocol (SIP) ping and checking SIP endpoint availability

- Pinging IP addresses for host availability

- Telneting to the host and running commands

- Logging in to remote servers/hosts and running commands

- Sample script to pull alarms from a Cisco BTS 10200 EMS

- An in-depth, tool-based case study that utilizes IP SLA data to perform bandwidth availability and performance assessment

Fully Qualified Domain Name (FQDN) Verification Tool

The FQDN verification tool takes a host name file, which contains a prepopulated FQDN list and runs a DNS query against the given DNS server for all the FQDNs to verify FQDN resolution to IP addresses. The script can be used to create an FQDN failure dashboard.

```
#!/bin/perl


use Data::Dumper;
```

```perl
use Getopt::Long;

GetOptions( "file=s"   => \$file_name,      # —file name
            "server=s"    => \$dns_server );          # — dns_server·

die " USAGE: ./check_dns.pl —file <fqdn_file_path> —server < dns_server to
use> "
unless defined $file_name;
die " USAGE: ./check_dns.pl —file <fqdn_file_path> —server < dns_server to
use> "
unless defined $dns_server;

#$fqdn_list="/tmp/dns_hosts.txt";
$fqdn_list=$file_name;

open FQDN_L,$fqdn_list
 or die "Cant open '$fqdn_list' : $!";

%nslookup_hash;
$count=0;
while ($fqdn = <FQDN_L>)
{
 #print "fqdn=$fqdn";
 chomp($fqdn);
 $nslookup_hash{$fqdn} = "FAIL";
    #&lookupaddress($fqdn,"68.87.72.180");                  # populates %results
    #&lookupaddress($fqdn,"10.89.42.1");                # populates %results
    &lookupaddress($fqdn,$dns_server);                # populates %results
$count++;
}
close(FQDN_L);

while ( ($tsap_addr,$check) = each %nslookup_hash) {
    if ($tsap_addr)
    {
       print "tsap_addr:$tsap_addr $check\n";
    }
}

print "count=$count\n";
$nslookup = "/usr/sbin/nslookup";                 # nslookup binary

# ask the server to look up the IP address for the host
# passed into this program on the command line, add info to
# the %nslookup_hash hash
sub lookupaddress {
```

```perl
    ($hostname,$server) = @_;

    #print "hostname = $hostname $server\n";
    open(NSLOOK,"/usr/sbin/nslookup $hostname $server 2>&1|") or
    #open(NSLOOK,"/usr/sbin/nslookup $hostname $server |") or
      die "Unable to start nslookup:$!\n";

    while (<NSLOOK>) {
        # ignore until we hit "Name: "
            next until (/^Name:/);
        # next line is Address: response
            $next_line = <NSLOOK>;
            #print "next_line = $next_line\n";
            if ( $next_line =~ /Address/ && ( $next_line !~ /Server failed/) )
            {
                $nslookup_hash{$hostname} = "PASS";
            }
            else
            {
                $nslookup_hash{$hostname} = "FAIL";
            }
            last;

    }
    close(NSLOOK);
}
```

SIP Ping Script

The SIP ping script performs an SIP ping for a given host and port to verify SIP endpoint availability.

```perl
#!/usr/bin/perl
use IO::Socket;
use POSIX 'strftime';
use Time::HiRes qw(gettimeofday tv_interval);
use Getopt::Long;
use strict;


my $USAGE = "Usage: sip_ping.pl [-v] [-t] [-s <src_host>] [-p <src_port]
<hostname>";

my $RECV_TIMEOUT = 5; # how long in seconds to wait for a response

my $sock = IO::Socket::INET->new(Proto => 'udp',
```

```
                                               LocalPort=>'6655',
                                               ReuseAddr=>1)

    or die "Could not make socket: $@";

# options
my ($verbose, $host, $my_ip, $my_port, $time);
GetOptions("verbose|v" => \$verbose,
                "source-ip|s=s" => \$my_ip,
                "source-port|p=n"=> \$my_port,
                "time|t" => \$time) or die "Invalid options:\n\n$USAGE\n";

# figure out who to ping
my $host = shift(@ARGV) or die $USAGE;
my $dst_addr = inet_aton($host) or die "Could not find host: $host";
my $dst_ip = inet_ntoa($dst_addr);
my $portaddr = sockaddr_in(5060, $dst_addr);

# figure out who we are
$my_ip = "127.0.0.1" unless defined($my_ip);
$my_port = "6655" unless defined($my_port);

# callid is just 32 random hex chars
my $callid = ""; $callid .= ('0'..'9', "a".."f")[int(rand(16))] for 1 .. 32;
# today's date
my $date = strftime('%a, %e %B %Y %I:%M:%S %Z',localtime());
# branch id - see rfc3261 for more info, using time() for uniqueness
my $branch="z9hG4bK" . time();

my $packet = qq(OPTIONS sip:$dst_ip SIP/2.0
Via: SIP/2.0/UDP $my_ip:$my_port;branch=$branch
From: <sip:ping\@$my_ip>
To: <sip:$host>
Contact: <sip:ping\@$my_ip>
Call-ID: $callid\@$my_ip>
```

Host Ping Script

The host ping script checks whether a host is responding to ping queries. This tool can be used to create a basic VoIP network host availability dashboard.

```
#!/usr/bin/perl
```

```perl
use Net::Ping;
$pinghost = 'server.onsight.com';
$p = Net::Ping->new("tcp", 2);
for (;;) {
        unless ($p->ping($pinghost)) {
                print "Fail: ", scalar(localtime), "\n";
        } else {
                print "Success: ", scalar(localtime), "\n";
        }
        sleep 10;
}
```

Telnet Script

This Telnet script is a Perl-based tool that can be used to telnet to a host and perform some metric collection tasks. It can be used to collect command-line interface (CLI)–based metrics. The tool includes a UNIX command 'who' as an example.

```perl
#!/usr/bin/perl -w

#Listing 13. telnet.pl

use strict;
use Net::Telnet;
my $host = shift || 'server.onsight.com';
my $user = shift || $ENV{USER};
die "no user!" unless $user;
my($pass, $command);
print 'Enter password: ';
system 'stty -echo';
chop($pass = <STDIN>);
system 'stty echo';
print "\n";
my $tn = new Net::Telnet(Host => $host)
    or die "connect failed: $!";
$tn->login($user, $pass)
    or die "login failed: $!";
print 'Hostname: ';
print $tn->cmd('/bin/hostname'), "\n";
my @who =  $tn->cmd('/usr/bin/who');
print "Here's who:\n", @who, "\n";
print "\nWhat is your command: ";
chop($command = <STDIN>);
print $tn->cmd($command), "\n";
$tn->close or die "close fail: $!";
```

Expect Script to Run Commands on Remote Hosts

Expect-based tools allow easy collection of data from remote hosts because the Expect language allows easy replication of the command responses that can be programmed in the scripts. This is shown in the Expect scripts that follow.

A command can replace the placeholder <put command here> that follows. The following sample script collects trunk information from a Cisco BTS 10200 EMS CLI.

```
#!/usr/bin/expect
```

```
set user_site [lindex $argv 0]
set timeout 30
spawn ssh root@$user_site
expect "password:"
send "cisco#x23\r"
expect "#"
set timeout 2000
send " <put command here>\r"
expect "#"
set timeout 2500
send "su - unixuser\r"
expect "CLI>"
#interact
set timeout 20000
set status_cmds [open "./trunk_status_$user_site"]
while {![eof $status_cmds]} {
  set buffer [read $status_cmds]
  send $buffer
}
#send  $status_cmds
expect "CLI"
#expect "CLI>"
set timeout 5
send "exit\r"
expect "#"
set timeout 5
send "exit\r"
#send \004
expect "#"
set timeout 5
send "exit\r"
close $status_cmds
```

Expect Script to Collect Alarms from a Cisco BTS 10200 EMS

The following script collects alarm information from a Cisco BTS 10200 EMS CLI. The UNIX **script** command is used to store the CLI alarm output on the remote box. The output can be retrieved by a periodic File Transfer Protocol (FTP)based transfer using a cron job.

```
#!/usr/bin/expect


set user_site [lindex $argv 0]
set timeout 30
spawn ssh $user_site
expect "password:"
send "opticall\r"
expect "#"
set timeout 2000
send "script /tmp/current_alarms\r"
expect "#"
set timeout 2500
send "su - optiuser\r"
expect "CLI>"
#interact
set timeout 20000
send "show alarm;\r"
expect "CLI"
#expect "CLI>"
set timeout 5
send "exit\r"
expect "#"
set timeout 5
send "exit\r"
#send \004
expect "#"
set timeout 5
send "exit\r"
```

Case Study: Network Path Analysis Tool Leveraging IP SLA to Assess Bandwidth Availability and Performance in IP Networks

As discussed in Chapter 6, "Managing VoIP Networks," the network must be ready to deliver the video and audio packets smoothly and reliably before the actual deployment of a service like VoIP. In this case study, the readiness of the network is analyzed by a network path analysis tool. This tool captures and analyzes network path characteristics (bandwidth, delay, jitter, and packet loss) and the version and configuration information of devices along that path.

The network path analysis tool is a sequence of scripts that gathers data to verify and report the following:

- Internet Operating System (IOS) version of WAN gateway routers involved in the path of audio and video traffic transmission
- Configuration of the WAN gateway routers including:
    - Egress and ingress WAN interfaces
    - Routing tables to determine the path(s) toward the destination WAN gateway
    - Routing process topologies
    - QoS configuration
    - Router capabilities (HW modules, interface types, CPU, and so on)
- Path confirmation, including intermediate hops in the transit network
- Analysis of network path bandwidth, delay, and jitter characteristics using the IOS-based IP service-level agreement (SLA) feature

Requirements to run this tool include the following:

- Two Cisco 2851 or any other IP SLA routers as probes at each end of a network path (total of two probes), or alternatively two routers running Cisco IOS Release 12.4(9)T or above
- PC for controlling probes and collecting data
- Permission to access the customer network, either through a VPN or directly on the network
- Permission to deploy and run a network path analysis tool in the customer network
- Network topology documentation
- Level 15 privileges (enable access) on the IOS routers in the path
- Opening conduits in firewall(s) and/or access control lists (ACL) on the devices present in the transit network

This tool has the following components:

- Cisco IOS routers with a software feature set containing IP SLA features
- Data-gathering unit
- Data-analysis unit
- Reporting unit

IP SLA Probes (Cisco IOS Routers/Switches)

This component is based on two IOS router-based probes (2851), although it can be any router or switch capable of running a Cisco IOS feature set that supports IP SLA and has a minimum of 1 Gigabit Ethernet (GE) interface. The script installs and executes the IP SLA–based tests on the router probes. The probes report on delay, jitter, and packet loss. The details of the IP SLA feature can be found at http://www.cisco.com/en/US/docs/switches/lan/catalyst4500/12.2/44sg/configuration/ guide/swipsla.html.

Essentially, one router generates packets, also known as a *source probe,* toward the other router, known as a *responder probe.* The responder returns these packets to the source with timestamps for analysis by the source probe. In this example, a source probe generates 1500-byte packets with prec 4 to simulate video packets and 80-byte packets with prec 5 to simulate the voice packets. The precedence values can be changed to reflect the values set by the TP codec. The probes are deployed at the two ends/sites that are being analyzed for TelePresence in this example. TelePresence is chosen because it has several video and audio streams, and hence the network path needs to be qualified for all of them. If more than two sites are involved, this test is repeated for every possible pair of communicating sites so that all the possibilities coinciding with a TelePresence call path are covered, as shown in Figure A-1.

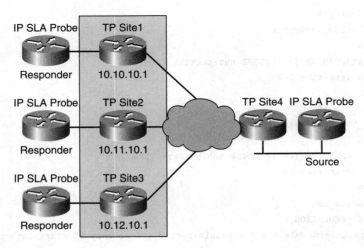

Figure A-1 *Source and Responder Probe Placement in the Network for Multiple-Path Analysis*

The source probe is capable of sending a data stream emulating media traffic to responder probes at multiple destinations, as shown in Figure A-1. The results of the test would be consolidated on the source probe by the **ip sla** test numbers corresponding to each site. In this example, ip lsa test 1 through 4 correspond to the first site, TP Site 1, where a responder with the IP address of 10.10.10.1 has been placed. The following commands are installed on the source probe to test destination TP Site 1:

Note   To test other sites simultaneously, the same commands can be repeated in the configuration with unique **ip sla** test identifiers, and the ip address after the **udp-jitter** command needs to correspond to the responder present at the site under test.

```
ip sla 1
 udp-jitter 10.10.10.1 16386 num-packets 1500 interval 30
 request-data-size 256
 tos 136
 owner Cisco_AS
 tag TP_VIDEO_256bytes
ip sla 2
 udp-jitter 10.10.10.1 16388 num-packets 3000 interval 15
 request-data-size 960
 tos 136
 owner Cisco_AS
 tag TP_VIDEO_960bytes
ip sla 3
 udp-jitter 10.10.10.1 16390 num-packets 5625 interval 12
 request-data-size 1400
 tos 136
 owner Cisco_AS
 tag TP_VIDEO_1400bytes
ip sla 4
 udp-jitter 10.10.10.1 16392 num-packets 1000
 request-data-size 200
 tos 136
 owner Cisco_AS
 tag TP_VOICE_200bytes
ip sla 5
 udp-jitter 10.16.126.17 17000 num-packets 1000
 request-data-size 80
 tos 184
 owner Cisco_AS
 tag VoIP_EMULATION
ip sla group schedule 1 1-5 schedule-period 1 frequency 60 start-time pending
```

The following command is installed on each responder probe:

```
ip sla responder
```

Data Gathering Unit

This unit is based on two Expect scripts that collect the data from WAN access gateway routers in the target network. WAN access gateway routers generally represent the anchor points for the slowest network links in the path. It is assumed that the LAN environment has enough bandwidth to support 10 Mbps or higher streams because most of the connections are 100 Mbps or higher. The LAN QoS is still needed. These scripts can be run remotely on a customer's network provided that customer facilitates VPN access. Alternatively, the scripts can be run through a PC residing on the customer's access. Secure Shell (SSH) or Telnet access to the routers is required.

The first script runs the following commands once, preferably on all the devices in the target network path:

```
show version
show  running-config
show mpls traffic-eng topology
show ip route
show ip interface brief
show ip eigrp topology
show policy-map
show policy-map class
show policy-map interface
show queuing
ping <destination ip address>
tracroute <destination ip address>
```

These commands gather information about the WAN access router because it is most likely to represent a network bottleneck. This set of commands is executed on the command-line interface in **enable** mode or at privilege level 15 to determine whether this router

- Has the right IOS version to support the QoS configuration needed for TelePresence

- Has multiple paths to the destination

- Has implemented a QoS policy

- Has enough bandwidth capacity to support the TelePresence requirements

- Can, if needed, support higher-speed links (slots, backplane capacity, and so on)

The output of these commands is saved in a file and sent to the analysis engine. The user of this tool might need to customize the following script to run it in a customer-specific environment. The possible script changes are shown in italics.

```
#!/opt/sfw/bin/expect
#####################  setup the variables
# $Author: tsiddiqu $
# $Log: expect-script,v $
#
#
# you need to reset the userID,password
# you need to reset the enable password
# Change telnet to ssh if the router only accepts ssh connection in
#  the Main application
# make sure that you have the right path to access expect
##############################################################
# How to run it:
# ./expect-script 1.1.1.1 where 1.1.1.1 is the router IP address
# for massive audit, you can have a main script and call each script
# ./expect-script 1.1.1.1
# ./expect-script 1.1.1.2
#
#
#Change the following variables to match the customer environment
set userID tsiddiqu
set password "Audit#03"
set prompt "\[A-Z,a-z\]#"
set enablepasswd "bf0Ic"

#This is the address of the target router or gateway being analyzed.
set destip "1.1.1.2"
set ROUTERIP [lindex $argv 0]
#
# The commandList can be modify to feed your needs
#
set CommandList {
 "term len 0"
 "show version"
 "show  running-config"
 "show mpls traffic-eng topology"
 "show ip route"
 "show ip interface brief"
 "show ip eigrp topology"
 "show policy-map"
 "show policy-map class"
 "show policy-map interface"
 "show queuing"
 "ping $destip"
 "tracroute $destip"
```

```
  "term len 24"
  ""
}


####################### Setup the procedures
proc processCommand { myCommand } {
global prompt
sleep 1
send "show clock\r"
expect "$prompt"
send "$myCommand\r"
send_log "\n|\n|\n|\n"
expect "$prompt"
expect *
}
####################### end of proc processCommand


####################### Main Application start
spawn telnet $argv
# change to ssh if applicable

log_file "gw-$argv"
sleep 3
expect ame:
send "$userID\r"
expect word:
send "$password\r"
expect ">"
send "en\r"
expect word:
send "$enablepasswd\r"
#
#
# start to send down all the commands
foreach c $CommandList  {
processCommand $c
}
sleep 2
```

The second script is run to periodically collect IP SLA data over time to gather network utilization statistics for the generated traffic that emulates TelePresence traffic in this example. These statistics mainly include packet propagation delay, loss, and jitter in both directions with minimum, average, and maximum values over time.

The possible script changes appear in italics.

```
#!/opt/local/bin/expect
#####################  setup the variables
# $Author: tsiddiqu $
# $Log: expect-script,v $
#
#
# you need to reset the userID,password
# you need to reset the enable password
# Change telnet to ssh if the router only accepts ssh connection in
# the Main application
# make sure that you have the right path to access expect
################################################################
# How to run it:
# For Unix
# ./IPSLA_check-v3.exp
# For MS-DOS using TCLSH
# TCLSH_Promtp% IPSLA_check-v3.exp
#
#
#
# Set path identifier and version for reference
set testnamever "Path1_v1"
set userID tsiddiqu
set password "cisco"
set prompt "\[A-Z,a-z\]#"
set enablepasswd "ciscoAS123"
# Set IP address of the IP SLA responder and source
set responderip "10.16.126.17"
set IPSLAROUTERIP "10.88.1.6"
# Set interval to 15 minutes run the IP Sla Monitor statistic command
set interval 900
# Set the number of iteration for the IP SLA Monitor command to run for
# 8 hours. It can be changed to run for longer. 72 is achived by running it 4 #
  times every hour, eight time.
set cmditeration 72


# The total duration of the test will be multiple of interval and duration
# Specify the first and last IP Sla Operation configured on the IP SLA source #
  router
# IMPORTANT: The IP SLA Operations must be in sequential order without
# skipping any number
# For example you have 5 monitors numbered 1 through 5


set SLAMONITORmin 1
set SLAMONITORmax 5
```

```
# The commandList can be modify to feed your needs#
set CommandList {
 "term len 0"
 "show version"
 "show clock"
 "show  running-config"
 "show ntp status"
 "conf t"
 "ip sla group schedule 1 1-5 schedule-period 1 frequency 60 start-time after
   00:01:00"
 "end"
 "clear counters"
 "y"
}


####################### Setup the procedures
proc processCommand { myCommand } {
global prompt
sleep 1
expect *
send "$myCommand\r"
send_log "\n|\n|\n|\n"
expect *
}
####################### end of proc processCommand

####################### Main Application start
spawn telnet $IPSLAROUTERIP
# change to ssh if applicable

log_file "gw-$testnamever"
sleep 1
expect ame:
send "$userID\r"
expect word:
send "$password\r"
expect ">"
send "en\r"
expect word:
send "$enablepasswd\r"
#
#
# start to send down all the commands
foreach c $CommandList  {
processCommand $c
```

```
}
sleep 2
set count 1
while { $count < $cmditeration } {
    send_log "Iteration number $count begins\n|\n|\n|\n"
                expect "*#"
                send "show clock\r"
                expect "*#"
                send "show ip sla statistics aggregated detail\r"
                sleep 1

    send_log "Iteration number $count ends\n|\n|\n|\n"
    incr count
    sleep $interval


}
expect "*#"
send "show clock"
expect "*#"
send "term len 24"
close
```

Data Analysis Module

The Analysis module is comprised of Awk and Perl scripts that analyze the configuration and IP SLA data.

The configuration data is formatted to show the routing table with the primary and secondary link (if applicable). It also shows the current version, existing link capacity, and potential to increase the link bandwidth by showing the empty slots and the available backplane bandwidth.

This script also analyzes the jitter, delay, and packet loss information using the results of IP SLA probes. A sample of **show ip sla monitor statistics aggregated 10 detail** follows:

```
router# show ip sla monitor statistics aggregated 10 detail


Round trip time (RTT) Index 10
Start Time Index: *08:27:07.920 PST Mon Dec 13 2004
Type of operation: jitter
Voice Scores:
MinOfICPIF: 0 MaxOfICPIF: 0 MinOfMOS: 0 MaxOfMOS: 0
RTT Values
Number Of RTT: 0
RTT Min/Avg/Max: 0/0/0 ms
Latency one-way time milliseconds
Number of Latency one-way Samples: 0
```

```
Source to Destination Latency one way Min/Avg/Max: 0/0/0 ms
Destination to Source Latency one way Min/Avg/Max: 0/0/0 ms
Source to Destination Latency one way Sum/Sum2: 0/0
Destination to Source Latency one way Sum/Sum2: 0/0
Jitter time milliseconds
Number of Jitter Samples: 0
Source to Destination Jitter Min/Avg/Max: 0/0/0 ms
Destination to Source Jitter Min/Avg/Max: 0/0/0 ms
Source to destination positive jitter Min/Avg/Max: 0/0/0 ms
Source to destination positive jitter Number/Sum/Sum2: 0/0/0
Source to destination negative jitter Min/Avg/Max: 0/0/0 ms
Source to destination negative jitter Number/Sum/Sum2: 0/0/0
Destination to Source positive jitter Min/Avg/Max: 0/0/0 ms
Destination to Source positive jitter Number/Sum/Sum2: 0/0/0
Destination to Source negative jitter Min/Avg/Max: 0/0/0 ms
Destination to Source negative jitter Number/Sum/Sum2: 0/0/0
Interarrival jitterout: 0 Interarrival jitterin: 0
Packet Loss Values
Loss Source to Destination: 0 Loss Destination to Source: 0
Out Of Sequence: 0 Tail Drop: 0 Packet Late Arrival: 0
Number of successes: 0
Number of failures: 1
Failed Operations due to over threshold: 0
Failed Operations due to Disconnect/TimeOut/Busy/No Connection: 0/0/0/1
Failed Operations due to Internal/Sequence/Verify Error: 0/0/0
Distribution Statistics:
Bucket Range: 0-19 ms
Avg. Latency: 0 ms
Percent of Total Completions for this Range: 0 %
Number of Completions/Sum of Latency: 0/0
Sum of RTT squared low 32 Bits/Sum of RTT squared high 32 Bits: 0/0
Operations completed over thresholds: 0
```

At the conclusion of IP SLA–based testing, the queues are reanalyzed using the following commands on the WAN access/gateway routers near the source and the responder probes:

```
show policy-map interface
show queuing
```

This information is parsed by the Perl script and reported in a CSV format. This data can be imported into an Excel spreadsheet for more analysis, such as plotting on a line graph, sorting, and filtering as needed.

The massive output file generated by the IP SLA scripts needs to be put into a flat file so that the data structure can be applied after it is imported from the customer's network. For this task, run the following Awk script. This can be run either using Cygwin (or equivalent) or directly on a UNIX workstation.

Awk Script

Awk script can help reduce the amount of information gathered by the Expect script from the IP SLA probes and restructures it for easier processing by the Perl script discussed in the next section. Change the source and the output filenames shown in italics to the desired value when executing the following commands:

```
#!/usr/bin/sh
grep -A 72 "Sunnyvale-ipsla#show clock" gw-Path1-4-29.txt  |  sed '/^$/d' | awk
'BEGIN { RS =" - "; FS ="\n"; ORS=""}  {x=1; while ( x < NF ) { print $x "," ; ++x
} print $NF "\n" }' |  sed 's/,$//' |sed 's/^,//' > flatten-4-29-.txt
```

This puts the output of the IP SLA file in a comma-and-space-delimited file, which is ready to be fed to the Perl script that applies the data structure to extract the useful information and put it in a comma-separated value (CSV) file to be imported to the Excel spreadsheet for charting and reporting.

Perl Script

Simply execute the following script after editing the input filename that was generated by the Awk script previously. The required changes are highlighted in italics.

```
#!/usr/bin/perl


# ->   1 Round Trip Time (RTT) for         Index 1
#      2 Start Time Index: 12:42:29.125 PST Thu Aug 17 2006
#      3 Type of operation: udp-jitter
#      4 Voice Scores:
#      5 MinOfICPIF: 0    MaxOfICPIF: 0    MinOfMOS: 0       MaxOfMOS: 0
#      6 RTT Values:
# ->   7         Number Of RTT: 90000              RTT Min/Avg/Max: 9/9/11
  milliseconds
#      8 Latency one-way time:
# ->   9         Number of Latency one-way Samples: 90000
# ->  10         Source to Destination Latency one way Min/Avg/Max: 3/4/9
  milliseconds
# -> 11          Destination to Source Latency one way Min/Avg/Max: 1/5/7
  milliseconds
#     12          Source to Destination Latency one way Sum/Sum2: 386059/1795349
#     13          Destination to Source Latency one way Sum/Sum2: 452646/2411532
#     14 Jitter Time:
# ->  15         Number of Jitter Samples: 89940
# ->  16         Source to Destination Jitter Min/Avg/Max: 1/1/2 milliseconds
# ->  17         Destination to Source Jitter Min/Avg/Max: 1/1/2 milliseconds
#     18        Source to destination positive jitter Min/Avg/Max: 1/1/2
  milliseconds
#     19        Source to destination positive jitter Number/Sum/Sum2:
  4333/4339/4351
```

```
#    20           Source to destination negative jitter Min/Avg/Max: 1/1/2
     milliseconds
#    21           Source to destination negative jitter Number/Sum/Sum2:
     4341/4351/4371
#    22           Destination to Source positive jitter Min/Avg/Max: 1/1/2
     milliseconds
#    23           Destination to Source positive jitter Number/Sum/Sum2:
     7495/7505/7525
#    24           Destination to Source negative jitter Min/Avg/Max: 1/1/2
     milliseconds
#    25           Destination to Source negative jitter Number/Sum/Sum2:
     7505/7516/7538
#    26           Interarrival jitterout: 0       Interarrival jitterin: 0
#    27 Packet Loss Values:
# -> 28           Loss Source to Destination: 0           Loss Destination to
     Source: 0
# -> 29           Out Of Sequence: 0     Tail Drop: 0     Packet Late Arrival: 0
#    30 Number of successes: 60
#    31 Number of failures: 0
#    32 Failed Operations due to over threshold: 0
#    33 Failed Operations due to Disconnect/TimeOut/Busy/No Connection: 0/0/0/0
#    34 Failed Operations due to Internal/Sequence/Verify Error: 0/0/0
#    35 Distribution Statistics:
#    36 Bucket Range: 0-19 ms
#    37 Avg. Latency: 9 ms
#    38 Percent of Total Completions for this Range: 100 %
#    39 Number of Completions/Sum of Latency: 60/540
#    40 Sum of RTT squared low 32 Bits/Sum of RTT squared high 32 Bits: 4860/0
#    41 Operations completed over thresholds: 0

$sla_file ="./flatten.txt";
open SLA,$sla_file
 or die "Cant open '$sla_file': $!";

# this array captures the set of Index, time based repeat
@Time_Array = ();
$index_count = 0;

while (<SLA>)
{
    #a count for the start of new record
    #a hash - array to store all the has information
    $INDEX_HASH  = {};
    $val1=0;$val2=0;$val3=0;$val4=0;$val5=0;$val6=0;$val7=0;

     $index_val = 0;
```

```perl
@fields = split /,/,$_;
#
# MUST do chomp here , otherwise, it keeps the newline, and the hash table
# ends up with a garbage entry
chomp(@fields);

foreach $entry (@fields)
{
    if ($entry =~ /Round Trip.*(Index (\d+))$/)
    {
        #print "$2\n";
        $index_val = $2;
        #
    }
    elsif ($entry =~ /.*Number Of RTT: (\d+).*RTT Min\/Avg\/Max:
       (\d+)\/(\d+)\/(\d+).*/)
    {
        #print "$1 $2 $3 $4\n";
        #
        #   [0] = Number Of RTT
        #   [1] = Min
        #   [2] = Avg
        #   [3] = Max
        #
        #$SLA_HASH{"RRT"} = [$1,$2.$3,$4];
        $INDEX_HASH{$index_val}{"RRT"} = [$1,$2.$3,$4];
    }
    elsif ($entry =~ /.*Number of Latency one-way Samples: (\d+).*/)
    {
        #
        $val1 = $1;
    }
    elsif ($entry =~ /.*Source to Destination Latency one way Min\/Avg\/Max:
       (\d+)\/(\d+)\/(\d+).*/)
    {
        #
        $val2 = $1;
        $val3 = $2;
        $val4 = $3;
        #print "source laten $1 $2 $3\n";
    }
    elsif ($entry =~ /.*Destination to Source Latency one way Min\/Avg\/Max:
       (\d+)\/(\d+)\/(\d+).*/)
    {
        #
```

```perl
    $val5 = $1;
    $val6 = $2;
    $val7 = $3;
    #print "destin laten $1 $2 $3\n";
    # Latency
    #   val1  = one-way Samples
    # Source
    #      val2  = Min
    #      val3  = Avg
    #      val4  = Max
    # Destination
    #      val5  = Min
    #      val6  = Avg
    #      val7  = Max
    #$SLA_HASH{"Latency"} = [$val1,$val2,$val3,$val4,$val5,$val6,$val7];
    $INDEX_HASH{$index_val}{"Latency"} =
      [$val1,$val2,$val3,$val4,$val5,$val6,$val7];
}
elsif ($entry =~ /.*Number of Jitter Samples: (\d+).*/)
{
    #
    $val1 = $1;
}
elsif ($entry =~ /.*Source to Destination Jitter Min\/Avg\/Max:
  (\d+)\/(\d+)\/(\d+).*/)
{
    #
    $val2 = $1;
    $val3 = $2;
    $val4 = $3;
    #print "source jitte $1 $2 $3\n";
}
elsif ($entry =~ /.*Destination to Source Jitter Min\/Avg\/Max:
  (\d+)\/(\d+)\/(\d+).*/)
{
    #
    $val5 = $1;
    $val6 = $2;
    $val7 = $3;
    #print "destin jitte $1 $2 $3\n";

    $INDEX_HASH{$index_val}{"Jitter"} =
      [$val1,$val2,$val3,$val4,$val5,$val6,$val7];
}
```

```perl
        elsif ($entry =~ /.*Loss Source to Destination: (\d+).*Loss Destination
          to Source:
(\d+).*/)
        {
            #
            $val1 = $1;
            $val2 = $2;
            #print "pkt loss $1 $2\n";
        }
        elsif ($entry =~ /.*Out Of Sequence: (\d+).*Tail Drop: (\d+).*Packet
          Late Arrival:
(\d+).*/)
        {
            #
            $val3 = $1;
            $val4 = $2;
            $val5 = $3;
            #print "out of seq $1 $2 $3\n";

            $INDEX_HASH{$index_val}{"Packet_Loss"} =
              [$val1,$val2,$val3,$val4,$val5];
        }
    }
    #print " jitter — hash:", $SLA_HASH{"Jitter"}[1],"\n";

    #$INDEX_HASH{$index_val} = $SLA_HASH;
    #print "index_val = $index_val\n";

    #print $INDEX_HASH{$index_val}{"Jitter"};
    #print " index — hash jitter — hash:", $INDEX_HASH{1}{"Jitter"},"\n";
    $index_count++;
    # there seem to be 10 Indexs
    if (  $index_count == 9 )
    {
        "print pushing entry\n";
        # SAVE RECORD
         push @Time_Array, $INDEX_HASH;
         $index_count = 0;
    }


}


&print_sla_report();


sub print_index_hash ()
```

```
{
    INDEX: for $index ( sort keys %INDEX_HASH )
    {
        #
        # SKIP INDEXES HERE
        #
        #print $index,"\n";

        if ( $index == 11 || $index == 12 || $index == 13 || $index == 14 ||
          $index == 15)
        {

            #print "SKIPPING INDEX",$index,"\n";
            next INDEX;
        }
        for $values ( keys %{ $INDEX_HASH{$index} }   )
        {
            #PRINTING THE RECORD
            print "[$index]$values\n";
            @sla_array = @{$INDEX_HASH{$index}{$values}};

            #
            # MATCH with the values from the HEADER
            #

        #   [5]Latency
#       60000   4       4       6       4       4       6
        #   [5]RRT
#               60000   99      11
        #   [5]Packet_Loss
#       0       0       0       0       0
        #   [5]Jitter
#       59940   1       1       2       1       1       2
        #

            foreach   ( @sla_array )
            {
                print "\t",$_;
            }
            print "\n";
        }
    }

}
```

```
sub print_sla_report ()
{
    foreach  ( @Time_Array )
    {
        $INDEX_HASH = $_;
        &print_index_hash();
    }
}
```

Data Reporting Module

The Date Reporting module essentially leverages the graphing capabilities of Microsoft Excel spreadsheets. The space-and-comma-delimited files generated by the Perl script in the Data Analysis module can be simply imported into Excel, and the graphs can be plotted.

Table A-1 shows how the static configuration information about the routers and switches in the path is displayed.

Table A-1   *Network Path Details*

Gateway/Switch Address	Ingress Interface(s)	Egress Interface	Redundancy	Delay (ms)	Comments
10.1.1.1	FE 2/30	GE 1/0	Yes	10	7200VXR with NPE-1GRunning 12.2
			No		
10.1.1.2	POS5/0	POS 7/0	Yes	12	To Sprint cloud
			No		
10.6.3.5	POS 4/3	GE1/2	Yes	15	GSR12000
			No		
10.6.48.2	GE4/1	GE4/24	Yes	9	CAT6507
			No		
172.12.18.72	GE3/2	GE4/2	Yes	9	CAT6507
			No		
172.12.90.1	GE 3/0	FE 2/42	Yes	1	CAT3650
			No		

Most of this information was retrieved using the **traceroute** command through the Expect script discussed in the section "Data Gathering Unit," earlier in this chapter. The report still needs manual input from the network engineer performing the test for validating redundancy of the device or the interface.

The IP SLA provides information on delay, jitter, latency, and packet drop from the source probe to the responder probe (S2D) and from the responder to the source probe (D2S) so that the path in both directions is verified. Figure A-2 shows a sample graph of packet jitter experienced by the stream of audio packets sent between the source and the responder probe over a period of two days. Both average and maximum values for the jitter are reported in both directions.

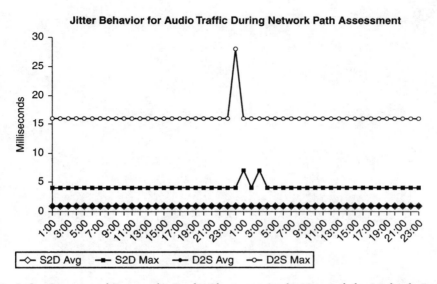

Figure A-2   *Source and Responder Probe Placement in the Network for Multiple-Path Analysis*

The graph shows that the audio test stream experienced an extraordinary amount of jitter between 1:00 a.m. and 3:00 a.m. It was determined that the IT department had scheduled massive system backups during this time frame, which is the potential cause of the spike in jitter value. With this knowledge, the network engineer is able to configure a proper QoS policy before the TelePresence units are deployed in this network, which otherwise meets the expected jitter value of 45–55 ms (or 10 ms peak to peak).

Appendix B

Detailed Call Flows

The following sections cover various call flows for the deployment models discussed in Chapter 3, "VoIP Deployment Models in Service Provider Networks," Chapter 4, "Internet Telephony," and Chapter 5, "VoIP Deployment Models in Enterprise Networks." These call flows explain the interaction between the devices, including Call Agent, Cisco Unified Communications Manager (CUCM), IP phones, gateways, and gatekeepers. They also provide context to performance and trend analysis discussion, which were discussed in Chapter 7, "Performance Analysis and Fault Isolation," and Chapter 8, "Trend Analysis and Optimization."

MGCP Call Flows

The PacketCable deployment model discussed in Chapter 3 uses Media Gateway Control Protocol (MGCP) for call control. In this section, we discuss two different call flows:

- On-net to On-net call flow

- On-net to Off-net call flow

Figures B-1 through B-3 illustrate an On-net to On-net MGCP call.

The callouts listed in the On-net to On-net call flow explain the various events during call setup and teardown.

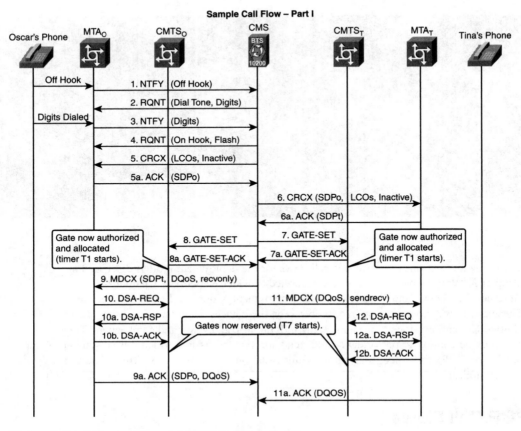

Figure B-1   *On-net to On-net MGCP Call Part 1 of 3*

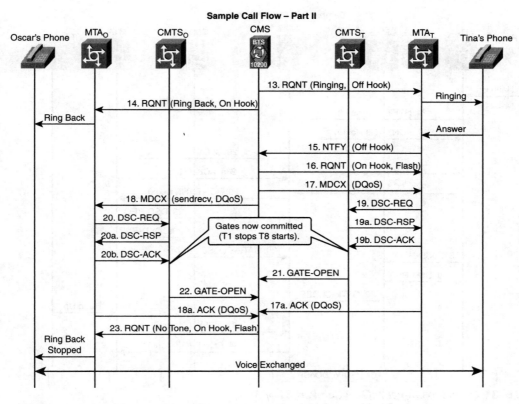

Figure B-2 *On-net to On-net MGCP Call Part 2 of 3*

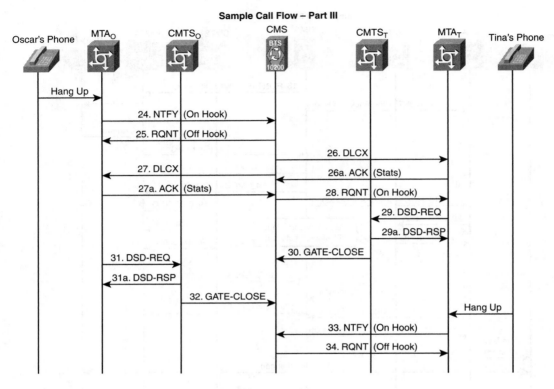

Figure B-3   *On-net to On-net MGCP Call Part 31 of 3*

Figures B-4 through B-8 illustrate an On-net to Off-net MGCP call.

The callouts listed in the On-net to Off-net call flow explain the various events during call setup and teardown.

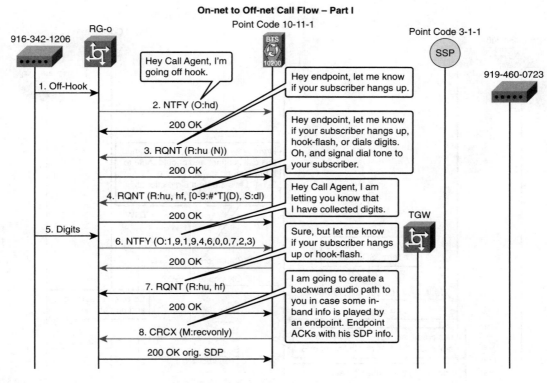

Figure B-4   *On-net to Off-net MGCP Call Part 1 of 5*

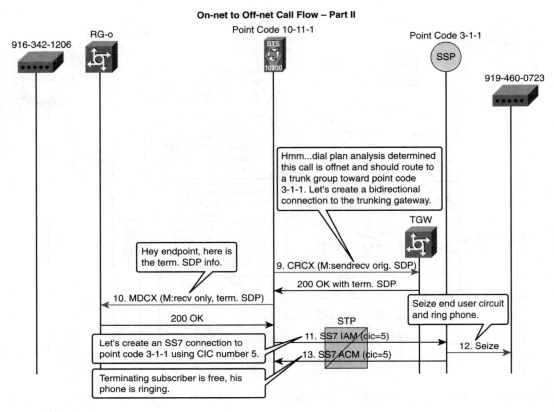

Figure B-5   *On-net to Off-net MGCP Call Part 2 of 5*

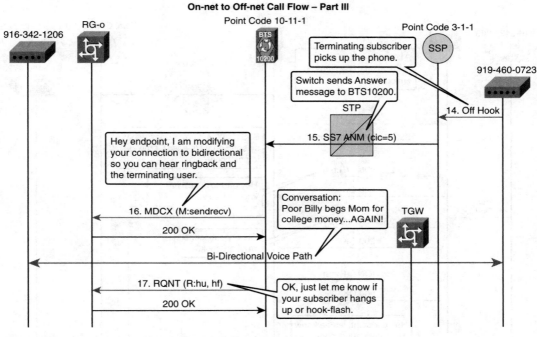

Figure B-6   *On-net to Off-net MGCP Call Part 3 of 5*

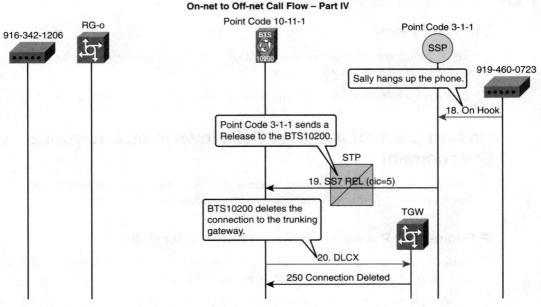

Figure B-7   *On-net to Off-net MGCP Call Part 4 of 5*

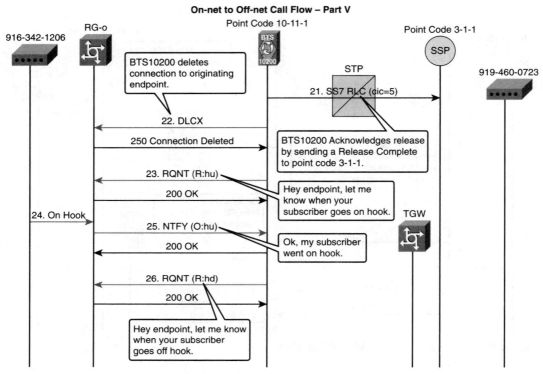

Figure B-8 *On-net to Off-net MGCP Call Part 5 of 5*

SIP Call Flow

The following is an example for Session Initiation Protocol (SIP)–based calls for the models discussed in Chapters 3 and 4.

Figure B-9 illustrates call flow for an SIP-based call.

Unified Communications Call Flows in an Enterprise Environment

This section discusses the common On-net and Off-net call scenarios in an enterprise environment with the help of call traces and flow diagrams.

IP Phone to IP Phone—Successful Intracluster Call

In this scenario, Phone A is registered to Cisco Unified Communication Manager X denoted by CUCM X. Phone B is registered to the same CUCM X.

A call is placed from Phone A (1000) to Phone B (1006).

Figure B-10 illustrates this basic setup for a call between two IP phones registered to the same CUCM.

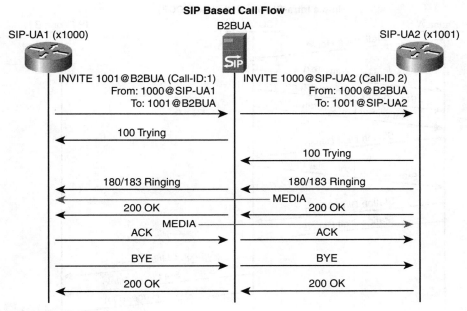

Figure B-9   *SIP-Based Call*

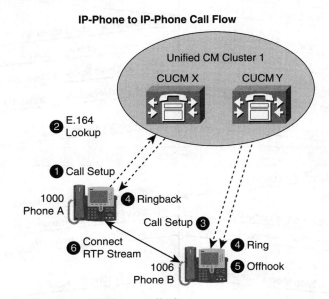

Figure B-10   *IP Phone–to–IP Phone Call Flow*

Figure B-11 illustrates the sequence of Skinny Call Control Protocol (SCCP) messages exchanged between the Unified CM (CUCM X) and the two IP phones described in Figure B-10.

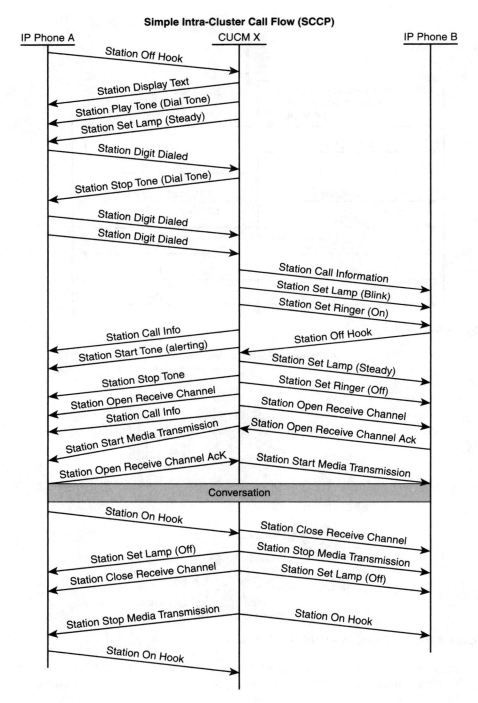

Figure B-11 *Simple Intracluster Call Flow (SCCP)*

The details of the call flow in Figure B-11 are as follows. Note that all traces are generated on CUCM X because both phones are registered to it. If Phone B were registered to Cisco Unified Communications Manager Y denoted by CUCM Y, we would have to collect traces from both servers.

Phone 1000 goes to OffHook:

```
03/20/2002 23:14:49.517 Cisco CallManager
|StationInit: 520443c OffHook.|
<CLID::WWCCM1-
Cluster><NID::172.21.51.216><CT::1,100,95,1.1243><IP::10.17.168.70><DEV::SEP0002
  FDAEFB9D>
```

Header is common to all the messages:

```
03/20/2002 23:14:49.517 Cisco CallManager
```

Trailer is common to all the messages to and from this phone:

```
<CLID::WWCCM1-
Cluster><NID::172.21.51.216><CT::1,100,95,1.1243><IP::10.17.168.70><DEV::SEP0002
  FDAEFB9D>
```

The 520443c is a unique ID for this phone. StationInit indicates that CUCM received a TCP message from a Skinny station (the IP phone).

CUCM sends a message to prompt the text message "Enter Number" on the phone and changes the softkeys on the phone.

```
-|Locations: reserve: cdccPID=(1.13.64) Orig=Dest=0 no need to reserve bw.|
-|StationD:     520443c DisplayText text='                        1000
  |StationD:       520443c SetLamp stimulus=9(Line) stimulusInstance=1
lampMode=2(LampOn).|
-|StationD:     520443c CallState callState=1(OffHook) lineInstance=1
callReference=16777239|
-|StationD:     520443c DisplayPromptStatus timeOutValue=0 promptStatus='Enter
  number'
    lineInstance=1 callReference=16777239.|
-|StationD:     520443c SelectSoftKeys instance=1 reference=16777239
  softKeySetIndex=4
    validKeyMask=-1.|
-|StationD:     520443c ActivateCallPlane lineInstance=1.|
```

StationD indicates that CUCM sends a message to the Skinny station. "lineInstance=1" means line 1 of the phone. For a Skinny call, each endpoint in a call is assigned a callReference, so for a two-way conversation, you will have two callReferences.

CUCM begins the routing process; the current set of dialed digits is empty. Every route pattern in the table is a potential match at this time. (These messages are an internal routing process within the CUCM, not sent to the phones.)

```
- |Insert an entry into CiCcp table, now this table has 1 entries|
-|Insert an entry into CiCcp table, now this table has 2 entries|
-|Digit analysis: match(fqcn="1000", cn="1000", pss="PA:Employee:CER", dd="")|
-|Digit analysis: potentialMatches=PotentialMatchesExist|
```

CUCM triggers the dial tone on the phone.

```
-StationD:      520443c StartTone tone=33(InsideDialTone).
```

CUCM sets the lamp state.

```
|StationD:      520443c SetLamp stimulus=9(Line) stimulusInstance=1
lampMode=2(LampOn).|
```

Note The following list explains some of the commonly occurring parameters for digit analysis:

- **Fqcn:** Fully Qualified Calling Party Number

- **Cn:** Calling Party Number

- **Pss:** Partition Search Space (the ordered list of partitions that make up the calling search space for this device and line)

- **dd:** Dialed Digits

- **Pss:** Indicates that Phone Search Space contains three partitions: PA, Employee, and CER

An indication of the type of dial tone is embedded within the dial tone. In this case, it is OutsideDialTone (tone 34). The user starts to enter the digits; the first digit, 1, is dialed:

```
|StationInit: 520443c KeypadButton kpButton=1.|
```

CUCM stops sending the dial tone to the phone 1000 and collects the other digits entered through the keypad (in the following trace, 1006 is dialed):

```
|StationD:      520443c StopTone.|
|StationD:      520443c SelectSoftKeys instance=1 reference=16777230
  softKeySetIndex=6 validKeyMask=-1.|
|Digit analysis: match(fqcn="1000", cn="1000", pss="PA:Employee:CER", dd="1")|
|Digit analysis: potentialMatches=PotentialMatchesExist|
|StationInit: 520443c KeypadButton kpButton=0.|
|Digit analysis: match(fqcn="1000", cn="1000", pss="PA:Employee:CER", dd="10")|
|Digit analysis: potentialMatches=PotentialMatchesExist|
|StationInit: 520443c KeypadButton kpButton=0.|
```

```
|Digit analysis: match(fqcn="1000", cn="1000", pss="PA:Employee:CER", dd="100")|
|Digit analysis: potentialMatches=PotentialMatchesExist|
|StationInit: 520443c KeypadButton kpButton=6.|
|Digit analysis: match(fqcn="1000", cn="1000", pss="PA:Employee:CER", dd="1006")|
```

Note that PotentialMatchesExist is put in the trace file for as long as there are still dial plan entries that *potentially* (that is, *partially*) match the accumulated digits.

The user starts to enter the digits; the first digit, 1, is dialed:

```
|StationInit: 520443c KeypadButton kpButton=1.|
```

CUCM stops sending the dial tone to the phone 1000 and collects the other digits entered through the keypad (in this example, 1006 is dialed):

```
|StationD:      520443c StopTone.|
|StationD:      520443c SelectSoftKeys instance=1 reference=16777230
  softKeySetIndex=6 validKeyMask=-1.|
|Digit analysis: match(fqcn="1000", cn="1000", pss="PA:Employee:CER", dd="1")|
|Digit analysis: potentialMatches=PotentialMatchesExist|
|StationInit: 520443c KeypadButton kpButton=0.|
|Digit analysis: match(fqcn="1000", cn="1000", pss="PA:Employee:CER", dd="10")|
|Digit analysis: potentialMatches=PotentialMatchesExist|
|StationInit: 520443c KeypadButton kpButton=0.|
|Digit analysis: match(fqcn="1000", cn="1000", pss="PA:Employee:CER", dd="100")|
|Digit analysis: potentialMatches=PotentialMatchesExist|
|StationInit: 520443c KeypadButton kpButton=6.|
|Digit analysis: match(fqcn="1000", cn="1000", pss="PA:Employee:CER", dd="1006")|
```

A match is found. Here are the CUCM digit analysis results:

```
|Digit analysis: analysis results|
|PretransformCallingPartyNumber=1000
|CallingPartyNumber=1000
|DialingPartition=Employee
|DialingPattern=1006
|DialingRoutePatternRegularExpression=(1006)
|DialingWhere=
|PatternType=Enterprise
|PotentialMatches=NoPotentialMatchesExist *
|DialingSdlProcessId=(1,34,20)
|PretransformDigitString=1006
|PretransformTagsList=SUBSCRIBER
|PretransformPositionalMatchList=1006
|CollectedDigits=1006
|UnconsumedDigits=
|TagsList=SUBSCRIBER
```

```
|PositionalMatchList=1006
|VoiceMailbox=1006
|DisplayName= Luc Bouchard
|RouteBlockFlag=RouteThisPattern
|InterceptPartition=Employee
|InterceptPattern=1006
|InterceptWhere=
|InterceptSdlProcessId=(1,25,1)
|InterceptSsType=16777220
|InterceptSsKey=16|
```

CUCM sends the call state information to indicate an incoming call on the called party phone:

```
|StationD:    52044a0 DisplayText text='                     1006        '.|
|StationD:    52044a0 CallState callState=4(RingIn) lineInstance=1
```

CUCM sends the calling party information to the called phone:

```
|StationD:    52044a0 CalInfo callingPartyName='Ramesh Kaza' callingParty=1000
 cgpnVoiceMailbox=1000   calledPartyName='Luc Bouchard' calledParty=1006
 cdpnVoiceMailbox=1006 originalCalledPartyName='Luc Bouchard'
originalCalledParty=1006
originalCdpnVoiceMailbox=1006 originalCdpnRedirectReason=0
lastRedirectingPartyName='Luc Bouchard' lastRedirectingParty=1006
lastRedirectingVoiceMailbox=1006 lastRedirectingReason=0
callType=1(InBound) lineInstance=1 callReference=16777240.|
```

CUCM sends the signal to the called party phone to blink the lamp:

```
|StationD:    52044a0 SetLamp stimulus=9(Line) stimulusInstance=1
lampMode=5(LampBlink).|
```

CUCM instructs Phone B to ring:

```
|StationD:    52044a0 SetRinger ringMode=2(InsideRing).|
```

CUCM sends the alerting tone (ringback) to the calling phone:

```
|StationD:    520443c StartTone tone=36(AlertingTone).|
```

CUCM received an OffHook message from the called party phone (1006); note that the IP address and its MAC address are shown in the traces:

```
|StationInit: 52044a0 OffHook.|<CLID::WWCCM1-
  Cluster><NID::172.21.51.216><CT::1,100,95,1.1248><IP::10.17.168.75><DEV::
  SEP00036B54BD01>
|StationD:    52044a0 ClearNotify.|
```

CUCM notifies the called phone to stop the ring tone:

```
|StationD:    52044a0 SetRinger ringMode=1(RingOff).|
```

Set the lamp on the called party phone to a steady state and update the CallState:

```
|StationD:     52044a0 SetLamp stimulus=9(Line) stimulusInstance=1
  lampMode=2(LampOn).|
|StationD:     52044a0 CallState callState=1(OffHook) lineInstance=1
  callReference=16777232|
|StationD:     52044a0 ActivateCallPlane lineInstance=1.|
```

CUCM instructs the calling party phone to stop the ringback tone:

```
|StationD: 520443c StopTone.|
```

CUCM asks the phones to give the Real-time Transfer Protocol (RTP) address from the phones:

```
|StationD: 520443c OpenReceiveChannel conferenceID=0 passThruPartyID=177
  millisecondPacketSize=20 compressionType=4(Media_Payload_G711Ulaw64k)
  qualifierIn=?. myIP: 46a8110a (10.17.168.70)|
|StationD: 52044a0 StopTone.|
|StationD: 52044a0 OpenReceiveChannel conferenceID=0 passThruPartyID=193
  millisecondPacketSize=20 compressionType=4(Media_Payload_G711Ulaw64k)
  qualifierIn=?. myIP: 4ba8110a (10.17.168.75)|
```

Some messages in the CUCM traces represent IP addresses in hexadecimal format. To convert the hexadecimal number to an IP address, first take each octet of the IP address starting from the end of string "0x4ba8110a" as it appears in the traces and arrange it like 0A 11 A8 4B. Then convert each piece to decimal:

0A = 10

11 = 17

A8 = 168

4B = 75

So, the IP address is 10.17.168.75.

The called party phone responds first:

```
StationInit: 52044a0 OpenReceiveChannelAck Status=0, IpAddr=0x4ba8110a,
Port=31016, PartyID=193|
```

Tell the calling party phone to start talking to the called party's RTP address; at this point, the called party can hear the calling party:

```
StationD:     520443c StartMediaTransmission conferenceID=0 passThruPartyID=193
  remoteIpAddress=4ba8110a(10.17.168.75) remotePortNumber=31016
  milliSecondPacketSize=20 compressType=4(Media_Payload_G711Ulaw64k)
  qualifierOut=?. myIP: 46a8110a (10.17.168.70)|
```

The calling party responds with the RTP details:

```
StationInit: 520443c OpenReceiveChannelAck Status=0, IpAddr=0x46a8110a,
  Port=28052, PartyID=177|
```

Inform the called party about the calling party's RTP information:

```
StationD:    52044a0 StartMediaTransmission conferenceID=0 passThruPartyID=193
  remoteIpAddress=46a8110a(10.17.168.70) remotePortNumber=28052
  milliSecondPacketSize=20 compressType=4(Media_Payload_G711Ulaw64k)
  qualifierOut=?.  myIP: 4ba8110a (10.17.168.75)|
```

Finally, the calling party phone goes OnHook:

```
|StationInit: 520443c OnHook.|
|StationD:    520443c SetSpeakerMode speakermode=2(Off).|
|StationD:    520443c ClearPromptStatus lineInstance=1 callReference=16777239.|
|StationD:    520443c CallState callState=2(OnHook) lineInstance=1
  callReference=16777239|
|StationD:    520443c SelectSoftKeys instance=0 reference=0 softKeySetIndex=0
  validKeyMask=7.|
|StationD:    520443c DisplayPromptStatus timeOutValue=0 promptStatus='Your
  current options' lineInstance=0 callReference=0.|
|StationD:    520443c ActivateCallPlane lineInstance=0.|
|StationD:    520443c SetLamp stimulus=9(Line) stimulusInstance=1
  lampMode=1(LampOff).|
|StationD:    520443c DefineTimeDate timeDateInfo=? systemTime=1016741367.|
|StationD:    520443c StopTone.|
```

CUCM instructs the phones to close the media channel and stop the media transmission:

```
|StationD:    520443c CloseReceiveChannel conferenceID=0 passThruPartyID=177.
  myIP: 46a8110a (10.17.168.70)|
|StationD:    520443c StopMediaTransmission conferenceID=0 passThruPartyID=177.
  myIP: 46a8110a (10.17.168.70)|
|StationD:    52044a0 CloseReceiveChannel conferenceID=0 passThruPartyID=193.
  myIP: 4ba8110a (10.17.168.75)|
|StationD:    52044a0 StopMediaTransmission conferenceID=0 passThruPartyID=193.
  myIP: 4ba8110a (10.17.168.75)|
```

CUCM updates the status of the called party phone to OnHook and changes the prompt status on the phone:

```
|StationD:    52044a0 DefineTimeDate timeDateInfo=? systemTime=1016741367.|
|StationD:    52044a0 SetSpeakerMode speakermode=2(Off).|
|StationD:    52044a0 ClearPromptStatus lineInstance=1 callReference=16777240.|
|StationD:    52044a0 CallState callState=2(OnHook) lineInstance=1
  callReference=16777240|
|StationD:    52044a0 SelectSoftKeys instance=0 reference=0 softKeySetIndex=0
  validKeyMask=7.|
|StationD:    52044a0 DisplayPromptStatus timeOutValue=0 promptStatus='Your
  current options' lineInstance=0 callReference=0.|
|StationD:    52044a0 ActivateCallPlane lineInstance=0.|
|StationD:    52044a0 SetLamp stimulus=9(Line) stimulusInstance=1
  lampMode=1(LampOff).|
|StationD:    52044a0 DefineTimeDate timeDateInfo=? systemTime=1016741367.|
|StationD:    52044a0 StopTone.|
```

IP Phone to Voice Gateway Using MGCP

In this scenario, Phone A is registered to the Unified CM (CUCM). Phone B is connected to a carrier's central office (CO) Switch. CUCM will access Phone B through the voice gateway.

Figure B-12 illustrates this basic setup for a call set up between an IP phone registered to the CUCM and another phone connected to a CO switch in the public switched telephone network (PSTN) through a voice gateway. This voice gateway is registered with the CUCM using Media Gateway Control Protocol (MGCP).

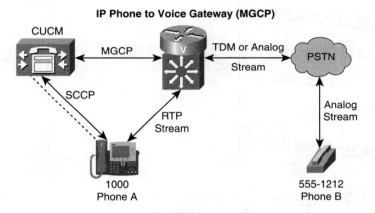

IP Phone to Voice Gateway (MGCP)

Figure B-12   *IP Phone–to–Voice Gateway Call Flow*

A call is placed from Phone A (1000) to Phone B (555-1212). Phone A will hang up first.

MGCP is used to relay ISDN messages that go between the CUCM and the actual CO switch.

Figures B-13 and B-14 include call flows that show ISDN messages going to the gateway (GW); they are really being sent to the CO switch through the gateway.

The details of the call flow in Figures B-13 and B-14 are as follows:

Phone 1000 goes to OffHook:

```
|StationInit: 51ca448 OffHook.|
```

CUCM instructs the phone to change the lamp status and changes the display on the phone:

```
|StationD:    51ca448 DisplayText text='                        1000        '.|
|StationD:    51ca448 SetLamp stimulus=9(Line) stimulusInstance=1
   lampMode=2(LampOn).|
|StationD:    51ca448 DisplayPromptStatus timeOutValue=0 promptStatus='Enter
   number' lineInstance=1 callReference=16777365.|
```

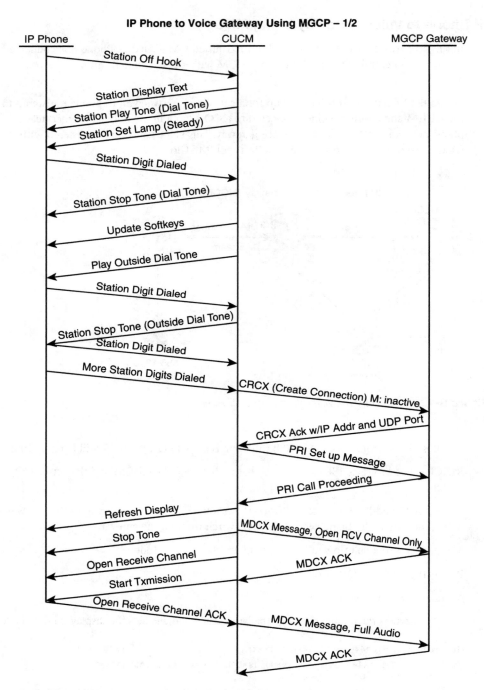

Figure B-13   *IP Phone to Voice Gateway—Part 1 of 2*

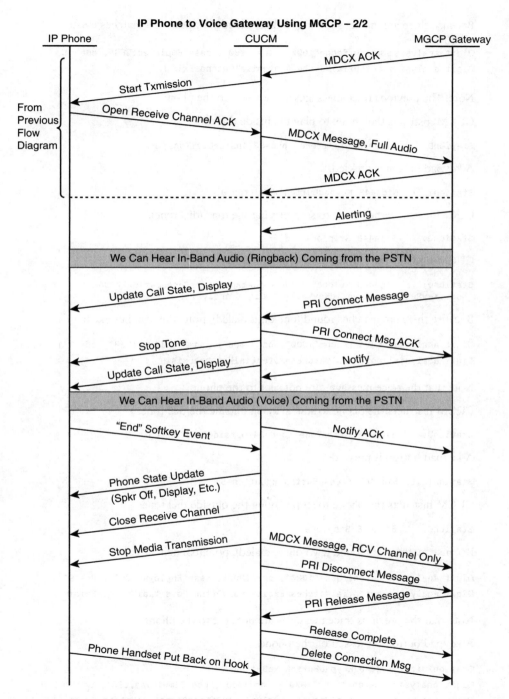

Figure B-14   *IP Phone to Voice Gateway—Part 2 of 2*

Because there are matches found (no digit dialed yet), potential matches exist:

```
|Digit analysis: match(fqcn="1000", cn="1000", pss="Employee:CER", dd="")|
|Digit analysis: potentialMatches=PotentialMatchesExist|
```

Note: The previous trace messages are not sent to the phone.

CUCM instructs the phone to play the inside dial tone:

```
|StationD:      51ca448 StartTone tone=33(InsideDialTone).|
```

A keypad button is pressed:

```
|StationInit: 51ca448 KeypadButton kpButton=9.|
```

CUCM instructs the phone to stop playing the tone (dial tone):

```
|StationD:      51ca448 StopTone.|
```

CUCM updates the softkeys on the phone display:

```
|StationD:      51ca448 SelectSoftKeys instance=1 reference=16777365
   softKeySetIndex=6 validKeyMask=-1.|Digit analysis: |
```

Because there are matches found (one digit dialed), potential matches exist:

```
|Digit analysis: match(fqcn="1000", cn="1000", pss="Employee:CER", dd="9")|
|Digit analysis: potentialMatches=PotentialMatchesExist|
```

Note that the trace messages are not sent to the phone.

CUCM instructs the phone to start playing outside the dial tone:

```
|StationD:      51ca448 StartTone tone=34(OutsideDialTone).|
```

A keypad button is pressed:

```
|StationInit: 51ca448 KeypadButton kpButton=5.|
```

CUCM instructs the phone to stop playing the outside dial tone:

```
|StationD:      51ca448 StopTone.|
```

Because matches are found (two digits dialed), potential matches exist:

```
|Digit analysis: match(fqcn="1000", cn="1000", pss="Employee:CER", dd="95")|
|Digit analysis: potentialMatches=ExclusivelyOffnetPotentialMatchesExist|
```

Note that the previous trace messages are not sent to the phone.

A keypad button is pressed on the phone:

```
|StationInit: 51ca448 KeypadButton kpButton=5.|
|Digit analysis: match(fqcn="1000", cn="1000", pss="Employee:CER", dd="955")|
|Digit analysis: potentialMatches=ExclusivelyOffnetPotentialMatchesExist|
```

Note that the previous trace messages are not sent to the phone.

Digit analysis is done:

```
|Digit analysis: match(fqcn="1000", cn="1000", pss="Employee:CER", dd="95551212")|
|Digit analysis: analysis results|<CT::1,100,95,1.561090>
|PretransformCallingPartyNumber=1000
|CallingPartyNumber=1000
|DialingPartition=Employee
|DialingPattern=9.
    */ the @ sign represents the North American Dial Plan
|DialingRoutePatternRegularExpression=(9)([2-9][02-9]X)(XXXX)
    */ this is the actual expression matched
|DialingWhere=
|PatternType=National
|PotentialMatches=NoPotentialMatchesExist
|DialingSdlProcessId=(1,58,7)
|PretransformDigitString=95551212
|PretransformTagsList=ACCESS-CODE:OFFICE-CODE:SUBSCRIBER
|PretransformPositionalMatchList=9:555:1212
|CollectedDigits=5551212
|UnconsumedDigits=
|TagsList=OFFICE-CODE:SUBSCRIBER
|PositionalMatchList=555:1212
|VoiceMailbox=
|DisplayName=
|RouteBlockFlag=RouteThisPattern
```

The @ sign is a symbolic representation of the 300+ dial plan entries that make up the North American Numbering Plan.

As you use the @ sign in partitions, you must be aware that more than 300 dial plan entries are invoked each time.

Call Manager sends a connection request to GW:

```
MGCPHandler send msg SUCCESSFULLY to: 172.21.57.243
CRCX 280 S0/DS1-0/23@SDA00033282D8E3 MGCP 0.1
C: D000000001000096000000000000000002
M: inactive|*
```

Inactive refers to the reservation of a DS0 channel.

GW sends an acknowledgment (ACK), including the port number:

```
MGCPHandler received msg from: 172.21.57.243
200 280
I: 23
v=0
```

```
o=- D00000000100009600000000000000002 0 IN IP4 172.21.57.243
s=Cisco SDP 0
c=IN IP4 172.21.57.243
t=0 0
m=audio 30656 RTP/AVP 0|*
```

Note the port number of GW's receive channel.

CUCM recognizes the ACK and port number to be used:

```
MGCPHandler recv CRCX Ack with RTP PortNum: 30656
```

CUCM sends the PRI setup message to the GW:

```
|Out Message — PriSetupMsg — Protocol= PriNi2Protocol|
|Ie - Ni2BearerCapabilityIe IEData= 04 03 80 90 A2 |
|Ie - Q931ChannelIdIe IEData= 18 03 A9 83 97 |
|Ie - Q931DisplayIe IEData= 28 0E 52 61 6A 65 73 68 20 52 61 6D 61 72 61 6F |
|Ie - Q931CallingPartyIe IEData= 6C 06 00 80 31 30 30 30 |
|Ie - Q931CalledPartyIe IEData= 70 08 A1 35 35 35 31 32 31 32 |
|MMan_Id= 0. (iep= 0 dsl= 0 sapi= 0 ces= 0 IpAddr=f33915ac IpPort=2427)|
|IsdnMsgData2= 08 02 00 02 05 04 03 80 90 A2 18 03 A9 83 97 28 0E 52 61 6A 65 73
    68 20 52 61 6D 61 72 61 6F 6C 06 00 80 31 30 30 30 70 08 A1 35 35 35 31 32 31
    32 |
```

> **Note**   The calling party's name is sent in hexadecimal digits. Here is the ASCII conversion of the hexadecimal digits in the previous trace.
>
> 52 61 6A 65 73 68 20 52 61 6D 61 72 61 6F = Rajesh Ramarao
>
> 31 30 30 30 = 1000
>
> 35 35 35 31 32 31 32 = 5551212
>
> Note that the PRI message is seen twice here: once in a "decoded" fashion and once in "raw" form (IsdnMsgData2).
>
> The ASCII conversion was done for this presentation. It is not included in the actual trace file.
>
> This message is intended for the other end of the ISDN link, meaning the CO switch. It is not, per se, an MGCP message.

GW sends the PRI a call proceeding message:

```
In  Message — PriCallProceedingMsg — Protocol= PriNi2Protocol|
|Ie - Q931ChannelIdIe — IEData= 18 03 A9 83 97 |
|MMan_Id= 0. (iep= 0 dsl= 0 sapi= 0 ces= 0 IpAddr=f33915ac IpPort=2427)|
|IsdnMsgData1= 08 02 80 02 02 18 03 A9 83 97 |
```

CUCM refreshes the phone display information:

```
IStationD:     51ca448 CallState callState=12(Proceed) lineInstance=1
callReference=16777365I
IStationD:     51ca448 CallInfo callingPartyName='Rajesh Ramarao'
   callingParty=1000 cgpnVoiceMailbox=1000 calledPartyName='' calledParty=5551212
   cdpnVoiceMailbox= originalCalledPartyName='' originalCalledParty=5551212
   originalCdpnVoiceMailbox= originalCdpnRedirectReason=0
   lastRedirectingPartyName='' lastRedirectingParty= lastRedirectingVoiceMailbox=
   lastRedirectingReason=0 callType=2(OutBound) lineInstance=1
   callReference=16777365.I
IStationD::star_StationOutputCallInfo(): callInfo: CI=16777365,
   CallingPartyName=Rajesh Ramarao,  CallingParty=1000, CalledPartyName=,
   CalledParty=5551212, OriginalCalledPartyName=, OriginalCalledParty=5551212,
   lastRedirectingPartyName=, lastRedirectingParty=I
IStationD:     51ca448 DialedNumber dialedNumber=95551212 lineInstance=1
   callReference=16777365.I
```

CUCM sends a stop tone message to phone:

```
IStationD:     51ca448 StopTone.I
```

CUCM sends an open receive channel to phone:

```
IStationD:     51ca448 OpenReceiveChannel conferenceID=0 passThruPartyID=513
   millisecondPacketSize=20 compressionType=4(Media_Payload_G711Ulaw64k)
   qualifierIn=?.
myIP: 46a8110a (10.17.168.70)I
```

CUCM sends a modify connection denoted by an MDCX message to the GW, opening a receive (rcv) channel only:

```
IMGCPHandler send msg SUCCESSFULLY to: 172.21.57.243
MDCX 281 S0/DS1-0/23@SDA00033282D8E3 MGCP 0.1
C: D000000001000096000000000000000002
I: 23
X: 17
L: p:20, a:PCMU, s:on
M: recvonly
R: D/[0-9ABCD*#]
Q: process,loop
```

GW sends an MDCX acknowledgment denoted by ack:

```
IMGCPHandler received msg from: 172.21.57.243
   200 281
```

CUCM identifies TransID:

```
IMGCPHandler received RESP header w/ transId= 281 FOUND a match for MDCXI
```

CUCM tells the phone to start transmitting.

```
|StationD:    51ca448 StartMediaTransmission conferenceID=0 passThruPartyID=513
  remoteIpAddress=f33915ac(172.21.57.243) remotePortNumber=30656
  milliSecondPacketSize=20 compressType=4(Media_Payload_G711Ulaw64k)
  qualifierOut=?.  myIP: 46a8110a (10.17.168.70)| *Note the port number of GW
  receive channel
```

Phone ACKs open receive channel:

```
|StationInit: 51ca448 OpenReceiveChannelAck Status=0, IpAddr=0x46a8110a,
  Port=27132, PartyID=513| *Note the port number of phone's receive channel
```

CUCM sends a GW MDCX to send/receive:

```
|MGCPHandler send msg SUCCESSFULLY to: 172.21.57.243
MDCX 282 S0/DS1-0/23@SDA00033282D8E3 MGCP 0.1
C: D0000000010000960000000000000002
I: 23
X: 17
L: p:20, a:PCMU, s:on
M: sendrecv
R: D/[0-9ABCD*#]
S:
Q: process,loop
v=0
o=- 35 0 IN EPN S0/DS1-0/23@SDA00033282D8E3
s=Cisco SDP 0
t=0 0
c=IN IP4 10.17.168.70
m=audio 27132 RTP/AVP 96 *Note the port number of phone's receive channel
  a=rtpmap:96 PCMU
```

GW sends an MDCX acknowledgment (ack):

```
|MGCPHandler received msg from: 172.21.57.243
200 282
GW sends alerting message
|In  Message — PriAlertingMsg — Protocol= PriNi2Protocol|
|Ie - Q931ProgressIndIe — IEData= 1E 02 84 88 |
|MMan_Id= 0. (iep=  0 dsl=  0 sapi=  0 ces= 0 IpAddr=f33915ac IpPort=2427)|
|IsdnMsgData1= 08 02 80 02 01 1E 02 84 88 |
```

*The last nibble (8h, or 1000b) indicates that "In-band information, or a
appropriate pattern is now available"; The ringback tone will be coming from
the PSTN in this case.

CUCM refreshes the IP phone call state and the display on the phone:

```
StationD:    51ca448 CallInfo callingPartyName='Rajesh Ramarao' callingParty=1000
  cgpnVoiceMailbox=1000 calledPartyName='' calledParty=5551212 cdpnVoiceMailbox=
  originalCalledPartyName='' originalCdpnVoiceMailbox=
  originalCdpnRedirectReason=0 lastRedirectingPartyName=''
  lastRedirectingParty=5551212 lastRedirectingVoiceMailbox=
  lastRedirectingReason=0 callType=2(OutBound) lineInstance=1
  callReference=16777365.|
StationD::star_StationOutputCallInfo(): callInfo: CI=16777365,
  CallingPartyName=Rajesh Ramarao,  CallingParty=1000, CalledPartyName=,
  CalledParty=5551212, OriginalCalledPartyName=, OriginalCalledParty=,
  lastRedirectingPartyName=, lastRedirectingParty=|
StationD:    51ca448 CallState callState=3(RingOut) lineInstance=1
  callReference=16777365|
StationD:    51ca448 SelectSoftKeys instance=1 reference=16777365
  softKeySetIndex=8 validKeyMask=-1.|
StationD:    51ca448 DisplayPromptStatus timeOutValue=0 promptStatus='Ring Out'
  lineInstance=1 callReference=16777365.|
```

Note In the CUCM environment, for public switched telephone network (PSTN) calls, the ringback audio is provided by the PSTN as in-band information. This is the case whether the destination station is ISDN or not.

In this case, the in-band information is ringback. It could have been an intercept of some sort or other audio information. Note that the phone now displays ringout to visually indicate that the other end is in Alerting mode.

GW sends the PRI a connect message:

```
In  Message — PriConnectMsg — Protocol= PriNi2Protocol|
Ie - Q931ProgressIndIe — IEData= 1E 02 84 82 |
MMan_Id= 0. (iep=  0 dsl=  0 sapi=  0 ces= 0 IpAddr=f33915ac IpPort=2427)|
IsdnMsgData1= 08 02 80 02 07 1E 02 84 82 |
```

The last nibble (2h, or 0010b) indicates that "Destination address is non-ISDN."

CUCM sends an ACK:

```
Out Message — PriConnectAcknowledgeMsg — Protocol= PriNi2Protocol|
MMan_Id= 0. (iep=  0 dsl=  0 sapi=  0 ces= 0 IpAddr=f33915ac IpPort=2427)|
IsdnMsgData2= 08 02 00 02 0F |
```

CUCM tells the phone to stop the tone:

```
StationD:    51ca448 StopTone.|
```

The call is connected. The PSTN's CO switch is the source of that message. CUCM tells the phone to stop the tone. In this case, it is redundant. CUCM updates the phone state and display:

```
|StationD:     51ca448 CallState callState=5(Connected) lineInstance=1
   callReference=16777365|
|StationD:     51ca448 CallInfo callingPartyName='Rajesh Ramarao'
   callingParty=1000 cgpnVoiceMailbox=1000 calledPartyName='' calledParty=5551212
   cdpnVoiceMailbox= originalCalledPartyName='' originalCalledParty=
   originalCdpnVoiceMailbox= originalCdpnRedirectReason=0
   lastRedirectingPartyName='' lastRedirectingParty=5551212
   lastRedirectingVoiceMailbox= lastRedirectingReason=0 callType=2(OutBound)
   lineInstance=1 callReference=16777365.|
|StationD::star_StationOutputCallInfo(): callInfo: CI=16777365,
   CallingPartyName=Rajesh Ramarao,  CallingParty=1000, CalledPartyName=,
   CalledParty=5551212, OriginalCalledPartyName=, OriginalCalledParty=,
   lastRedirectingPartyName=, lastRedirectingParty=|
|StationD:     51ca448 SelectSoftKeys instance=1 reference=16777365
   softKeySetIndex=1 validKeyMask=-1.|
|StationD:     51ca448 DisplayPromptStatus timeOutValue=0 promptStatus='Connected'
   lineInstance=1 callReference=16777365.|
```

This illustrates the phone softkey event where the "end" key is pressed:

```
|StationInit: 51ca448 SoftKeyEvent softKeyEvent=9(EndCall) lineInstance=1
   callReference=16777365.|
```

CUCM updates the IP phone state (tone, spkr off, display, and so on):

```
|StationD:     51ca448 SetSpeakerMode speakermode=2(Off).|
|StationD:     51ca448 SetSpeakerMode speakermode=2(Off).|
|StationD:     51ca448 ClearPromptStatus lineInstance=1 callReference=16777365.|
|StationD:     51ca448 CallState callState=2(OnHook) lineInstance=1
   callReference=16777365|
|StationD:     51ca448 SelectSoftKeys instance=0 reference=0 softKeySetIndex=0
   validKeyMask=7.|
|StationD:     51ca448 DisplayPromptStatus timeOutValue=0 promptStatus='Your
   current options' lineInstance=0 callReference=0.|
|StationD:     51ca448 ActivateCallPlane lineInstance=0.|
|StationD:     51ca448 SetLamp stimulus=9(Line) stimulusInstance=1
   lampMode=1(LampOff).|
|StationD:     51ca448 DefineTimeDate timeDateInfo=? systemTime=1020105057.|
|StationD:     51ca448 StopTone.|
```

CUCM tells the phone to close the receive channel:

```
|StationD:     51ca448 CloseReceiveChannel conferenceID=0 passThruPartyID=513.
   myIP: 46a8110a (10.17.168.70)|
```

CUCM tells the phone to stop transmitting:

```
|StationD:     51ca448 StopMediaTransmission conferenceID=0 passThruPartyID=513.
   myIP: 46a8110a (10.17.168.70)|
```

CUCM sends an MDCX receive-only message to the GW:

```
MGCPHandler send msg SUCCESSFULLY to: 172.21.57.243
MDCX 283 S0/DS1-0/23@SDA00033282D8E3 MGCP 0.1
C: D0000000010000960000000000000002
I: 23
X: 17
M: recvonly
R: D/[0-9ABCD*#]
Q: process,loopl
```

CUCM sends a PRI disconnect message (msg) to the GW:

```
lOut Message — PriDisconnectMsg — Protocol= PriNi2Protocoll
lIe - Q931CauseIe IEData= 08 02 80 90 l
lMMan_Id= 0. (iep= 0 dsl= 0 sapi= 0 ces= 0 IpAddr=f33915ac IpPort=2427)l
lIsdnMsgData2= 08 02 00 02 45 08 02 80 90 l
```

GW sends a PRI release message to the CUCM:

```
lIn  Message — PriReleaseMsg — Protocol= PriNi2Protocoll
lMMan_Id= 0. (iep= 0 dsl= 0 sapi= 0 ces= 0 IpAddr=f33915ac IpPort=2427)l
lIsdnMsgData1= 08 02 80 02 4D l
```

CUCM sends a PRI release message complete to the GW:

```
lOut Message — PriReleaseCompleteMsg — Protocol= PriNi2Protocoll
lMMan_Id= 0. (iep= 0 dsl= 0 sapi= 0 ces= 0 IpAddr=f33915ac IpPort=2427)l
lIsdnMsgData2= 08 02 00 02 5A l
```

CUCM sends a delete connection message to the GW:

```
MGCPHandler send msg SUCCESSFULLY to: 172.21.57.243
DLCX 284 S0/DS1-0/23@SDA00033282D8E3 MGCP 0.1
C: D0000000010000960000000000000002
I: 23
X: 17
S: l
```

The phone handset is put back on hook:

```
lStationInit: 51ca448 OnHook.l
```

CUCM refreshes the IP phone call state and its display:

```
|StationD:    51ca448 DefineTimeDate timeDateInfo=? systemTime=1020105059.|
|StationD:    51ca448 StopTone.|
|StationD:    51ca448 SelectSoftKeys instance=0 reference=0 softKeySetIndex=0
  validKeyMask=7.|
```

IP Phone to H.323 Voice Gateway with Gatekeeper

In this scenario, Phone A is registered to the CUCM. Phone B is connected to a carrier's CO switch. Figure B-15 illustrates this call setup between two IP phones registered to the same CUCM.

A call is placed from Phone A (555-2001) to Phone B (555-2001). Phone B hangs up first. The CUCM must register with the gatekeeper prior to placing any calls, and the gatekeeper will police the bandwidth by implementing Call Admission Control (CAC).

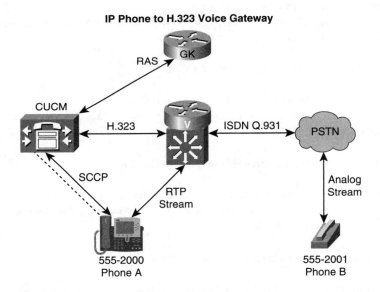

IP Phone to H.323 Voice Gateway

Figure B-15 *IP Phone to H.323 Voice Gateway Call Flow*

Because the initial call setup phases between the IP phone and the CUCM have been covered earlier, they are not repeated in this call flow. Figure B-16 includes all the steps involving CUCM registration with the gatekeeper, skips the initial IP Phone A interaction with the CUCM when it goes off-hook to call Phone B in the PSTN, and then illustrates the CUCM and H.323 gateway interaction, including the ISDN Q.931 messaging exchange with the CO in the PSTN to set up the call with Phone B. The communication beyond the CO, which would include SS7 messages, with the CO serving Phone B is not part of this overall example.

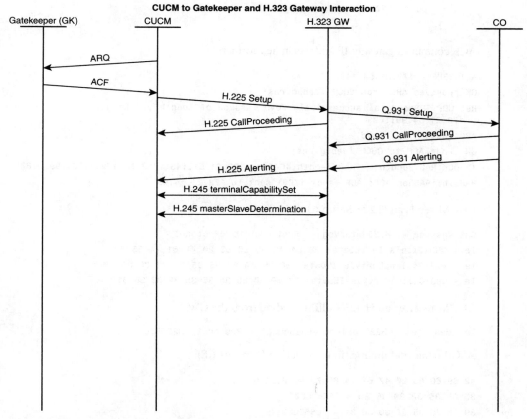

CUCM to Gatekeeper and H.323 Gateway Interaction

Figure B-16   *CUCM to Gatekeeper and H.323 Gateway Interaction*

The details of the call flow in Figure B-16 are as follows. CUCM sends an Admission Request (ARQ) to the gatekeeper (GK):

```
value V2Message ::= admissionRequest :
  {
    destinationInfo
    {
      e164 : "95552001"
    }
```

CUCM receives an Admission Confirm (ACF) from the GK:

```
value V2Message ::= admissionConfirm :
  {
    destCallSignalAddress ipAddress :
      {
        ip 'AC15338F'H,
```

```
      port 1720
   }
```

ACF contains a gateway IP address in hex format:

```
AC15338F = 172.21.54.143
GK receives ARQ from CUCM (debug ras)
RecvUDP_IPSockData  successfully rcvd message of length 118 from
   172.21.54.99:1603
ARQ (seq# 1706) rcvd
GK sends ACF to CUCM (debug ras)
IPSOCK_RAS_sendto:   msg length 46 from 172.21.51.143:1719 to 172.21.54.99: 1603
RASLib::RASSendACF: ACF (seq# 1706) sent to 172.21.54.99
```

CUCM sends an H.225 Setup to the GW:

```
Out Message — H225SetupMsg — Protocol= H225Protocol
Ie - Q931DisplayIe IEData= 28 0A 42 69 6C 6C 20 47 61 74 65 73
Ie - H225CallingPartyIe IEData= 6C 09 00 81 35 35 35 32 30 30 30
Ie - Q931CalledPartyIe IEData= 70 09 80 39 35 35 35 32 30 30 31
```

CUCM receives an H.225 CallProceeding from the GW:

```
In  Message — H225CallProceedingMsg — Protocol= H225Protocol
```

ASCII name and number in Information Elements (IE):

```
42 69 6C 6C 20 47 61 74 65 73 = "Bill Gates"
35 35 35 32 30 30 30 = "5552000"
39 35 35 35 32 30 30 31 = "95552001"
```

GW receives an H.225 Setup from the CUCM (debug h225 asn1):

```
value H323_UserInformation ::=
    {
      h323-uu-pdu
      {
        h323-message-body setup :
        {
          sourceAddress
          {
            h323-ID : {"Bill Gates"}
          }
          destinationAddress
          {
            dialedDigits : "95552001"
          }
```

GW sends a Q931 Setup to the CO (debug isdn q931):

```
ISDN Se1/0:23 Q931: TX -> SETUP pd = 8  callref = 0x002A
    Display i = 'Bill Gates'
    Calling Party Number i = 0x0081, '5552000'
    Called Party Number i = 0x80, '5552001'
GW receives  "proceeding" from CO
(debug isdn q931)
ISDN Se1/0:23 Q931: RX <- CALL_PROC pd = 8  callref = 0x802A
    Channel ID i = 0xA98386
        Exclusive, Channel 6
```

GW receives a Q.931 Alerting from the CO (debug isdn q931):

```
ISDN Se1/0:23 Q931: RX <- ALERTING pd = 8  callref = 0x802A
    Progress Ind i = 0x8088 - In-band info or appropriate now available
```

GW sends an H.225 Alerting to the CUCM (debug h225 asn1):

```
value H323_UserInformation ::=      {
      h323-uu-pdu
      {
         h323-message-body alerting :
value H323_UU_NonStdInfo ::=      {
      protoParam qsigNonStdInfo :
      {
        iei 30
        rawMesg '1E028088'H
```

CUCM receives an H.225 Alerting from the GW (ccm trace). In this case: progress_ind alert enable 8:

```
In  Message — H225AlertMsg — Protocol= H225Protocol
Ie - Q931ProgressIndIe — IEData= 1E 02 80 88
Progress Indicators values
ISDN Se1/0:23 Q931: RX <- ALERTING pd = 8  callref = 0x802A
Progress Ind i = 0x8088 - In-band info or appropriate now available
1:  not end-to-end ISDN, further call prog info may be available
    in-band
2:  destination address is non-ISDN
3:  origination address is non-ISDN
8:  in-band info now available
Force PI to make Call Progress tones work
dial-peer voice 1 pots
  progress_ind [setup | alert] enable [1 | 2 | 3 | 8]
```

GW advertises the preferred codecs (debug h245 asn1):

```
value MultimediaSystemControlMessage ::= request : terminalCapabilitySet : {
capabilityTable    {
 {
          capabilityTableEntryNumber 4
          capability receiveAudioCapability : g729AnnexA : 2        },
        {
          capabilityTableEntryNumber 3
          capability receiveAudioCapability : g729 : 2        },
        {
          capabilityTableEntryNumber 1
         capability receiveAudioCapability : g711Ulaw64k : 20
```

CUCM advertises preferred codecs in "terminalCapabilitySet" message.

```
H245ASN MultimediaSystemControlMessage ::= request : terminalCapabilitySet    {
        capabilityTable    {
         {
          capabilityTableEntryNumber 1,
          capability receiveAudioCapability : g711Ulaw64k : 40
```

GW announces a chosen codec (debug h245 asn1):

```
value MultimediaSystemControlMessage ::= response : terminalCapabilitySetAck :
    {
      sequenceNumber 1
    }
```

CUCM announces the chosen codec in terminalCapabilitySetAck:

```
H245ASN - TtPid=(1,100,108,42) -Outgoing -value MultimediaSystemControlMessage
  ::= response : terminalCapabilitySetAck :
    {
      sequenceNumber 1
    }
```

Codec is controlled through Region on CUCM and voice-class on the GW:

```
voice class codec 10
  codec preference 1 g711ulaw
  codec preference 2 g729r8
dial-peer voice 10000 voip
  incoming called-number.
  voice-class codec 10
```

At this point, the CUCM has received all the information along with confirmation including the bandwidth availability and the codec to proceed with the call. Now the CUCM is ready to respond to Phone A, which originated the call. The following interaction is now among Phone A, the CUCM, and the gateway, as shown in Figure B-17. The call flows from this point onward are described in Figure B-18.

IP Phone to CUCM and H.323 Voice Gateway Interaction

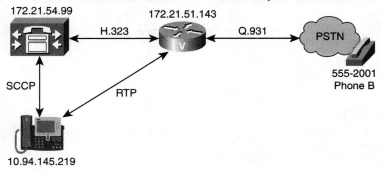

Figure B-17   *IP Phone, CUCM, and H.323 Gateway Interaction*

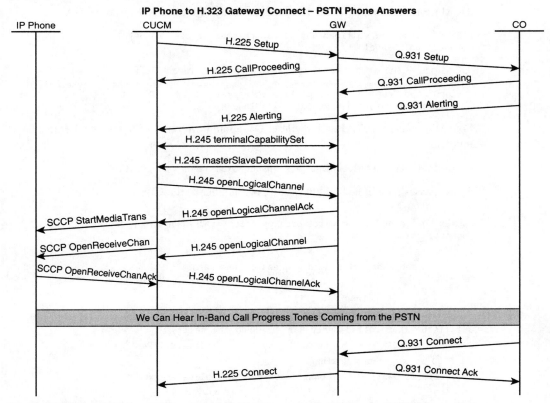

Figure B-18   *CUCM to IP Phone and H.323 Gateway Call Flows—PSTN Phone B Answers*

Here is the detailed explanation of the transactions shown in Figure B-18. CUCM tells the GW to get ready to receive an RTP from the IP phone:

```
H245ASN - Outgoing MultimediaSystemControlMessage ::= request :
  openLogicalChannel :
```

GW responds to the CUCM with the UDP port number it is listening on:

```
value MultimediaSystemControlMessage ::= response : openLogicalChannelAck :
    {
        mediaChannel unicastAddress : iPAddress :
        {
          network 'AC15338F'H
        tsapIdentifier 17256
```

CUCM directs the phone to start transmitting RTP:

```
StationD:   StartMediaTransmission remoteIpAddress=8f3315ac(172.21.51.143)
  remotePortNumber=17256 milliSecondPacketSize=20
  compressType=4(Media_Payload_G711Ulaw64k)
```

IP address format: AC15338F = 172.21.51.143:

CUCM hears from the GW that it should get ready to receive RTP from the GW.

```
H245ASN - Incoming MultimediaSystemControlMessage ::= request :
  openLogicalChannel :
```

CUCM passes the word on to the phone: Get ready to receive RTP:

```
StationD: OpenReceiveChannel millisecondPacketSize=20
  compressionType=4(Media_Payload_G711Ulaw64k)
```

Phone responds to CUCM with the User Datagram Protocol (UDP) port number it is listening on:

```
StationInit: OpenReceiveChannelAck IpAddr=0xdb915e0a, Port=24614
```

CUCM passes the phone IP address and port number to the GW:

```
H245ASN -Outgoing -value MultimediaSystemControlMessage ::= response :
  openLogicalChannelAck :
    {
        mediaChannel unicastAddress : iPAddress :
            {
                network '0A5E91DB'H,
                tsapIdentifier 24614
CO tells GW that called party has answered the call
ISDN Se1/0:23 Q931: RX <- CONNECT pd = 8  callref = 0x802A
    Display i = 'John Chambers'
ISDN Se1/0:23 Q931: TX -> CONNECT_ACK pd = 8  callref = 0x002A
```

GW tells CUCM that the call has been answered

```
value H323_UserInformation ::=
    {
        h323-uu-pdu
        {
          h323-message-body connect :
```

Note that IP address format 0A5E91DB and db915e0a translate to 10.94.145.219.

CUCM hears from the GW that Phone B has answered:

```
In  Message — H225ConnectMsg — Protocol= H225Protocol
```

The PSTN phone will hang up. The CO switch alerts the gateway through which the call was established using ISDN Q.931, as shown in Figure B-19. This process includes the call teardown, including notification to the gatekeeper so that it can relinquish the bandwidth used for this call. Because the call teardown between the IP phone and the CUCM has been previously discussed, it has been omitted from this call flow illustration.

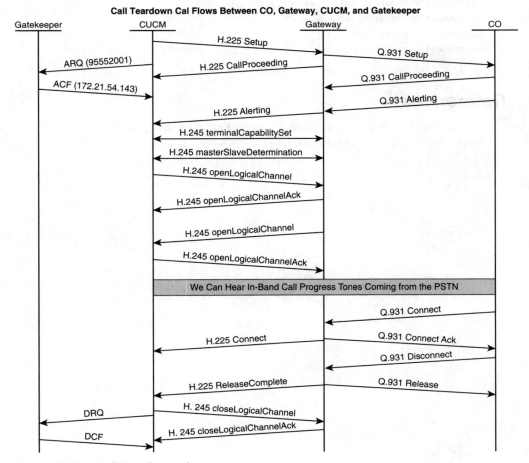

Figure B-19 *Call Teardown Flows*

CO tells the GW that Phone B has hung up:

```
ISDN Se1/0:23 Q931: RX <- DISCONNECT pd = 8  callref = 0x802A
    Cause i = 0x8090 - Normal call clearing
ISDN Se1/0:23 Q931: TX -> RELEASE pd = 8  callref = 0x002A
```

CUCM hears from the GW that Phone B has hung up:

```
In  Message — H225ReleaseCompleteMsg — Protocol= H225Protocol
Ie - Q931CauseIe — IEData= 08 02 80 90
```

Disconnect cause codes assist in troubleshooting:

```
90h = 10010000b > 0010000b = 16
gateway#show call history voice | include DisconnectText
DisconnectText=normal call clearing (16)
DisconnectText=unassigned number (1)
DisconnectText=user busy (17)
DisconnectText=no circuit (34)
```

References

1. Bouchard, Luc, Rajesh Ramarao, and Ramesh Kaza. "Troubleshooting IP Telephony Networks." Cisco Networkers 2001.

VoIP Dashboard

This appendix describes a tool case study. It is based on the Cisco BTS 10200 product. A set of open-source tools are also listed that can be used to implement a dashboard.

Case Study: Dashboard for BTS 10200

Trend analysis and visual monitoring of VoIP performance metrics are performed for counters from a BTS 10200 to create a VoIP dashboard. The VoIP dashboard is built from metrics collected from various VoIP network components such as the Signaling System 7 (SS7), public switched telephone network (PSTN), and others, thus providing visibility into the overall health and capacity of the VoIP network.

The reports were collected at 15-minute intervals for 24 hours and dumped to flat files (CSV); 3 months worth of data was collected.

The data was sent through FTP to a Linux server that had Perl, Round Robin Database (RRD), public tools, and DRRAW (a CGI script) installed on them. The RRD was created for the counters and updated on a periodic basis. Perl was used to parse the CSV files and update the RRDs. We cover some of these tools later to provide more context.

Figure C-1 shows the architecture of the tools-based dashboard.

The Cisco BTS offers a wide set of private Simple Network Management Protocol (SNMP) Management Information Base (MIB)–based performance counters. The following set of BTS counters captures system health across various VoIP call segments and is presented in the VoIP dashboard:

- ISDN user part (SS7/PSTN) signaling protocol-related information

- Media Gateway Control Protocol (MGCP) signaling protocol-related information

- Session Initiation Protocol (SIP) interface adapter-related information

- Call-processing specific information

- Trunk Group (TG) usage information

- Dynamic quality of service–related information

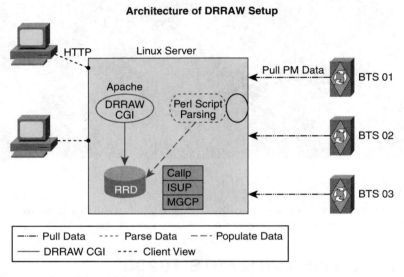

Figure C-1   *Tool Dashboard Architecture*

The following code shows an example of the performance counter raw data for a BTS 10200 Trunk Group. The raw data comes from a trunk group performance counter file. The switch collects and stores various trunk group–related Performance Management (PM) counters: SS7, ANNC, CAS, ISDN, and SIP.

```
TIMESTAMP, call_agent_id, CONDITION, TRKGRP_TYPE,tgn_id, TRKGRP_INCOM_ATTMP,
TRKGRP_OUTG_ATTMP, TRKGRP_OUTBOUND_FAIL, TRKGRP_TOTAL_OVERFLOW, TRKGRP_TOTAL_TRK,
TRKGRP_INCOM_BUSY_TRK, TRKGRP_OUTG_BUSY_TRK, TRKGRP_TOTAL_OOS_TRK,
  TRKGRP_INCOM_USAGE,
TRKGRP_OUTG_USAGE, TRKGRP_TOTAL_USAGE, TRKGRP_AVERAGE_USAGE
2005-08-21 00:15:00,CA101,Normal,SS7,1002,0,0,0,0,24,0,0,4,0,0,0,0
2005-08-21 19:50:00,CA101,Normal,ANNC,9802,0,0,0,0,96,0,0,6,0,0,0,0
2005-08-21 19:40:00,CA101,Normal,ISDN,9950,0,0,0,0,23,0,0,0,0,0,0,0
2005-08-21 19:45:00,CA101,Normal,CAS,9401,0,0,0,0,4,0,0,0,0,0,0,0
19:45:00,CA101,Normal,SOFTSW,9601,0,1,0,0,0,0,0,0,0,0,0,0
```

Performance counter information reflects how many Circuit Identifier (CIC)s are Out of Service (OOS) along with other trunk information. The DRRAW tool was customized to create a dashboard for some of these counters. The DRRAW script is driven by the performance counter RRDs.

A pseudo-real-time display of the performance stats is created. This dashboard displays the past 28 hours, 1 week, 1 month, and 1 year of data. It can be used to create a systemwide view of the VoIP call flow, do capacity planning, and a track Service Level Agreement (SLA).

The RRD tool is a database and a set of tools that store, retrieve, and display time-series-based data. It stores data in a very efficient way and provides a mechanism to generate nice time-series-based graphs. RRD is effective for short-term trending. The data sources would get wrapped around after its set limit is reached, because the RRD database size for a data source is determined at DB creation time.

The following code is an RRD creation example for Trunk Group performance counters:

```
rrdtool create tg_ss7.rrd \
        —start 1124600400 \   UTC time in ticks
        —step 900 \    15 min. intervals
        DS:TRKGRP_OFAIL:ABSOLUTE:900:0:100000 \  min, max values
        DS:TRKGRP_TOTAL_OFLOW:ABSOLUTE:900:0:100000 \
        DS:TRKGRP_INCOM_BSYTRK:ABSOLUTE:900:0:100000 \
        DS:TRKGRP_OUTG_BSYTRK:ABSOLUTE:900:0:100000 \
        DS:TRKGRP_TUSAGE:ABSOLUTE:900:0:100000 \
       RRA:AVERAGE:0.5:1:57600   600 days worth of data
```

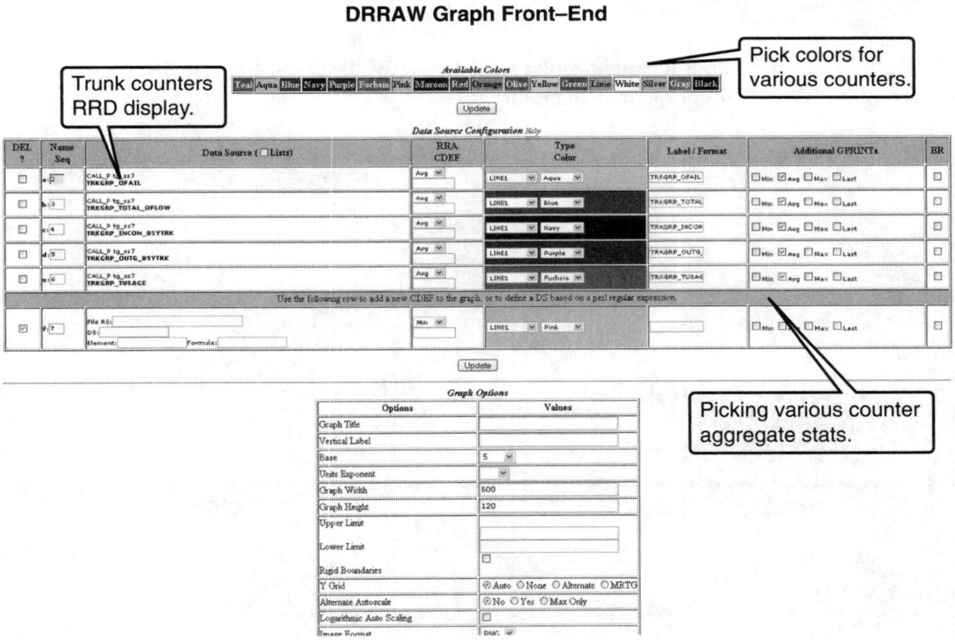

Figure C-2 *DRRAW CGI Tool*

DRRAW is an open source graphing tool—based on rrdtool and is freely available on the internet. It is comprised of a set of CGI scripts written in Perl, which enables the creation of real-time graphs through a web front end. It also enables the creation of consolidated dashboard views of these graphs. Figure C-2 shows the DRRAW dashboard configuration screen. Refer to the DRRAW website link for details on configuring DRRAW.

The following text summarizes the DRRAW-based VoIP dashboard. It also provides a high-level readout of the various VoIP segment graphs presented in the dashboard:

- Call stats dashboard captures the number of call attempts of all types, call originating failures, and call successes on the reporting BTS 10200.

- The Dynamic Quality of Service (DQoS) dashboard shows no issues, which reflects the Cable Modem Termination System (CMTS) leg. It shows that the Gate SET attempts are equal to the Gate SET successes.

- The MGCP dashboard shows the number of MGCP attempt successes, failures, or abandons.

- The ISDN User Part (ISUP) dashboard shows the SS7 signaling pattern, which includes the patterns for the Initial Answer Message (IAM), Answer Message (ANM), and Release (REL) signaling message.

- The SIP dashboard shows the number of SIP messages going through the switch.

- The Trunk dashboard shows the utilization of trunks and overflows, which might help in capacity planning.

Figure C-3 shows an implementation of the DRRAW dashboard-based RRDs created from the BTS 10200 performance counters.

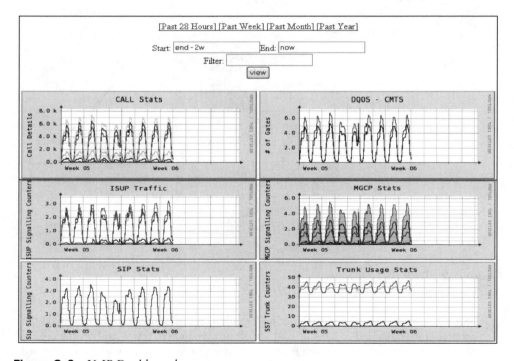

Figure C-3   *VoIP Dashboard*

The following links represent open-source tool download sites for the key tools used in the VoIP dashboard creation:

■ RRDtool: http://oss.oetiker.ch/rrdtool/download.en.html

■ DRRAW tool: http://web.taranis.org/drraw

Open-Source Trending and Capacity-Planning Tools

The following list covers a set of tools for metric collection and event notification:

■ Zabbix: http://zabbix.com

■ Hyperic HQ: http://hyperic.com

■ Munin: http://munin.projects.linpro.no

■ ZenOSS: http://www.zenoss.com

■ OpenNMS: http://opennms.org

■ GroundWork: http://www.groundworkopensource.com

■ Ganglia: http://ganglia.info

■ Nagios: http://nagios.org

■ Cacti: http://cacti.net

The following set of tools can be used to generate graphs for the collected metrics:

■ RRDTool: http://oss.oetiker.ch/rrdtool

■ Collectd: http://collectd.org

■ Dstat: http://dag.wieers.com/home-made/dstat

■ GraphClick: http://www.arizona-software.ch/graphclick

Debugs, Traces, and Logs

Ths appendix covers debugs, traces, and logs that can be collected from various VoIP network components for the following purposes:

- Collecting and analyzing data for monitoring the health of a VoIP network

- Isolating network problems and identifying the root causes of those problems

- Trending data for identifying resource utilization issues and capacity-planning exercises

CMTS Debugs for a PacketCable Call

The PacketCable deployment model was discussed in Chapter 3, "VoIP Deployment Models in Service Provider Networks." This section illustrates the debugs and logs that can be collected from the Cable Modem Termination System (CMTS) to view GATE messaging between the CMTS and Call Management Server (CMS). This can help verify proper resource allocation on the CMTS for a PacketCable voice call.

Figure D-1 illustrates an On-net–to–On-net PacketCable call between two CMTSs.

The following debugs have been collected from a Cisco CMTS for an On-net–to–On-net PacketCable call that originated and terminated on the same CMTS. You can see pairs of GATE messages for GATE-SET, GATE-SET-ACK, GATE-OPEN, and GATE-CLOSE.

You can also see the Data Over Cable Service Interface Specification (DOCSIS) layer messaging between the Multimedia Terminal Adapter (MTA) and the CMTS. To view these messages on the CMTS, turn on the following debugs:

```
CMTS# debug packetcable subscriber MTA IP Address [verbose]
CMTS# debug packetcable gate docsis-mapping
CMTS# debug packetcable gate control
CMTS# debug packetcable gate events
```

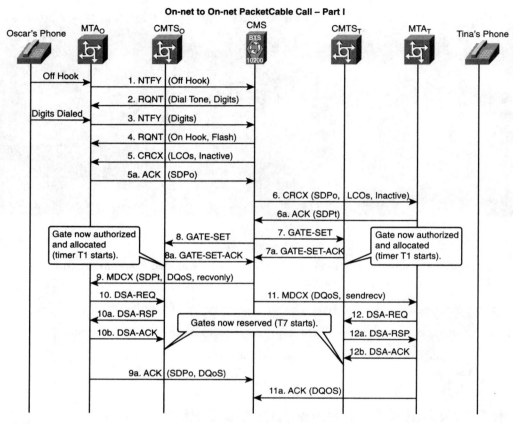

Figure D-1 *On-net–to–On-net PacketCable Call*

On-net to On-net PacketCable Call – Part II

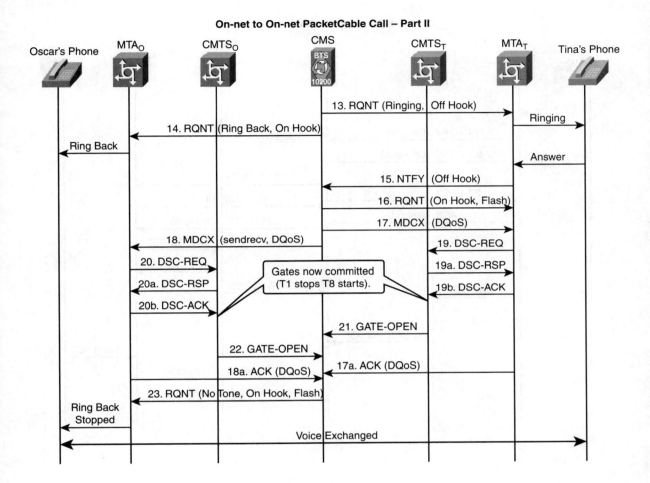

On-net to On-net PacketCable Call – Part III

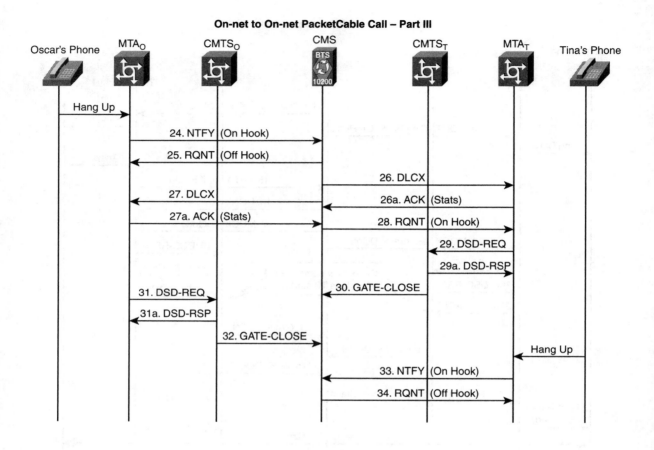

Note that the following debugs represent a PacketCable call originating and terminating on the same CMTS. Both MTAs are connected to the same CMTS. This is different from what is illustrated in Figure D-1, where two MTAs (originating and terminating) are connected to two different CMTSs. The debugs that follow are still applicable to Figure D-1 because the GATE messages are the same, even if the call originates and terminates on the same CMTS or two different CMTSs.

```
089650: Jan 20 17:16:49.975: Pktcbl(gcp): Received GATE SET message, tid=0xA5
089651: Jan 20 17:16:49.975: -- Pktcbl(gcp): Received GCP message ------
089652: Jan 20 17:16:49.975: TRANSACTION ID        : Object.[snum/stype/len 1/1/8]
089653: Jan 20 17:16:49.975:      transaction id   : 0xA5
089654: Jan 20 17:16:49.975:      gcp cmd          : 4 (GATE SET)
[7] CMTS receives a GATE SET message from the CMS for the first call leg
089655: Jan 20 17:16:49.975: SUBSCRIBER ID (IPV4) : Object.[snum/stype/len 2/1/8]
089656: Jan 20 17:16:49.975:      Addr             : 24.31.240.28
089657: Jan 20 17:16:49.975: ACTIVITY COUNT       : Object.[snum/stype/len 4/1/8]
```

```
089658: Jan 20 17:16:49.975:       Count              : 28
089659: Jan 20 17:16:49.975: GATE SPEC              : Object.[snum/stype/len
   5/1/60]
089660: Jan 20 17:16:49.975:       direction          : 1 (UPSTREAM)
089661: Jan 20 17:16:49.975:       protocol id        : 17
089662: Jan 20 17:16:49.975:       commit flag        : 0x0
089663: Jan 20 17:16:49.975:       session class      : 1
089664: Jan 20 17:16:49.975:       source             : 24.31.240.28
089665: Jan 20 17:16:49.975:       dest               : 24.31.240.36
089666: Jan 20 17:16:49.975:       src port           : 0 (0x0)
089667: Jan 20 17:16:49.975:       dest port          : 53456 (0xD0D0)
089668: Jan 20 17:16:49.975:       dscp               : 0xB8
089669: Jan 20 17:16:49.975:       timer t1           : 200
089670: Jan 20 17:16:49.975:       timer t7           : 300
089671: Jan 20 17:16:49.975:       timer t8           : 600
089672: Jan 20 17:16:49.975:       flowspec # 1       : [r/b/p/m/M
   10000/200/10000/200/200] [R/S: 10000/800]
089673: Jan 20 17:16:49.975: GATE SPEC              : Object.[snum/stype/len
   5/1/60]
089674: Jan 20 17:16:49.975:       direction          : 0 (DOWNSTREAM)
089675: Jan 20 17:16:49.975:       protocol id        : 17
089676: Jan 20 17:16:49.975:       commit flag        : 0x0
089677: Jan 20 17:16:49.975:       session class      : 1
089678: Jan 20 17:16:49.975:       source             : 24.31.240.36
089679: Jan 20 17:16:49.975:       dest               : 24.31.240.28
089680: Jan 20 17:16:49.975:       src port           : 0 (0x0)
089681: Jan 20 17:16:49.975:       dest port          : 53456 (0xD0D0)
089682: Jan 20 17:16:49.975:       dscp               : 0xB8
089683: Jan 20 17:16:49.975:       timer t1           : 200
089684: Jan 20 17:16:49.975:       timer t7           : 300
089685: Jan 20 17:16:49.975:       timer t8           : 600
089686: Jan 20 17:16:49.975:       flowspec # 1       : [r/b/p/m/M
   10000/200/10000/200/200] [R/S: 10000/0]
089687: Jan 20 17:16:49.975: -------------------------------------
089688: Jan 20 17:16:49.987: Pktcbl(gcp): Building GCP message, added obj
   TRANSACTION ID      ; len:8 padding:0
089689: Jan 20 17:16:49.987: Pktcbl(gcp): Building GCP message, added obj
   SUBSCRIBER ID (IPV4); len:8 padding:0
089690: Jan 20 17:16:49.987: Pktcbl(gcp): Building GCP message, added obj GATE
   ID            ; len:8 padding:0
089691: Jan 20 17:16:49.987: Pktcbl(gcp): Building GCP message, added obj
   ACTIVITY COUNT      ; len:8 padding:0
089692: Jan 20 17:16:49.987: Pktcbl(gcp): Built GCP message, GATE SET ACK
   , length: 32, copsLen 60
089693: Jan 20 17:16:49.987:  ---- Pktcbl: Sending GCP message ------
089694: Jan 20 17:16:49.987: TRANSACTION ID      : Object.[snum/stype/len 1/1/8]
089695: Jan 20 17:16:49.987:       transaction id   : 0xA5
```

```
089696: Jan 20 17:16:49.987:      gcp cmd          : 5 (GATE SET ACK)
```

[7a] CMTS allocates a GATE ID of 74557, sets the ACTIVITY COUNT to 2 which means the gates have been allocated for both upstream and downstream directions, and responds back to the CMS.

```
089697: Jan 20 17:16:49.987: SUBSCRIBER ID (IPV4) : Object.[snum/stype/len 2/1/8]
089698: Jan 20 17:16:49.987:     Addr             : 24.31.240.28
089699: Jan 20 17:16:49.987: GATE ID              : Object.[snum/stype/len 3/1/8]
089700: Jan 20 17:16:49.987:     GateID           : 74557 (0x1233D)
089701: Jan 20 17:16:49.987: ACTIVITY COUNT       : Object.[snum/stype/len 4/1/8]
089702: Jan 20 17:16:49.987:     Count            : 2
089703: Jan 20 17:16:49.987:  ----------------------------------------
089704: Jan 20 17:16:50.051: Pktcbl(gcp): Received GATE SET message, tid=0xA6
089705: Jan 20 17:16:50.051: ---- Pktcbl(gcp): Received GCP message ------
089706: Jan 20 17:16:50.051: TRANSACTION ID       : Object.[snum/stype/len 1/1/8]
089707: Jan 20 17:16:50.051:     transaction id   : 0xA6
089708: Jan 20 17:16:50.051:      gcp cmd          : 4 (GATE SET)
```

[8] CMTS receives a GATE SET message from the CMS for the second call leg

```
089709: Jan 20 17:16:50.051: SUBSCRIBER ID (IPV4) : Object.[snum/stype/len 2/1/8]
089710: Jan 20 17:16:50.051:     Addr             : 24.31.240.36
089711: Jan 20 17:16:50.051: ACTIVITY COUNT       : Object.[snum/stype/len 4/1/8]
089712: Jan 20 17:16:50.051:     Count            : 28
089713: Jan 20 17:16:50.051: GATE SPEC            : Object.[snum/stype/len
    5/1/60]
089714: Jan 20 17:16:50.051:     direction        : 1 (UPSTREAM)
089715: Jan 20 17:16:50.051:     protocol id      : 17
089716: Jan 20 17:16:50.051:     commit flag      : 0x0
089717: Jan 20 17:16:50.051:     session class    : 1
089718: Jan 20 17:16:50.051:     source           : 24.31.240.36
089719: Jan 20 17:16:50.051:     dest             : 24.31.240.28
089720: Jan 20 17:16:50.051:     src port         : 0 (0x0)
089721: Jan 20 17:16:50.051:     dest port        : 53456 (0xD0D0)
089722: Jan 20 17:16:50.051:     dscp             : 0xB8
089723: Jan 20 17:16:50.051:     timer t1         : 200
089724: Jan 20 17:16:50.051:     timer t7         : 300
089725: Jan 20 17:16:50.051:     timer t8         : 600
089726: Jan 20 17:16:50.051:     flowspec # 1     : [r/b/p/m/M
    10000/200/10000/200/200] [R/S: 10000/800]
089727: Jan 20 17:16:50.051: GATE SPEC            : Object.[snum/stype/
    len 5/1/60]
089728: Jan 20 17:16:50.051:     direction        : 0 (DOWNSTREAM)
089729: Jan 20 17:16:50.051:     protocol id      : 17
089730: Jan 20 17:16:50.051:     commit flag      : 0x0
089731: Jan 20 17:16:50.051:     session class    : 1
089732: Jan 20 17:16:50.051:     source           : 24.31.240.28
089733: Jan 20 17:16:50.051:     dest             : 24.31.240.36
089734: Jan 20 17:16:50.051:     src port         : 0 (0x0)
```

```
089735: Jan 20 17:16:50.051:     dest port        : 53456 (0xD0D0)
089736: Jan 20 17:16:50.055:     dscp             : 0xB8
089737: Jan 20 17:16:50.055:     timer t1         : 200
089738: Jan 20 17:16:50.055:     timer t7         : 300
089739: Jan 20 17:16:50.055:     timer t8         : 600
089740: Jan 20 17:16:50.055:     flowspec # 1     : [r/b/p/m/M
   10000/200/10000/200/200] [R/S: 10000/0]
089741: Jan 20 17:16:50.055: ---------------------------------------
089742: Jan 20 17:16:50.171: Pktcbl(gcp): Building GCP message, added obj
   TRANSACTION ID     ; len:8 padding:0
089743: Jan 20 17:16:50.171: Pktcbl(gcp): Building GCP message, added obj
   SUBSCRIBER ID (IPV4); len:8 padding:0
089744: Jan 20 17:16:50.171: Pktcbl(gcp): Building GCP message, added obj GATE
   ID                 ; len:8 padding:0
089745: Jan 20 17:16:50.171: Pktcbl(gcp): Building GCP message, added obj
   ACTIVITY COUNT     ; len:8 padding:0
089746: Jan 20 17:16:50.171: Pktcbl(gcp): Built GCP message, GATE SET ACK,
   length: 32, copsLen 60
089747: Jan 20 17:16:50.171:  ---- Pktcbl: Sending GCP message ------
089748: Jan 20 17:16:50.171: TRANSACTION ID        : Object.[snum/stype/len 1/1/8]
089749: Jan 20 17:16:50.171:     transaction id    : 0xA6
089750: Jan 20 17:16:50.171:     gcp cmd           : 5 (GATE SET ACK)
```

[8a] CMTS allocates a GATE ID of 9033, sets the ACTIVITY COUNT to 2 which means the gates have been allocated for both upstream and downstream directions, and responds back to the CMS.

```
089751: Jan 20 17:16:50.171: SUBSCRIBER ID (IPV4) : Object.[snum/stype/len 2/1/8]
089752: Jan 20 17:16:50.171:     Addr              : 24.31.240.36
089753: Jan 20 17:16:50.171: GATE ID               : Object.[snum/stype/len 3/1/8]
089754: Jan 20 17:16:50.171:     GateID            : 9033 (0x2349)
089755: Jan 20 17:16:50.171: ACTIVITY COUNT        : Object.[snum/stype/len 4/1/8]
089756: Jan 20 17:16:50.171:     Count             : 2
089757: Jan 20 17:16:50.171: ---------------------------------------
Jan 20 17:16:50.278: PktCbl(d2r): extract id: gate=9033, resource=0
Jan 20 17:16:50.278: PktCbl(d2r): extract id: gate=9033, resource=0
Jan 20 17:16:50.278: PktCbl(d2r): extract id: gate=9033, resource=0
Jan 20 17:16:50.278: PktCbl(d2r): extract id: gate=9033, resource=0
Jan 20 17:16:50.278: PktCbl(d2r): extract id: gate=9033, resource=0
Jan 20 17:16:50.278: PktCbl(d2r): extract id: gate=9033, resource=0
Jan 20 17:16:50.278: PktCbl(d2r): extract id: gate=9033, resource=0
Jan 20 17:16:50.278: PktCbl(d2r): DSA-REQ received, gateid: 9033
Jan 20 17:16:50.278: PktCbl(d2r): US flowspec cfr: proto=17 src=24.31.240.36
   dst=24.31.240.28 sport=0 dport=53456
Jan 20 17:16:50.278: PktCbl(r2d): cfr: proto=17 src=24.31.240.36 dst=24.31.240.28
   sport=53456 dport=53456
089758: Jan 20 17:16:50.307: PktCbl(d2r): extract id: gate=9033, resource=789
089759: Jan 20 17:16:50.307: PktCbl(d2r): extract id: gate=9033, resource=789
089760: Jan 20 17:16:50.311: PktCbl(d2r): extract id: gate=9033, resource=789
```

```
089761: Jan 20 17:16:50.311: PktCbl(d2r): extract id: gate=9033, resource=789
089762: Jan 20 17:16:50.311: PktCbl(d2r): DSA-ACK notification received on RP,
  gateid : 9033
Jan 20 17:16:50.278: PktCbl(r2d): REQ - r/b/p/m/M 1174515712 1126563840
  1174515712 166 166 (R/S) 1174515712 800
Jan 20 17:16:50.278: PktCbl(d2r): DS flowspec cfr: proto=17 src=24.31.240.28
  dst=24.31.240.36 sport=0 dport=53456
Jan 20 17:16:50.278: PktCbl(r2d): cfr: proto=17 src=24.31.240.28 dst=24.31.240.36
  sport=0 dport=53456
Jan 20 17:16:50.278: PktCbl(r2d): REQ - r/b/p/m/M 1175539712 1127874560
  1175539712 186 186 (R/S) 1175539712 0
Jan 20 17:16:50.310: PktCbl(d2r): DSA-ACK received, gateid: 9033
Jan 20 17:16:50.378: PktCbl(d2r): extract id: gate=74557, resource=0
Jan 20 17:16:50.378: PktCbl(d2r): extract id: gate=74557, resource=0
Jan 20 17:16:50.378: PktCbl(d2r): extract id: gate=74557, resource=0
Jan 20 17:16:50.378: PktCbl(d2r): extract id: gate=74557, resource=0
Jan 20 17:16:50.378: PktCbl(d2r): extract id: gate=74557, resource=0
Jan 20 17:16:50.378: PktCbl(d2r): extract id: gate=74557, resource=0
Jan 20 17:16:50.378: PktCbl(d2r): extract id: gate=74557, resource=0
Jan 20 17:16:50.378: PktCbl(d2r): DSA-REQ received, gateid: 74557
Jan 20 17:16:50.378: PktCbl(d2r): US flowspec cfr: proto=17 src=24.31.240.28
  dst=24.31.240.36 sport=0 dport=53456
089763: Jan 20 17:16:50.455: PktCbl(d2r): extract id: gate=74557, resource=791
089764: Jan 20 17:16:50.455: PktCbl(d2r): extract id: gate=74557, resource=791
089765: Jan 20 17:16:50.455: PktCbl(d2r): extract id: gate=74557, resource=791
089766: Jan 20 17:16:50.455: PktCbl(d2r): extract id: gate=74557, resource=791
089767: Jan 20 17:16:50.455: PktCbl(d2r): DSA-ACK notification received on RP,
  gateid : 74557
Jan 20 17:16:50.378: PktCbl(r2d): cfr: proto=17 src=24.31.240.28 dst=24.31.240.36
  sport=53456 dport=53456
Jan 20 17:16:50.378: PktCbl(r2d): REQ - r/b/p/m/M 1174515712 1126563840
  1174515712 166 166 (R/S) 1174515712 800
Jan 20 17:16:50.378: PktCbl(r2d): DS flowspec cfr: proto=17 src=24.31.240.36
  dst=24.31.240.28 sport=0 dport=53456
Jan 20 17:16:50.378: PktCbl(r2d): cfr: proto=17 src=24.31.240.36 dst=24.31.240.28
  sport=0 dport=53456
Jan 20 17:16:50.378: PktCbl(r2d): REQ - r/b/p/m/M 1175539712 1127874560
  1175539712 186 186 (R/S) 1175539712 0
Jan 20 17:16:50.422: PktCbl(d2r): DSA-ACK received, gateid: 74557
Jan 20 17:16:51.154: PktCbl(d2r): extract id: gate=9033, resource=789
Jan 20 17:16:51.154: PktCbl(d2r): extract id: gate=9033, resource=789
Jan 20 17:16:51.154: PktCbl(d2r): extract id: gate=9033, resource=789
Jan 20 17:16:51.154: PktCbl(d2r): extract id: gate=9033, resource=789
Jan 20 17:16:51.154: PktCbl(d2r): extract id: gate=9033, resource=789
Jan 20 17:16:51.158: PktCbl(d2r): extract id: gate=9033, resource=789
Jan 20 17:16:51.158: PktCbl(d2r): DSC-REQ received, gateid(9033) flow handle(9033)
Jan 20 17:16:51.158: PktCbl(d2r): max_fs_type us=2, ds=2 qos_set_type us=6 ds=6
Jan 20 17:16:51.158: PktCbl(d2r): US flowspec cfr: proto=17 src=24.31.240.36
  dst=24.31.240.28 sport=0 dport=53456
```

```
Jan 20 17:16:51.158: PktCbl(r2d): cfr: proto=17 src=24.31.240.36 dst=24.31.240.28
   sport=53456 dport=53456

Jan 20 17:16:51.158: PktCbl(r2d): REQ - r/b/p/m/M 1174515712 1126563840
   1174515712 166 166 (R/S) 1174515712 800

Jan 20 17:16:51.158: PktCbl(d2r): DS flowspec cfr: proto=17 src=24.31.240.28
   dst=24.31.240.36 sport=0 dport=53456

Jan 20 17:16:51.158: PktCbl(r2d): cfr: proto=17 src=24.31.240.28 dst=24.31.240.36
   sport=0 dport=53456

Jan 20 17:16:51.158: PktCbl(r2d): REQ - r/b/p/m/M 1175539712 1127874560
   1175539712 186 186 (R/S) 1175539712 0

Jan 20 17:16:51.158: PktCbl(d2r): extract id: gate=74557, resource=791

Jan 20 17:16:51.158: PktCbl(d2r): extract id: gate=74557, resource=791

Jan 20 17:16:51.158: PktCbl(d2r): extract id: gate=74557, resource=791

Jan 20 17:16:51.158: PktCbl(d2r): extract id: gate=74557, resource=791

Jan 20 17:16:51.158: PktCbl(d2r): extract id: gate=74557, resource=791

Jan 20 17:16:51.158: PktCbl(d2r): extract id: gate=74557, resource=791

Jan 20 17:16:51.158: PktCbl(d2r): DSC-REQ received, gateid(74557) flow
   handle(74557)

Jan 20 17:16:51.158: PktCbl(d2r): max_fs_type us=2, ds=2 qos_set_type us=6 ds=6

Jan 20 17:16:51.158: PktCbl(d2r): US flowspec cfr: proto=17 src=24.31.240.28
   dst=24.31.240.36 sport=0 dport=53456

089768: Jan 20 17:16:51.279: PktCbl(d2r): extract id: gate=9033, resource=789

089769: Jan 20 17:16:51.279: PktCbl(d2r): extract id: gate=9033, resource=789

089770: Jan 20 17:16:51.283: PktCbl(d2r): extract id: gate=9033, resource=789

089771: Jan 20 17:16:51.283: PktCbl(d2r): extract id: gate=9033, resource=789

089772: Jan 20 17:16:51.283: PktCbl(d2r): DSC-ACK notification received on RP,
   gateid : 9033

089773: Jan 20 17:16:51.283: Pktcbl(gcp): Building GCP message, added obj
   TRANSACTION ID      ; len:8 padding:0

089774: Jan 20 17:16:51.283: Pktcbl(gcp): Building GCP message, added obj GATE
   ID              ; len:8 padding:0

089775: Jan 20 17:16:51.283: Pktcbl(gcp): Built GCP message, GATE OPEN,
   length: 16, copsLen 44

089776: Jan 20 17:16:51.283:  ---- Pktcbl: Sending GCP message ------

089777: Jan 20 17:16:51.283: TRANSACTION ID       : Object.[snum/stype/len 1/1/8]

089778: Jan 20 17:16:51.283:      transaction id   : 0x0

089779: Jan 20 17:16:51.283:      gcp cmd          : 13 (GATE OPEN)
```

[21] CMTS activates the gate with GATE ID 9033 and notifies the CMS.

```
089780: Jan 20 17:16:51.283: GATE ID            : Object.[snum/stype/len 3/1/8]

089781: Jan 20 17:16:51.283:      GateID         : 9033 (0x2349)

089782: Jan 20 17:16:51.283: -------------------------------------flow
   handle(9033)

089783: Jan 20 17:16:51.435: PktCbl(d2r): extract id: gate=74557, resource=791

089784: Jan 20 17:16:51.435: PktCbl(d2r): extract id: gate=74557, resource=791

089785: Jan 20 17:16:51.439: PktCbl(d2r): extract id: gate=74557, resource=791

089786: Jan 20 17:16:51.439: PktCbl(d2r): extract id: gate=74557, resource=791

089787: Jan 20 17:16:51.439: PktCbl(d2r): DSC-ACK notification received on RP,
   gateid : 74557

089788: Jan 20 17:16:51.439: Pktcbl(gcp): Building GCP message, added obj
   TRANSACTION ID      ; len:8 padding:0
```

```
089789: Jan 20 17:16:51.439: Pktcbl(gcp): Building GCP message, added obj GATE
  ID              ; len:8 padding:0
089790: Jan 20 17:16:51.439: Pktcbl(gcp): Built GCP message, GATE OPEN,
  length: 16, copsLen 44
089791: Jan 20 17:16:51.439:   ---- Pktcbl: Sending GCP message ------
089792: Jan 20 17:16:51.439: TRANSACTION ID      : Object.[snum/stype/len 1/1/8]
089793: Jan 20 17:16:51.439:     transaction id  : 0x0
089794: Jan 20 17:16:51.439:     gcp cmd         : 13 (GATE OPEN)
```
[22] CMTS activates the gate with GATE ID 74557 and notifies the CMS.
```
089795: Jan 20 17:16:51.439: GATE ID             : Object.[snum/stype/len 3/1/8]
089796: Jan 20 17:16:51.439:     GateID          : 74557 (0x1233D)
089797: Jan 20 17:16:51.439: ------------------------------------flow
  handle(74557)
Jan 20 17:16:51.158: PktCbl(r2d): cfr: proto=17 src=24.31.240.28 dst=24.31.240.36
  sport=53456 dport=53456
Jan 20 17:16:51.158: PktCbl(r2d): REQ - r/b/p/m/M 1174515712 1126563840
  1174515712 166 166 (R/S) 1174515712 800
Jan 20 17:16:51.158: PktCbl(d2r): DS flowspec cfr: proto=17 src=24.31.240.36
  dst=24.31.240.28 sport=0 dport=53456
Jan 20 17:16:51.158: PktCbl(r2d): cfr: proto=17 src=24.31.240.36 dst=24.31.240.28
  sport=0 dport=53456
Jan 20 17:16:51.158: PktCbl(r2d): REQ - r/b/p/m/M 1175539712 1127874560
  1175539712 186 186 (R/S) 1175539712 0
089800: Jan 20 17:18:52.138: Pktcbl(gcp): Building GCP message, added obj
  TRANSACTION ID       ; len:8 padding:0
089801: Jan 20 17:18:52.142: Pktcbl(gcp): Building GCP message, added obj GATE
  ID              ; len:8 padding:0
089802: Jan 20 17:18:52.142: Pktcbl(gcp): Building GCP message, added obj PKTCBL
  REASON          ; len:8 padding:0
089803: Jan 20 17:18:52.142: Pktcbl(gcp): Built GCP message, GATE CLOSE,
  length: 24, copsLen 52
089804: Jan 20 17:18:52.142:   ---- Pktcbl: Sending GCP message ------
089805: Jan 20 17:18:52.142: TRANSACTION ID      : Object.[snum/stype/len 1/1/8]
089806: Jan 20 17:18:52.142:     transaction id  : 0x0
089807: Jan 20 17:18:52.142:     gcp cmd         : 14 (GATE CLOSE)
```
[30] CMTS sends a GATE CLOSE message to CMS for the gate with GATE ID 74557.
This releases the resources for the first call leg.
```
089808: Jan 20 17:18:52.142: GATE ID             : Object.[snum/stype/len 3/1/8]
089809: Jan 20 17:18:52.142:     GateID          : 74557 (0x1233D)
089810: Jan 20 17:18:52.142: PKTCBL REASON       : Object.[snum/stype/len 13/1/8]
089811: Jan 20 17:18:52.142:     reason          : [code/subcode: 0x1/0x0]
089812: Jan 20 17:18:52.142: ------------------------------------
089813: Jan 20 17:18:52.178: Pktcbl(gcp): Building GCP message, added obj
  TRANSACTION ID       ; len:8 padding:0
089814: Jan 20 17:18:52.178: Pktcbl(gcp): Building GCP message, added obj GATE
  ID              ; len:8 padding:0
089815: Jan 20 17:18:52.178: Pktcbl(gcp): Building GCP message, added obj PKTCBL
  REASON          ; len:8 padding:0
089816: Jan 20 17:18:52.178: Pktcbl(gcp): Built GCP message, GATE CLOSE,
  length: 24, copsLen 52
```

```
089817: Jan 20 17:18:52.178:   ---- Pktcbl: Sending GCP message ------
089818: Jan 20 17:18:52.178: TRANSACTION ID        : Object.[snum/stype/len 1/1/8]
089819: Jan 20 17:18:52.178:   transaction id    : 0x0
089820: Jan 20 17:18:52.178:      gcp cmd         : 14 (GATE CLOSE)
[32] CMTS sends a GATE CLOSE message to CMS for the gate with GATE ID 9033. This
   releases the resources for the second call leg.
089821: Jan 20 17:18:52.178: GATE ID                : Object.[snum/stype/len 3/1/8]
089822: Jan 20 17:18:52.178:     GateID             : 9033 (0x2349)
089823: Jan 20 17:18:52.178: PKTCBL REASON          : Object.[snum/stype/len 13/1/8]
089824: Jan 20 17:18:52.178:      reason            : [code/subcode: 0x1/0x0]
089825: Jan 20 17:18:52.178: ------------------------------------
Jan 20 17:18:52.142: PktCbl(d2r): DSD-REQ received, gateid: 74557
Jan 20 17:18:52.178: PktCbl(d2r): DSD-REQ received, gateid: 9033
```

Session Initiation Protocol (SIP) Call Traces

This section illustrates call traces for the following use cases. These are collected via Unified CM (CUCM) logs and debugs on Cisco IOS devices. The Unified CM System Diagnostic Interface (SDI) trace logs are at debug trace level 12, which is the default setting. These were collected using the device-specific filters on CUCM version 7.0.1 with the SIP stack traced enabled. Instructions for setting this trace log can be found at http://www.cisco.com/en/US/docs/voice_ip_comm/cucm/service/7_0_1/admin_master/satrace.html. Cisco IOS gateway used the **debug ccsip all** command on IOS Release 12.4(8)T. The purpose is to use these as reference for fault isolation and establishing a pattern for trend analysis, as discussed in Chapter 7, "Performance Analysis and Fault Isolation," and Chapter 8, "Trend Analysis and Optimization."

The following call flows and traces are discussed in this section:

- Basic Early Media Call through an Inter-Cluster SIP trunk

- SIP trunk passes through T38 Fax Call between SIP and H.323 networks

- Call between Cisco Unified Communications Manager (CUCM) Skinny Client Control Protocol (SCCP)–controlled phone and SIP Gateway using KeyPad Markup Language (KPML)

- Call admission control using Resource Reservation Protocol (RSVP)

Basic Early Media Call Through an Inter-Cluster SIP Trunk

The following call flow illustrates the SIP messaging that takes place between two Unified CM (CUCM) via an inter-cluster SIP trunk. Early media is referred to the exchange of information before the connection is established. Cisco Unified CM1 sent out the initial INVITE with Session Description Protocol (SDP). SDP contains the information about the capability of the calling party's device that originated the call. SDP might include the protocol version, supported codecs, bandwidth, encryption, and other session attributes. Early media calls have the SDP as part of the first SIP INVITE message, even if the called party has not responded.

Configuration is as follows:

Node = Unified CM1, IP = 10.10.201.228

Node = Unified CM2, IP = 10.10.201.227

```
 [1] INVITE sip:90025060 SIP/2.0
Via: SIP/2.0/TCP 10.10.201.228;branch=z9hG4bK44ebed7d
Remote-Party-ID: <sip:9309@10.10.201.228>;party=calling;screen=yes;privacy=off
From: <sip:9309@10.10.201.228>;tag=69a5de24-d471-4214-a73b-7c3e594b36b1-29935926
To: <sip:9002@10.10.201.227>
Date: Tue, 14 Feb 2006 19:42:14 GMT
Call-ID: fe167480-3f213296-2-e4c912ac@10.10.201.228
Supported: timer,replaces
Min-SE: 1800
User-Agent: Cisco-CCM7.0
Allow: INVITE, OPTIONS, INFO, BYE, CANCEL, ACK, PRACK, UPDATE, REFER,
  SUBSCRIBE, NOTIFY
CSeq: 101 INVITE
Contact: <sip:93095060;transport=tcp>
Expires: 180
Allow-Events: presence
Max-Forwards: 70
Content-Type: application/sdp
Content-Length: 216
v=0
o=CiscoSystemsCCM-SIP 2000 1 IN IP4 10.10.201.228
s=SIP Call
c=IN IP4 10.10.201.228
t=0 0
m=audio 24598 RTP/AVP 0 101
a=rtpmap:0 PCMU/8000
a=ptime:20
a=rtpmap:101 telephone-event/8000
a=fmtp:101 0-15

[2] SIP/2.0 100 Trying
Via: SIP/2.0/TCP 10.10.201.228;branch=z9hG4bK44ebed7d
From: <sip:9309@10.10.201.228>;tag=69a5de24-d471-4214-a73b-7c3e594b36b1-29935926
To: <sip:9002@10.10.201.227>
Date: Tue, 14 Feb 2006 19:42:14 GMT
Call-ID: fe167480-3f213296-2-e4c912ac@10.10.201.228
CSeq: 101 INVITE
```

```
Allow-Events: presence
Content-Length: 0

[3] SIP/2.0 180 Ringing
Via: SIP/2.0/TCP 10.10.201.228;branch=z9hG4bK44ebed7d
From: <sip:9309@10.10.201.228>;tag=69a5de24-d471-4214-a73b-7c3e594b36b1-29935926
To: <sip:9002@10.10.201.227>;tag=f5356241-600b-496b-91af-e076425f07d3-28540947
Date: Tue, 14 Feb 2006 19:42:14 GMT
Call-ID: fe167480-3f213296-2-e4c912ac@10.10.201.228
CSeq: 101 INVITE
Allow: INVITE, OPTIONS, INFO, BYE, CANCEL, ACK, PRACK, UPDATE, REFER,
  SUBSCRIBE, NOTIFY
Allow-Events: presence
Remote-Party-ID: <sip:9002@10.10.201.227>;party=called;screen=yes;privacy=off
Contact: <sip:90025060;transport=tcp>
Content-Length: 0

[4] SIP/2.0 200 OK
Via: SIP/2.0/TCP 10.10.201.228;branch=z9hG4bK44ebed7d
From: <sip:9309@10.10.201.228>;tag=69a5de24-d471-4214-a73b-7c3e594b36b1-29935926
To: <sip:9002@10.10.201.227>;tag=f5356241-600b-496b-91af-e076425f07d3-28540947
Date: Tue, 14 Feb 2006 19:42:14 GMT
Call-ID: fe167480-3f213296-2-e4c912ac@10.10.201.228
CSeq: 101 INVITE
Allow: INVITE, OPTIONS, INFO, BYE, CANCEL, ACK, PRACK, UPDATE, REFER,
  SUBSCRIBE, NOTIFY
Allow-Events: presence
Remote-Party-ID: <sip:9002@10.10.201.227>;party=called;screen=yes;privacy=off
Contact: <sip:90025060;transport=tcp>
Session-Expires: 1800;refresher=uas
Require: timer
Content-Type: application/sdp
Content-Length: 216
v=0
o=CiscoSystemsCCM-SIP 2000 1 IN IP4 10.10.201.227
s=SIP Call
c=IN IP4 10.10.201.227
t=0 0
m=audio 24588 RTP/AVP 0 101
a=rtpmap:0 PCMU/8000
a=ptime:20
a=rtpmap:101 telephone-event/8000
a=fmtp:101 0-15
```

```
[5] ACK sip:90025060;transport=tcp SIP/2.0
Via: SIP/2.0/TCP 10.10.201.228;branch=z9hG4bK29bffd76
From: <sip:9309@10.10.201.228>;tag=69a5de24-d471-4214-a73b-7c3e594b36b1-29935926
To: <sip:9002@10.10.201.227>;tag=f5356241-600b-496b-91af-e076425f07d3-28540947
Date: Tue, 14 Feb 2006 19:42:14 GMT
Call-ID: fe167480-3f213296-2-e4c912ac@10.10.201.228
Max-Forwards: 70
CSeq: 101 ACK
Allow-Events: presence
Content-Length: 0


[6] BYE sip:90025060;transport=tcp SIP/2.0
Via: SIP/2.0/TCP 10.10.201.228;branch=z9hG4bK3db4b620
From: <sip:9309@10.10.201.228>;tag=69a5de24-d471-4214-a73b-7c3e594b36b1-29935926
To: <sip:9002@10.10.201.227>;tag=f5356241-600b-496b-91af-e076425f07d3-28540947
Date: Tue, 14 Feb 2006 19:42:14 GMT
Call-ID: fe167480-3f213296-2-e4c912ac@10.10.201.228
User-Agent: Cisco-CCM7.0
Max-Forwards: 70
CSeq: 102 BYE
Content-Length: 0


[7] SIP/2.0 200 OK
Via: SIP/2.0/TCP 10.10.201.228;branch=z9hG4bK3db4b620
From: <sip:9309@10.10.201.228>;tag=69a5de24-d471-4214-a73b-7c3e594b36b1-29935926
To: <sip:9002@10.10.201.227>;tag=f5356241-600b-496b-91af-e076425f07d3-28540947
Date: Tue, 14 Feb 2006 19:42:16 GMT
Call-ID: fe167480-3f213296-2-e4c912ac@10.10.201.228
CSeq: 102 BYE
Content-Length: 0
```

SIP Trunk Passes Through T.38 Fax Call Between SIP and H.323 Networks

Cisco Unified CM can pass through T.38 fax calls using an SIP trunk. This call flow shows how an SIP trunk lets a fax call pass the Unified CM between SIP and H.323 networks. Fax calls have notoriously high failure rates, especially when protocol interworking is involved. Chapter 8 discusses this, and Figure 8-9 illustrates it.

The SIP gateway is the side that initiates the fax transition. The receiving end is controlled through the H.323 protocol. The CUCM acts like a protocol translator in this transaction. The trace shows that in the final stage, denoted by [12], the originating side (SIP gateway) indicates that it cannot support the requested media type, hence terminating the call.

Configuration is as follows:

Node = Unified CM, IP = 10.10.136.15

Node = SIP Gateway, IP = 10.10.136.61

```
[1] INVITE sip:456885060 SIP/2.0
Via: SIP/2.0/TCP 10.10.136.15;branch=z9hG4bK48426491
From: <sip:10.10.136.15>;tag=5ddbd6a3-63b3-4ec9-832f-f5e1c961600e-23767069
To: <sip:45688@10.10.136.61>
Date: Sat, 28 Jan 2006 02:18:47 GMT
Call-ID: 6862a300-3da1d487-6-f8813ac@10.10.136.15
Supported: timer,replaces
Min-SE: 1800
User-Agent: Cisco-CCM7.0
Allow: INVITE, OPTIONS, INFO, BYE, CANCEL, ACK, PRACK, UPDATE, REFER,
  SUBSCRIBE, NOTIFY
CSeq: 101 INVITE
Contact: <sip:10.10.136.15:5060;transport=tcp>
Expires: 180
Allow-Events: presence, dialog, kpml
Call-Info: <sip:10.10.136.15:5060>;method="NOTIFY;Event=telephone-
  event;Duration=500"
Max-Forwards: 70
Content-Length: 0


[2] SIP/2.0 100 Trying
Via: SIP/2.0/TCP 10.10.136.15;branch=z9hG4bK48426491
From: <sip:10.10.136.15>;tag=5ddbd6a3-63b3-4ec9-832f-f5e1c961600e-23767069
To: <sip:45688@10.10.136.61>;tag=295AB278-1B1B
Date: Wed, 30 Jul 2003 20:04:16 GMT
Call-ID: 6862a300-3da1d487-6-f8813ac@10.10.136.15
Server: Cisco-SIPGateway/IOS-12.x
CSeq: 101 INVITE
Allow-Events: telephone-event
Content-Length: 0


[3] SIP/2.0 183 Session Progress
Via: SIP/2.0/TCP 10.10.136.15;branch=z9hG4bK48426491
From: <sip:10.10.136.15>;tag=5ddbd6a3-63b3-4ec9-832f-f5e1c961600e-23767069
To: <sip:45688@10.10.136.61>;tag=295AB278-1B1B
Date: Wed, 30 Jul 2003 20:04:16 GMT
Call-ID: 6862a300-3da1d487-6-f8813ac@10.10.136.15
Server: Cisco-SIPGateway/IOS-12.x
CSeq: 101 INVITE
```

```
Allow: INVITE, OPTIONS, BYE, CANCEL, ACK, PRACK, UPDATE, REFER, SUBSCRIBE,
  NOTIFY, INFO, REGISTER
Allow-Events: telephone-event

Contact: <sip:456885060;transport=tcp>
Content-Type: application/sdp
Content-Disposition: session;handling=required
Content-Length: 230
v=0
o=CiscoSystemsSIP-GW-UserAgent 7024 4914 IN IP4 10.10.136.61
s=SIP Call
c=IN IP4 10.10.136.61
t=0 0
m=audio 17308 RTP/AVP 0 8 19
c=IN IP4 10.10.136.61
a=rtpmap:0 PCMU/8000
a=rtpmap:8 PCMA/8000
a=rtpmap:19 CN/8000


[4] SIP/2.0 200 OK
Via: SIP/2.0/TCP 10.10.136.15;branch=z9hG4bK48426491
From: <sip:10.10.136.15>;tag=5ddbd6a3-63b3-4ec9-832f-f5e1c961600e-23767069
To: <sip:45688@10.10.136.61>;tag=295AB278-1B1B
Date: Wed, 30 Jul 2003 20:04:16 GMT
Call-ID: 6862a300-3da1d487-6-f8813ac@10.10.136.15
Server: Cisco-SIPGateway/IOS-12.x
CSeq: 101 INVITE
Allow: INVITE, OPTIONS, BYE, CANCEL, ACK, PRACK, UPDATE, REFER, SUBSCRIBE,
  NOTIFY, INFO, REGISTER
Allow-Events: telephone-event

Contact: <sip:456885060;transport=tcp>
Content-Type: application/sdp
Content-Disposition: session;handling=required
Content-Length: 230
v=0
o=CiscoSystemsSIP-GW-UserAgent 7024 4914 IN IP4 10.10.136.61
s=SIP Call
c=IN IP4 10.10.136.61
t=0 0
m=audio 17308 RTP/AVP 0 8 19
c=IN IP4 10.10.136.61
a=rtpmap:0 PCMU/8000
a=rtpmap:8 PCMA/8000
a=rtpmap:19 CN/8000


[5] ACK sip:456885060;transport=tcp SIP/2.0
Via: SIP/2.0/TCP 10.10.136.15;branch=z9hG4bK293d6d6c
```

```
From: <sip:10.10.136.15>;tag=5ddbd6a3-63b3-4ec9-832f-f5e1c961600e-23767069
To: <sip:45688@10.10.136.61>;tag=295AB278-1B1B
Date: Sat, 28 Jan 2006 02:18:47 GMT
Call-ID: 6862a300-3da1d487-6-f8813ac@10.10.136.15
Max-Forwards: 70
CSeq: 101 ACK
Allow-Events: presence, dialog, kpml
Content-Type: application/sdp
Content-Length: 157
v=0
o=CiscoSystemsCCM-SIP 2000 1 IN IP4 10.10.136.15
s=SIP Call
c=IN IP4 10.10.13.36
t=0 0
m=audio 17264 RTP/AVP 0
a=rtpmap:0 PCMU/8000
a=ptime:20


[6] INVITE sip:10.10.136.15:5060;transport=tcp SIP/2.0
Via: SIP/2.0/TCP 10.10.136.61;branch=z9hG4bK2DFBA
From: <sip:45688@10.10.136.61>;tag=295AB278-1B1B
To: <sip:10.10.136.15>;tag=5ddbd6a3-63b3-4ec9-832f-f5e1c961600e-23767069
Date: Wed, 30 Jul 2003 20:04:39 GMT
Call-ID: 6862a300-3da1d487-6-f8813ac@10.10.136.15
Supported: 100rel,timer,resource-priority,replaces
Min-SE: 1800
Cisco-Guid: 3572621395-3254718935-2153631767-1618905505
User-Agent: Cisco-SIPGateway/IOS-12.x
Allow: INVITE, OPTIONS, BYE, CANCEL, ACK, PRACK, UPDATE, REFER, SUBSCRIBE,
  NOTIFY, INFO, REGISTER
CSeq: 101 INVITE
Max-Forwards: 70
Timestamp: 1059595479

Contact: <sip:456885060;transport=tcp>
Expires: 180
Allow-Events: telephone-event
Content-Type: application/sdp
Content-Length: 399
v=0
o=CiscoSystemsSIP-GW-UserAgent 7024 4916 IN IP4 10.10.136.61
s=SIP Call
c=IN IP4 10.10.136.61
t=0 0
m=image 17308 udptl t38
c=IN IP4 10.10.136.61
```

```
a=T38FaxVersion:0
a=T38MaxBitRate:14400
a=T38FaxFillBitRemoval:0
a=T38FaxTranscodingMMR:0
a=T38FaxTranscodingJBIG:0
a=T38FaxRateManagement:transferredTCF
a=T38FaxMaxBuffer:200
a=T38FaxMaxDatagram:72
a=T38FaxUdpEC:t38UDPRedundancy


[7] SIP/2.0 100 Trying
Via: SIP/2.0/TCP 10.10.136.61;branch=z9hG4bK2DFBA
From: <sip:45688@10.10.136.61>;tag=295AB278-1B1B
To: <sip:10.10.136.15>;tag=5ddbd6a3-63b3-4ec9-832f-f5e1c961600e-23767069
Date: Sat, 28 Jan 2006 02:19:10 GMT
Call-ID: 6862a300-3da1d487-6-f8813ac@10.10.136.15
CSeq: 101 INVITE
Allow-Events: presence, dialog, kpml
Content-Length: 0


[8] SIP/2.0 200 OK
Via: SIP/2.0/TCP 10.10.136.61;branch=z9hG4bK2DFBA
From: <sip:45688@10.10.136.61>;tag=295AB278-1B1B
To: <sip:10.10.136.15>;tag=5ddbd6a3-63b3-4ec9-832f-f5e1c961600e-23767069
Date: Sat, 28 Jan 2006 02:19:10 GMT
Call-ID: 6862a300-3da1d487-6-f8813ac@10.10.136.15
CSeq: 101 INVITE
Allow: INVITE, OPTIONS, INFO, BYE, CANCEL, ACK, PRACK, UPDATE, REFER,
   SUBSCRIBE, NOTIFY
Allow-Events: presence, dialog, kpml
Contact: <sip:10.10.136.15:5060;transport=tcp>
Content-Type: application/sdp
Content-Length: 123
v=0
o=CiscoSystemsCCM-SIP 2000 2 IN IP4 10.10.136.15
s=SIP Call
t=0 0
m=image 17264 udptl t38
c=IN IP4 10.10.13.36


[9] ACK sip:10.10.136.15:5060;transport=tcp SIP/2.0
Via: SIP/2.0/TCP 10.10.136.61;branch=z9hG4bK2E18E2
From: <sip:45688@10.10.136.61>;tag=295AB278-1B1B
To: <sip:10.10.136.15>;tag=5ddbd6a3-63b3-4ec9-832f-f5e1c961600e-23767069
Date: Wed, 30 Jul 2003 20:04:39 GMT
```

```
Call-ID: 6862a300-3da1d487-6-f8813ac@10.10.136.15
Max-Forwards: 70
CSeq: 101 ACK
Content-Length: 0


[10] INVITE sip:10.10.136.15:5060;transport=tcp SIP/2.0
Via: SIP/2.0/TCP 10.10.136.61;branch=z9hG4bK2F1D17
From: <sip:45688@10.10.136.61>;tag=295AB278-1B1B
To: <sip:10.10.136.15>;tag=5ddbd6a3-63b3-4ec9-832f-f5e1c961600e-23767069
Date: Wed, 30 Jul 2003 20:05:05 GMT
Call-ID: 6862a300-3da1d487-6-f8813ac@10.10.136.15
Supported: 100rel,timer,resource-priority,replaces
Min-SE: 1800
Cisco-Guid: 3572621395-3254718935-2153631767-1618905505
User-Agent: Cisco-SIPGateway/IOS-12.x
Allow: INVITE, OPTIONS, BYE, CANCEL, ACK, PRACK, UPDATE, REFER, SUBSCRIBE,
   NOTIFY, INFO, REGISTER
CSeq: 102 INVITE
Max-Forwards: 70
Timestamp: 1059595505

Contact: <sip:456885060;transport=tcp>
Expires: 180
Allow-Events: telephone-event
Content-Type: application/sdp
Content-Length: 230
v=0
o=CiscoSystemsSIP-GW-UserAgent 7024 4917 IN IP4 10.10.136.61
s=SIP Call
c=IN IP4 10.10.136.61
t=0 0
m=audio 17308 RTP/AVP 0 8 19
c=IN IP4 10.10.136.61
a=rtpmap:0 PCMU/8000
a=rtpmap:8 PCMA/8000
a=rtpmap:19 CN/8000


[11] SIP/2.0 100 Trying
Via: SIP/2.0/TCP 10.10.136.61;branch=z9hG4bK2F1D17
From: <sip:45688@10.10.136.61>;tag=295AB278-1B1B
To: <sip:10.10.136.15>;tag=5ddbd6a3-63b3-4ec9-832f-f5e1c961600e-23767069
Date: Sat, 28 Jan 2006 02:19:36 GMT
Call-ID: 6862a300-3da1d487-6-f8813ac@10.10.136.15
CSeq: 102 INVITE
Allow-Events: presence, dialog, kpml
Content-Length: 0
```

```
[12] SIP/2.0 415 Unsupported Media Type
Via: SIP/2.0/TCP 10.10.136.61;branch=z9hG4bK2F1D17
From: <sip:45688@10.10.136.61>;tag=295AB278-1B1B
To: <sip:10.10.136.15>;tag=5ddbd6a3-63b3-4ec9-832f-f5e1c961600e-23767069
Date: Sat, 28 Jan 2006 02:19:36 GMT
Call-ID: 6862a300-3da1d487-6-f8813ac@10.10.136.15
CSeq: 102 INVITE
Allow-Events: presence, dialog, kpml
Content-Length: 0


[13] BYE sip:456885060;transport=tcp SIP/2.0
Via: SIP/2.0/TCP 10.10.136.15;branch=z9hG4bK7ce13ea
From: <sip:10.10.136.15>;tag=5ddbd6a3-63b3-4ec9-832f-f5e1c961600e-23767069
To: <sip:45688@10.10.136.61>;tag=295AB278-1B1B
Date: Sat, 28 Jan 2006 02:19:36 GMT
Call-ID: 6862a300-3da1d487-6-f8813ac@10.10.136.15
User-Agent: Cisco-CCM7.0
Max-Forwards: 70
CSeq: 102 BYE
Reason: Q.850;cause=47
Content-Length: 0


[14] ACK sip:10.10.136.15:5060;transport=tcp SIP/2.0
Via: SIP/2.0/TCP 10.10.136.61;branch=z9hG4bK2F1D17
From: <sip:45688@10.10.136.61>;tag=295AB278-1B1B
To: <sip:10.10.136.15>;tag=5ddbd6a3-63b3-4ec9-832f-f5e1c961600e-23767069
Date: Wed, 30 Jul 2003 20:05:05 GMT
Call-ID: 6862a300-3da1d487-6-f8813ac@10.10.136.15
Max-Forwards: 70
CSeq: 102 ACK
Content-Length: 0


[15] SIP/2.0 200 OK
Via: SIP/2.0/TCP 10.10.136.15;branch=z9hG4bK7ce13ea
From: <sip:10.10.136.15>;tag=5ddbd6a3-63b3-4ec9-832f-f5e1c961600e-23767069
To: <sip:45688@10.10.136.61>;tag=295AB278-1B1B
Date: Wed, 30 Jul 2003 20:05:05 GMT
Call-ID: 6862a300-3da1d487-6-f8813ac@10.10.136.15
Server: Cisco-SIPGateway/IOS-12.x
CSeq: 102 BYE
Reason: Q.850;cause=16
Content-Length: 0
```

Call Between Cisco Unified CM SCCP Phone and SIP Gateway Using KPML

The collection of dialed digits by the call-originating device and propagating it in the call-signaling message are critical for a successful call. SIP is designed to send the entire address of the called party in the initial call setup message, although it might create a user experience issue in which the user has to change habits by dialing # after the last dialed digit or wait for a timer to expire before the call is initiated and the user hears a ringback tone. To rectify this potential issue with SIP call signaling, the Keypress Stimulus Protocol is leveraged that uses message bodies from KeyPad Markup Language (KPML) for transport using SIP. The following call flow illustrates the SIP messaging that takes place between an SCCP phone hosted by a Cisco Unified CM and an SIP Gateway through an SIP trunk. The SIP trunk dual-tone multifrequency (DTMF) signaling method is configured for NoPreference, and the SIP Gateway (GW) dtmf-relay for the associated dial peer is configured for sip-kpml.

These traces can be useful in isolating a potential issue in a device's capability to send the digits through KPML, dial plan issue, or protocol compatibility. Traces to note are the bidirectional subscriptions for DTMF notification between the CUCM and the Gateway, and the NOTIFY methods exchanged when the digit 1 is pressed from the called and calling party phones.

Note that after the call is released, the CUCM and the SIP GW terminate both local and remote KPML subscriptions. Therefore, it is normal to see the CUCM either send or receive a 489 Bad Event. This signifies that the subscription has already been removed.

Configuration is as follows:

Node = Unified CM, IP = 10.10.199.61

Node = Gateway, IP = 10.10.199.93

```
[1] INVITE sip:36015060 SIP/2.0
Via: SIP/2.0/UDP 10.10.199.61:5060;branch=z9hG4bKe40d0d
Remote-Party-ID: "sccp_3000"
<sip:3000@10.10.199.61>;party=calling;screen=yes;privacy=off
From: "sccp_3000" <sip:3000@10.10.199.61>;tag=4afadcdc-0169-479d-ba6a-
    855a1847684f-30228326
To: <sip:3601@10.10.199.93>
Date: Wed, 15 Feb 2006 16:00:35 GMT
Call-ID: 31add900-3f315023-4-3dc712ac@10.10.199.61
Supported: timer,replaces
Min-SE: 1800
User-Agent: Cisco-CCM7.0
Allow: INVITE, OPTIONS, INFO, BYE, CANCEL, ACK, PRACK, UPDATE, REFER,
    SUBSCRIBE, NOTIFY
CSeq: 101 INVITE
Contact: <sip:30005060>
Expires: 180
```

```
Allow-Events: presence, kpml
Call-Info: <sip:10.10.199.61:5060>;method="NOTIFY;Event=telephone-
  event;Duration=500"
Max-Forwards: 70
Content-Length: 0


[2] SIP/2.0 100 Trying
Via: SIP/2.0/UDP 10.10.199.61:5060;branch=z9hG4bKe40d0d
From: "sccp_3000" <sip:3000@10.10.199.61>;tag=4afadcdc-0169-479d-ba6a-
  855a1847684f-30228326
To: <sip:3601@10.10.199.93>;tag=43A6301C-1EA
Date: Wed, 15 Feb 2006 16:00:35 GMT
Call-ID: 31add900-3f315023-4-3dc712ac@10.10.199.61
Server: Cisco-SIPGateway/IOS-12.x
CSeq: 101 INVITE
Allow-Events: kpml, telephone-event
Content-Length: 0


[3] SIP/2.0 183 Session Progress
Via: SIP/2.0/UDP 10.10.199.61:5060;branch=z9hG4bKe40d0d
From: "sccp_3000" <sip:3000@10.10.199.61>;tag=4afadcdc-0169-479d-ba6a-
  855a1847684f-30228326
To: <sip:3601@10.10.199.93>;tag=43A6301C-1EA
Date: Wed, 15 Feb 2006 16:00:35 GMT
Call-ID: 31add900-3f315023-4-3dc712ac@10.10.199.61
Server: Cisco-SIPGateway/IOS-12.x
CSeq: 101 INVITE
Allow: INVITE, OPTIONS, BYE, CANCEL, ACK, PRACK, UPDATE, REFER, SUBSCRIBE,
  NOTIFY, INFO, REGISTER
Allow-Events: kpml, telephone-event
Contact: <sip:36015060>
Content-Type: application/sdp
Content-Disposition: session;handling=required
Content-Length: 218
v=0
o=CiscoSystemsSIP-GW-UserAgent 5775 9505 IN IP4 10.10.199.93
s=SIP Call
c=IN IP4 10.10.199.93
t=0 0
m=audio 16636 RTP/AVP 0 19
c=IN IP4 10.10.199.93
a=rtpmap:0 PCMU/8000
a=rtpmap:19 CN/8000
a=ptime:20


[4] SIP/2.0 200 OK
```

```
Via: SIP/2.0/UDP 10.10.199.61:5060;branch=z9hG4bKe40d0d
From: "sccp_3000" <sip:3000@10.10.199.61>;tag=4afadcdc-0169-479d-ba6a-
   855a1847684f-30228326
To: <sip:3601@10.10.199.93>;tag=43A6301C-1EA
Date: Wed, 15 Feb 2006 16:00:35 GMT
Call-ID: 31add900-3f315023-4-3dc712ac@10.10.199.61
Server: Cisco-SIPGateway/IOS-12.x
CSeq: 101 INVITE
Allow: INVITE, OPTIONS, BYE, CANCEL, ACK, PRACK, UPDATE, REFER, SUBSCRIBE,
   NOTIFY, INFO, REGISTER
Allow-Events: kpml, telephone-event
Contact: <sip:36015060>
Content-Type: application/sdp
Content-Disposition: session;handling=required
Content-Length: 218
v=0
o=CiscoSystemsSIP-GW-UserAgent 5775 9505 IN IP4 10.10.199.93
s=SIP Call
c=IN IP4 10.10.199.93
t=0 0
m=audio 16636 RTP/AVP 0 19
c=IN IP4 10.10.199.93
a=rtpmap:0 PCMU/8000
a=rtpmap:19 CN/8000
a=ptime:20


[5] ACK sip:36015060 SIP/2.0
Via: SIP/2.0/UDP 10.10.199.61:5060;branch=z9hG4bK5cd72c7c
From: "sccp_3000" <sip:3000@10.10.199.61>;tag=4afadcdc-0169-479d-ba6a-
   855a1847684f-30228326
To: <sip:3601@10.10.199.93>;tag=43A6301C-1EA
Date: Wed, 15 Feb 2006 16:00:35 GMT
Call-ID: 31add900-3f315023-4-3dc712ac@10.10.199.61
Max-Forwards: 70
CSeq: 101 ACK
Allow-Events: presence, kpml
Content-Type: application/sdp
Content-Length: 158
v=0
o=CiscoSystemsCCM-SIP 2000 1 IN IP4 10.10.199.61
s=SIP Call
c=IN IP4 10.10.199.57
t=0 0
m=audio 20186 RTP/AVP 0
a=rtpmap:0 PCMU/8000
```

```
a=ptime:20

[6] SUBSCRIBE sip:36015060 SIP/2.0
Via: SIP/2.0/UDP 10.10.199.61:5060;branch=z9hG4bK44ba8143
From: "sccp_3000" <sip:3000@10.10.199.61>;tag=4afadcdc-0169-479d-ba6a-
  855a1847684f-30228326
To: <sip:3601@10.10.199.93>;tag=43A6301C-1EA
Call-ID: 31add900-3f315023-4-3dc712ac@10.10.199.61
CSeq: 102 SUBSCRIBE
Max-Forwards: 70
Date: Wed, 15 Feb 2006 16:00:38 GMT
User-Agent: Cisco-CCM7.0
Event: kpml
Expires: 7200
Contact: <sip:10.10.199.61:5060>
Accept: application/kpml-response+xml
Content-Type: application/kpml-request+xml
Content-Length: 213
<?xml version="1.0" encoding="UTF-8" standalone="no" ?>
<kpml-request version="1.0">
<pattern interdigittimer="500000" persist="persist">
<regex tag="dtmf"> [x*#ABCD]</regex>
</pattern>
</kpml-request>


[7] SUBSCRIBE sip:30005060 SIP/2.0
Via: SIP/2.0/UDP 10.10.199.93:5060;branch=z9hG4bK7C1148
From: <sip:3601@10.10.199.93>;tag=43A6301C-1EA
To: "sccp_3000" <sip:3000@10.10.199.61>;tag=4afadcdc-0169-479d-ba6a-855a1847684f-
  30228326
Call-ID: 31add900-3f315023-4-3dc712ac@10.10.199.61
CSeq: 101 SUBSCRIBE
Max-Forwards: 70
Date: Wed, 15 Feb 2006 16:00:38 GMT
User-Agent: Cisco-SIPGateway/IOS-12.x
Event: kpml
Expires: 7200
Contact: <sip:10.10.199.93:5060>
Content-Type: application/kpml-request+xml
Content-Length: 327
<?xml version="1.0" encoding="UTF-8"?><kpml-request
  xmlns="urn:ietf:params:xml:ns:kpml-request"
xmlns:xsi="http://www.w3.org/2001/XMLSchema-instance"
  xsi:schemaLocation="urn:ietf:params:xml:ns:kpmlrequest
kpml-request.xsd" version="1.0"><pattern persist="persist"><regex
tag="dtmf"> [x*#ABCD]</regex></pattern></kpml-request>
```

```
[8] SIP/2.0 200 OK
Via: SIP/2.0/UDP 10.10.199.61:5060;branch=z9hG4bK44ba8143
From: "sccp_3000" <sip:3000@10.10.199.61>;tag=4afadcdc-0169-479d-ba6a-
   855a1847684f-30228326
To: <sip:3601@10.10.199.93>;tag=43A6301C-1EA
Call-ID: 31add900-3f315023-4-3dc712ac@10.10.199.61
CSeq: 102 SUBSCRIBE
Content-Length: 0
Contact: <sip:10.10.199.93:5060>
Expires: 7200


[9] NOTIFY sip:10.10.199.61:5060 SIP/2.0
Via: SIP/2.0/UDP 10.10.199.93:5060;branch=z9hG4bK7DBB5
From: <sip:3601@10.10.199.93>;tag=43A6301C-1EA
To: "sccp_3000" <sip:3000@10.10.199.61>;tag=4afadcdc-0169-479d-ba6a-855a1847684f-
   30228326
Call-ID: 31add900-3f315023-4-3dc712ac@10.10.199.61
CSeq: 102 NOTIFY
Max-Forwards: 70
Date: Wed, 15 Feb 2006 16:00:38 GMT
User-Agent: Cisco-SIPGateway/IOS-12.x
Event: kpml
Subscription-State: active
Contact: <sip:10.10.199.93:5060>
Content-Length: 0


[10] SIP/2.0 200 OK
Via: SIP/2.0/UDP 10.10.199.93:5060;branch=z9hG4bK7C1148
From: <sip:3601@10.10.199.93>;tag=43A6301C-1EA
To: "sccp_3000" <sip:3000@10.10.199.61>;tag=4afadcdc-0169-479d-ba6a-855a1847684f-
   30228326
Date: Wed, 15 Feb 2006 16:00:39 GMT
Call-ID: 31add900-3f315023-4-3dc712ac@10.10.199.61
CSeq: 101 SUBSCRIBE
Content-Length: 0
Contact: <sip:10.10.199.61:5060>
Expires: 3600


[11] NOTIFY sip:10.10.199.93:5060 SIP/2.0
Via: SIP/2.0/UDP 10.10.199.61:5060;branch=z9hG4bK5ba303c0
From: "sccp_3000" <sip:3000@10.10.199.61>;tag=4afadcdc-0169-479d-ba6a-
   855a1847684f-30228326
To: <sip:3601@10.10.199.93>;tag=43A6301C-1EA
Call-ID: 31add900-3f315023-4-3dc712ac@10.10.199.61
CSeq: 103 NOTIFY
```

```
Max-Forwards: 70
Date: Wed, 15 Feb 2006 16:00:39 GMT
User-Agent: Cisco-CCM7.0
Event: kpml
Subscription-State: active;expires=3600
Contact: <sip:10.10.199.61:5060>
Content-Length: 0


[12] SIP/2.0 200 OK
Via: SIP/2.0/UDP 10.10.199.93:5060;branch=z9hG4bK7DBB5
From: <sip:3601@10.10.199.93>;tag=43A6301C-1EA
To: "sccp_3000" <sip:3000@10.10.199.61>;tag=4afadcdc-0169-479d-ba6a-855a1847684f-
   30228326
Date: Wed, 15 Feb 2006 16:00:39 GMT
Call-ID: 31add900-3f315023-4-3dc712ac@10.10.199.61
CSeq: 102 NOTIFY
Content-Length: 0


[13] SIP/2.0 200 OK
Via: SIP/2.0/UDP 10.10.199.61:5060;branch=z9hG4bK5ba303c0
From: "sccp_3000" <sip:3000@10.10.199.61>;tag=4afadcdc-0169-479d-ba6a-
   855a1847684f-30228326
To: <sip:3601@10.10.199.93>;tag=43A6301C-1EA
Call-ID: 31add900-3f315023-4-3dc712ac@10.10.199.61
CSeq: 103 NOTIFY
Content-Length: 0


[14] NOTIFY sip:10.10.199.93:5060 SIP/2.0
Via: SIP/2.0/UDP 10.10.199.61:5060;branch=z9hG4bK7e715b4
From: "sccp_3000" <sip:3000@10.10.199.61>;tag=4afadcdc-0169-479d-ba6a-
   855a1847684f-30228326
To: <sip:3601@10.10.199.93>;tag=43A6301C-1EA
Call-ID: 31add900-3f315023-4-3dc712ac@10.10.199.61
CSeq: 104 NOTIFY
Max-Forwards: 70
Date: Wed, 15 Feb 2006 16:00:48 GMT
User-Agent: Cisco-CCM7.0
Event: kpml
Subscription-State: active;expires=3591
Contact: <sip:10.10.199.61:5060>
Content-Type: application/kpml-response+xml
Content-Length: 177
<?xml version="1.0" encoding="UTF-8" standalone="no" ?>
<kpml-response code="200" digits="1" forced_flush="false" suppressed="false"
   tag="dtmf" text="Success"
version="1.0"/>
```

```
[15] SIP/2.0 200 OK
Via: SIP/2.0/UDP 10.10.199.61:5060;branch=z9hG4bK7e715b4
From: "sccp_3000" <sip:3000@10.10.199.61>;tag=4afadcdc-0169-479d-ba6a-
   855a1847684f-30228326
To: <sip:3601@10.10.199.93>;tag=43A6301C-1EA
Call-ID: 31add900-3f315023-4-3dc712ac@10.10.199.61
CSeq: 104 NOTIFY
Content-Length: 0


[16] NOTIFY sip:10.10.199.61:5060 SIP/2.0
Via: SIP/2.0/UDP 10.10.199.93:5060;branch=z9hG4bK7E5F0
From: <sip:3601@10.10.199.93>;tag=43A6301C-1EA
To: "sccp_3000" <sip:3000@10.10.199.61>;tag=4afadcdc-0169-479d-ba6a-855a1847684f-
   30228326
Call-ID: 31add900-3f315023-4-3dc712ac@10.10.199.61
CSeq: 103 NOTIFY
Max-Forwards: 70
Date: Wed, 15 Feb 2006 16:00:58 GMT
User-Agent: Cisco-SIPGateway/IOS-12.x
Event: kpml
Subscription-State: active
Contact: <sip:10.10.199.93:5060>
Content-Type: application/kpml-response+xml
Content-Length: 113
<?xml version="1.0" encoding="UTF-8"?><kpml-response version="1.0" code="200"
   text="OK" digits="1"
tag="dtmf"/>


[17] SIP/2.0 200 OK
Via: SIP/2.0/UDP 10.10.199.93:5060;branch=z9hG4bK7E5F0
From: <sip:3601@10.10.199.93>;tag=43A6301C-1EA
To: "sccp_3000" <sip:3000@10.10.199.61>;tag=4afadcdc-0169-479d-ba6a-855a1847684f-
   30228326
Date: Wed, 15 Feb 2006 16:00:58 GMT
Call-ID: 31add900-3f315023-4-3dc712ac@10.10.199.61
CSeq: 103 NOTIFY
Content-Length: 0


[18] BYE sip:30005060 SIP/2.0
Via: SIP/2.0/UDP 10.10.199.93:5060;branch=z9hG4bK7F44E
From: <sip:3601@10.10.199.93>;tag=43A6301C-1EA
To: "sccp_3000" <sip:3000@10.10.199.61>;tag=4afadcdc-0169-479d-ba6a-855a1847684f-
   30228326
Date: Wed, 15 Feb 2006 16:00:38 GMT
Call-ID: 31add900-3f315023-4-3dc712ac@10.10.199.61
User-Agent: Cisco-SIPGateway/IOS-12.x
```

```
Max-Forwards: 70
Timestamp: 1140019264
CSeq: 104 BYE
Reason: Q.850;cause=16
Content-Length: 0


[19] SIP/2.0 200 OK
Via: SIP/2.0/UDP 10.10.199.93:5060;branch=z9hG4bK7F44E
From: <sip:3601@10.10.199.93>;tag=43A6301C-1EA
To: "sccp_3000" <sip:3000@10.10.199.61>;tag=4afadcdc-0169-479d-ba6a-855a1847684f-
   30228326
Date: Wed, 15 Feb 2006 16:01:04 GMT
Call-ID: 31add900-3f315023-4-3dc712ac@10.10.199.61
Timestamp: 1140019264
CSeq: 104 BYE
Content-Length: 0


[20] SUBSCRIBE sip:10.10.199.93:5060 SIP/2.0
Via: SIP/2.0/UDP 10.10.199.61:5060;branch=z9hG4bK1dc235ee
From: "sccp_3000" <sip:3000@10.10.199.61>;tag=4afadcdc-0169-479d-ba6a-
   855a1847684f-30228326
To: <sip:3601@10.10.199.93>;tag=43A6301C-1EA
Call-ID: 31add900-3f315023-4-3dc712ac@10.10.199.61
CSeq: 105 SUBSCRIBE
Max-Forwards: 70
Date: Wed, 15 Feb 2006 16:01:04 GMT
User-Agent: Cisco-CCM7.0
Event: kpml
Expires: 0
Contact: <sip:10.10.199.61:5060>
Content-Length: 0


[21] NOTIFY sip:10.10.199.93:5060 SIP/2.0
Via: SIP/2.0/UDP 10.10.199.61:5060;branch=z9hG4bKce2da53
From: "sccp_3000" <sip:3000@10.10.199.61>;tag=4afadcdc-0169-479d-ba6a-
   855a1847684f-30228326
To: <sip:3601@10.10.199.93>;tag=43A6301C-1EA
Call-ID: 31add900-3f315023-4-3dc712ac@10.10.199.61
CSeq: 106 NOTIFY
Max-Forwards: 70
Date: Wed, 15 Feb 2006 16:01:04 GMT
User-Agent: Cisco-CCM7.0
Event: kpml
Subscription-State: terminated
Contact: <sip:10.10.199.61:5060>
Content-Length: 0
```

```
[22] SUBSCRIBE sip:10.10.199.61:5060 SIP/2.0
Via: SIP/2.0/UDP 10.10.199.93:5060;branch=z9hG4bK80232D
From: <sip:3601@10.10.199.93>;tag=43A6301C-1EA
To: "sccp_3000" <sip:3000@10.10.199.61>;tag=4afadcdc-0169-479d-ba6a-855a1847684f-
  30228326
Call-ID: 31add900-3f315023-4-3dc712ac@10.10.199.61
CSeq: 105 SUBSCRIBE
Max-Forwards: 70
Date: Wed, 15 Feb 2006 16:01:04 GMT
User-Agent: Cisco-SIPGateway/IOS-12.x
Event: kpml
Expires: 0
Contact: <sip:10.10.199.93:5060>
Content-Length: 0


[23] SIP/2.0 200 OK
Via: SIP/2.0/UDP 10.10.199.93:5060;branch=z9hG4bK80232D
From: <sip:3601@10.10.199.93>;tag=43A6301C-1EA
To: "sccp_3000" <sip:3000@10.10.199.61>;tag=4afadcdc-0169-479d-ba6a-855a1847684f-
  30228326
Date: Wed, 15 Feb 2006 16:01:04 GMT
Call-ID: 31add900-3f315023-4-3dc712ac@10.10.199.61
CSeq: 105 SUBSCRIBE
Content-Length: 0
Contact: <sip:10.10.199.61:5060>
Expires: 0


[24] NOTIFY sip:10.10.199.93:5060 SIP/2.0
Via: SIP/2.0/UDP 10.10.199.61:5060;branch=z9hG4bK2ce6c7d2
From: "sccp_3000" <sip:3000@10.10.199.61>;tag=4afadcdc-0169-479d-ba6a-
  855a1847684f-30228326
To: <sip:3601@10.10.199.93>;tag=43A6301C-1EA
Call-ID: 31add900-3f315023-4-3dc712ac@10.10.199.61
CSeq: 107 NOTIFY
Max-Forwards: 70
Date: Wed, 15 Feb 2006 16:01:04 GMT
User-Agent: Cisco-CCM7.0
Event: kpml
Subscription-State: terminated;reason=timeout
Contact: <sip:10.10.199.61:5060>
Content-Type: application/kpml-response+xml
Content-Length: 189
<?xml version="1.0" encoding="UTF-8" standalone="no" ?>
<kpml-response code="487" digits="" forced_flush="false" suppressed="false"
  tag="dtmf" text="Subscription
```

```
Expired" version="1.0"/>


[25] SIP/2.0 200 OK
Via: SIP/2.0/UDP 10.10.199.61:5060;branch=z9hG4bK1dc235ee
From: "sccp_3000" <sip:3000@10.10.199.61>;tag=4afadcdc-0169-479d-ba6a-
   855a1847684f-30228326
To: <sip:3601@10.10.199.93>;tag=43A6301C-1EA
Call-ID: 31add900-3f315023-4-3dc712ac@10.10.199.61
CSeq: 105 SUBSCRIBE
Content-Length: 0
Contact: <sip:10.10.199.93:5060>
Expires: 0


[26] NOTIFY sip:10.10.199.61:5060 SIP/2.0
Via: SIP/2.0/UDP 10.10.199.93:5060;branch=z9hG4bK811043
From: <sip:3601@10.10.199.93>;tag=43A6301C-1EA
To: "sccp_3000" <sip:3000@10.10.199.61>;tag=4afadcdc-0169-479d-ba6a-855a1847684f-
   30228326
Call-ID: 31add900-3f315023-4-3dc712ac@10.10.199.61
CSeq: 106 NOTIFY
Max-Forwards: 70
Date: Wed, 15 Feb 2006 16:01:04 GMT
User-Agent: Cisco-SIPGateway/IOS-12.x
Event: kpml
Subscription-State: terminated
Contact: <sip:10.10.199.93:5060>
Content-Type: application/kpml-response+xml
Content-Length: 109
<?xml version="1.0" encoding="UTF-8"?><kpml-response version="1.0" code="487"
   text="Subscription Expired"/>


[27] SIP/2.0 200 OK
Via: SIP/2.0/UDP 10.10.199.61:5060;branch=z9hG4bKce2da53
From: "sccp_3000" <sip:3000@10.10.199.61>;tag=4afadcdc-0169-479d-ba6a-
   855a1847684f-30228326
To: <sip:3601@10.10.199.93>;tag=43A6301C-1EA
Call-ID: 31add900-3f315023-4-3dc712ac@10.10.199.61
CSeq: 106 NOTIFY
Content-Length: 0


[28] SIP/2.0 200 OK
Via: SIP/2.0/UDP 10.10.199.93:5060;branch=z9hG4bK811043
From: <sip:3601@10.10.199.93>;tag=43A6301C-1EA
To: "sccp_3000" <sip:3000@10.10.199.61>;tag=4afadcdc-0169-479d-ba6a-855a1847684f-
   30228326
Date: Wed, 15 Feb 2006 16:01:04 GMT
Call-ID: 31add900-3f315023-4-3dc712ac@10.10.199.61
```

```
CSeq: 106 NOTIFY
Content-Length: 0


[29] SIP/2.0 489 Bad Event - 'Malformed/Unsupported Event'
Via: SIP/2.0/UDP 10.10.199.61:5060;branch=z9hG4bK2ce6c7d2
From: "sccp_3000" <sip:3000@10.10.199.61>;tag=4afadcdc-0169-479d-ba6a-
  855a1847684f-30228326
To: <sip:3601@10.10.199.93>;tag=43A6301C-1EA
Call-ID: 31add900-3f315023-4-3dc712ac@10.10.199.61
CSeq: 107 NOTIFY
Allow-Events: telephone-event
Content-Length: 0
```

Call Admission Control Using RSVP

This section illustrates the scenario for an RSVP outgoing call from a Cisco Unified IP
Phone to a UserAgent reachable through Cisco Unified CallManager's SIP Trunk.

The Cisco Unified IP Phone and the SIP Trunk devices are in different locations and have
RSVP policy=Mandatory between them. This is to ensure that bandwidth availability is
confirmed before the call path is established.

During call setup, an RSVP agent is allocated for each device in a different location,
which provides an RSVP media path as follows:

```
Cisco Unified IP Phone<--unreserved media->RSVPAgent1(10.10.195.249)<--reserved
  media
path-->RSVPAgent2(10.10.195.250)<--unreserved media-->remoteUA
```

The RSVPAgent1 caters to the Cisco Unified IP Phone's location. The RSVPAgent2 caters
to the Cisco Unified CM's SIP Trunk's location.

The RSVP reservation is relevant only between the location of the IP Phone and outgoing
SIP Trunk device. The standard rfc3312 for negotiation of the resources through precon-
ditions is not supported through the SIP trunk. Thus, the remote is not made aware of the
reservation.

Configuration is as follows:

Node = Unified CM, IP = 10.10.253.4
Node = Remote UA, IP = 10.10.253.3

```
[1] INVITE sip:45205060 SIP/2.0
Via: SIP/2.0/UDP 10.10.253.4:5060;branch=z9hG4bK59f140f7
Remote-Party-ID: <sip:4501@10.10.253.4>;party=calling;screen=yes;privacy=off
From: <sip:4501@10.10.253.4>;tag=1b15671c-e6a2-4e16-bb03-7cd50862f491-30129245
To: <sip:4520@10.10.253.3>
```

```
Date: Wed, 15 Feb 2006 10:00:55 GMT
Call-ID: f2ff2b00-3f21fbd7-81-4fd4c0a@10.10.253.4
Supported: timer,replaces
Min-SE: 1800
User-Agent: Cisco-CCM7.0
Allow: INVITE, OPTIONS, INFO, BYE, CANCEL, ACK, PRACK, UPDATE, REFER,
   SUBSCRIBE, NOTIFY
CSeq: 101 INVITE
Contact: <sip:45015060>
Expires: 180
Allow-Events: presence, kpml
Call-Info: <sip:10.10.253.4:5060>;method="NOTIFY;Event=telephone-
   event;Duration=500"
Max-Forwards: 70
Content-Length: 0


[2] SIP/2.0 100 Trying
Via: SIP/2.0/UDP 10.10.253.4:5060;branch=z9hG4bK59f140f7
From: <sip:4501@10.10.253.4>;tag=1b15671c-e6a2-4e16-bb03-7cd50862f491-30129245
To: <sip:4520@10.10.253.3>
Date: Wed, 15 Feb 2006 10:00:55 GMT
Call-ID: f2ff2b00-3f21fbd7-81-4fd4c0a@10.10.253.4
CSeq: 101 INVITE
Allow-Events: presence, dialog
Content-Length: 0


[3] SIP/2.0 180 Ringing
Via: SIP/2.0/UDP 10.10.253.4:5060;branch=z9hG4bK59f140f7
From: <sip:4501@10.10.253.4>;tag=1b15671c-e6a2-4e16-bb03-7cd50862f491-30129245
To: <sip:4520@10.10.253.3>;tag=7b42a936-ba2d-4980-b674-fb7d79cc9280-23513586
Date: Wed, 15 Feb 2006 10:00:55 GMT
Call-ID: f2ff2b00-3f21fbd7-81-4fd4c0a@10.10.253.4
CSeq: 101 INVITE
Allow: INVITE, OPTIONS, INFO, BYE, CANCEL, ACK, PRACK, UPDATE, REFER,
   SUBSCRIBE, NOTIFY
Allow-Events: presence, dialog
Remote-Party-ID: "Alert4520" <sip:4520@10.10.253.3>;party=called;screen=yes;
   privacy=off
Contact: <sip:45205060>
Content-Length: 0


[4] SIP/2.0 183 Session Progress
Via: SIP/2.0/UDP 10.10.253.4:5060;branch=z9hG4bK59f140f7
From: <sip:4501@10.10.253.4>;tag=1b15671c-e6a2-4e16-bb03-7cd50862f491-30129245
To: <sip:4520@10.10.253.3>;tag=7b42a936-ba2d-4980-b674-fb7d79cc9280-23513586
Date: Wed, 15 Feb 2006 10:00:55 GMT
```

```
Call-ID: f2ff2b00-3f21fbd7-81-4fd4c0a@10.10.253.4
CSeq: 101 INVITE
Allow: INVITE, OPTIONS, INFO, BYE, CANCEL, ACK, PRACK, UPDATE, REFER,
  SUBSCRIBE, NOTIFY
Allow-Events: presence, dialog, kpml
Remote-Party-ID: "Alert4520" <sip:4520@10.10.253.3>;party=called;screen=yes;
  privacy=off
Contact: <sip:45205060>
Content-Type: application/sdp
Content-Length: 295
v=0
o=CiscoSystemsCCM-SIP 2000 1 IN IP4 10.10.253.3
s=SIP Call
c=IN IP4 10.10.253.3
t=0 0
m=audio 4000 RTP/AVP 0 8 18 101
a=rtpmap:0 PCMU/8000
a=ptime:20
a=rtpmap:8 PCMA/8000
a=ptime:20
a=rtpmap:18 G729/8000
a=ptime:20
a=sendonly
a=rtpmap:101 telephone-event/8000
a=fmtp:101 0-15


[5] SIP/2.0 200 OK
Via: SIP/2.0/UDP 10.10.253.4:5060;branch=z9hG4bK59f140f7
From: <sip:4501@10.10.253.4>;tag=1b15671c-e6a2-4e16-bb03-7cd50862f491-30129245
To: <sip:4520@10.10.253.3>;tag=7b42a936-ba2d-4980-b674-fb7d79cc9280-23513586
Date: Wed, 15 Feb 2006 10:00:55 GMT
Call-ID: f2ff2b00-3f21fbd7-81-4fd4c0a@10.10.253.4
CSeq: 101 INVITE
Allow: INVITE, OPTIONS, INFO, BYE, CANCEL, ACK, PRACK, UPDATE, REFER,
  SUBSCRIBE, NOTIFY
Allow-Events: presence, dialog, kpml
Remote-Party-ID: "Caller4520"
  <sip:4520@10.10.253.3>;party=called;screen=yes;privacy=off
Contact: <sip:45205060>
Session-Expires: 1800;refresher=uas
Require: timer
Content-Type: application/sdp
Content-Length: 295
v=0
o=CiscoSystemsCCM-SIP 2000 1 IN IP4 10.10.253.3
s=SIP Call
```

```
c=IN IP4 10.10.253.3
t=0 0
m=audio 4000 RTP/AVP 0 8 18 101
a=rtpmap:0 PCMU/8000
a=ptime:20
a=rtpmap:8 PCMA/8000
a=ptime:20
a=rtpmap:18 G729/8000
a=ptime:20
a=sendonly
a=rtpmap:101 telephone-event/8000
a=fmtp:101 0-15


[6] ACK sip:45205060 SIP/2.0
Via: SIP/2.0/UDP 10.10.253.4:5060;branch=z9hG4bK5694d66a
From: <sip:4501@10.10.253.4>;tag=1b15671c-e6a2-4e16-bb03-7cd50862f491-30129245
To: <sip:4520@10.10.253.3>;tag=7b42a936-ba2d-4980-b674-fb7d79cc9280-23513586
Date: Wed, 15 Feb 2006 10:00:55 GMT
Call-ID: f2ff2b00-3f21fbd7-81-4fd4c0a@10.10.253.4
Max-Forwards: 70
CSeq: 101 ACK
Allow-Events: presence, kpml
Content-Type: application/sdp
Content-Length: 225
v=0
o=CiscoSystemsCCM-SIP 2000 1 IN IP4 10.10.253.4
s=SIP Call
c=IN IP4 10.10.195.250
t=0 0
m=audio 16768 RTP/AVP 0 101
a=rtpmap:0 PCMU/8000
a=ptime:20
a=recvonly
a=rtpmap:101 telephone-event/8000
a=fmtp:101 0-15


[7] INVITE sip:45015060 SIP/2.0
Via: SIP/2.0/UDP 10.10.253.3:5060;branch=z9hG4bK8964d39
Remote-Party-ID: "Caller4520"
  <sip:4520@10.10.253.3>;party=calling;screen=yes;privacy=off
From: <sip:4520@10.10.253.3>;tag=7b42a936-ba2d-4980-b674-fb7d79cc9280-23513586
To: <sip:4501@10.10.253.4>;tag=1b15671c-e6a2-4e16-bb03-7cd50862f491-30129245
Date: Wed, 15 Feb 2006 10:00:56 GMT
Call-ID: f2ff2b00-3f21fbd7-81-4fd4c0a@10.10.253.4
Supported: timer,replaces
```

```
Min-SE: 1800
Cisco-Guid: 4076808960-1059191767-28861-66931722
User-Agent: Cisco-CCM7.0
Allow: INVITE, OPTIONS, INFO, BYE, CANCEL, ACK, PRACK, UPDATE, REFER,
  SUBSCRIBE, NOTIFY
CSeq: 103 INVITE
Max-Forwards: 70
Contact: <sip:45205060>
Expires: 180
Allow-Events: presence, dialog, kpml
Session-Expires: 1800;refresher=uac
Content-Type: application/sdp
Content-Length: 211
v=0
o=CiscoSystemsCCM-SIP 2000 2 IN IP4 10.10.253.3
s=SIP Call
c=IN IP4 10.10.253.11
t=0 0
m=audio 23698 RTP/AVP 0 101
a=rtpmap:0 PCMU/8000
a=ptime:20
a=rtpmap:101 telephone-event/8000
a=fmtp:101 0-15


[8] SIP/2.0 100 Trying
Via: SIP/2.0/UDP 10.10.253.3:5060;branch=z9hG4bK8964d39
From: <sip:4520@10.10.253.3>;tag=7b42a936-ba2d-4980-b674-fb7d79cc9280-23513586
To: <sip:4501@10.10.253.4>;tag=1b15671c-e6a2-4e16-bb03-7cd50862f491-30129245
Date: Wed, 15 Feb 2006 10:00:56 GMT
Call-ID: f2ff2b00-3f21fbd7-81-4fd4c0a@10.10.253.4
CSeq: 103 INVITE
Allow-Events: presence, kpml
Content-Length: 0


[9] SIP/2.0 200 OK
Via: SIP/2.0/UDP 10.10.253.3:5060;branch=z9hG4bK8964d39
From: <sip:4520@10.10.253.3>;tag=7b42a936-ba2d-4980-b674-fb7d79cc9280-23513586
To: <sip:4501@10.10.253.4>;tag=1b15671c-e6a2-4e16-bb03-7cd50862f491-30129245
Date: Wed, 15 Feb 2006 10:00:56 GMT
Call-ID: f2ff2b00-3f21fbd7-81-4fd4c0a@10.10.253.4
CSeq: 103 INVITE
Allow: INVITE, OPTIONS, INFO, BYE, CANCEL, ACK, PRACK, UPDATE, REFER,
  SUBSCRIBE, NOTIFY
Allow-Events: presence, kpml
Remote-Party-ID: <sip:4501@10.10.253.4>;party=called;screen=yes;privacy=off
```

```
Contact: <sip:45205060>
Session-Expires: 1800;refresher=uac
Require: timer
Content-Type: application/sdp
Content-Length: 213
v=0
o=CiscoSystemsCCM-SIP 2000 2 IN IP4 10.10.253.4
s=SIP Call
c=IN IP4 10.10.195.250
t=0 0
m=audio 16768 RTP/AVP 0 101
a=rtpmap:0 PCMU/8000
a=ptime:20
a=rtpmap:101 telephone-event/8000
a=fmtp:101 0-15


[10] ACK sip:45205060 SIP/2.0
Via: SIP/2.0/UDP 10.10.253.3:5060;branch=z9hG4bK291c7d95
From: <sip:4520@10.10.253.3>;tag=7b42a936-ba2d-4980-b674-fb7d79cc9280-23513586
To: <sip:4501@10.10.253.4>;tag=1b15671c-e6a2-4e16-bb03-7cd50862f491-30129245
Date: Wed, 15 Feb 2006 10:00:56 GMT
```

Index

T

cisco
cisco

ciscopress.com: Your Cisco Certification and Networking Learning Resource

Subscribe to the monthly Cisco Press newsletter to be the first to learn about new releases and special promotions.

Visit **ciscopress.com/newsletters.**

While you are visiting, check out the offerings available at your finger tips.

– Free Podcasts from experts:
  - OnNetworking
  - OnCertification
  - OnSecurity

Podcasts

View them at **ciscopress.com/podcasts.**

– Read the latest author **articles** and **sample chapters** at **ciscopress.com/articles.**

– Bookmark the Certification Reference Guide available through our partner site at **informit.com/certguide.**

Connect with Cisco Press authors and editors via Facebook and Twitter, visit
informit.com/socialconnect.

FREE Online Edition

Your purchase of **VoIP Performance Management and Optimization** includes access to a free online edition for 45 days through the Safari Books Online subscription service. Nearly every Cisco Press book is available online through Safari Books Online, along with more than 5,000 other technical books and videos from publishers such as Addison-Wesley Professional, Exam Cram, IBM Press, O'Reilly, Prentice Hall, Que, and Sams.

SAFARI BOOKS ONLINE allows you to search for a specific answer, cut and paste code, download chapters, and stay current with emerging technologies.

Activate your FREE Online Edition at www.informit.com/safarifree

> **STEP 1:** Enter the coupon code: YENIAZG.

> **STEP 2:** New Safari users, complete the brief registration form.
> Safari subscribers, just log in.

If you have difficulty registering on Safari or accessing the online edition, please e-mail customer-service@safaribooksonline.com

  Addison Wesley  Adobe Press   ALPHA Cisco Press   FT Press FINANCIAL TIMES   IBM Press   lynda.com   Microsoft Press New Riders

O'REILLY Peachpit Press PRENTICE HALL Que Redbooks SAMS SAS Publishing Sun microsystems WILEY